5G IoT and Edge Computing for Smart Healthcare

Intelligent Data-centric Systems

5G IoT and Edge Computing for Smart Healthcare

Edited by

Akash Kumar Bhoi

KIET Group of Institutions, Delhi-NCR, Ghaziabad, India

Victor Hugo Costa de Albuquerque

Federal University of Ceará, Fortaleza, Graduate Program on Teleinformatics Engineering, Fortaleza/CE, Brazil

Samarendra Nath Sur

Department of Electronics and Communication Engineering, Sikkim Manipal Institute of Technology, Majitar, Sikkim, India

Paolo Barsocchi

Information Science and Technologies Institute (ISTI), National Research Council (CNR), Pisa, Italy

Series Editor

Fatos Xhafa

UPC-BarcelonaTech, Barcelona, Spain

ELSEVIER

ACADEMIC PRESS
An imprint of Elsevier

Academic Press is an imprint of Elsevier
125 London Wall, London EC2Y 5AS, United Kingdom
525 B Street, Suite 1650, San Diego, CA 92101, United States
50 Hampshire Street, 5th Floor, Cambridge, MA 02139, United States
The Boulevard, Langford Lane, Kidlington, Oxford OX5 1GB, United Kingdom

Notices

Knowledge and best practice in this field are constantly changing. As new research and experience broaden our
understanding, changes in research methods, professional practices, or medical treatment may become necessary.

Practitioners and researchers must always rely on their own experience and knowledge in evaluating and using any
information, methods, compounds, or experiments described herein. In using such information or methods they
should be mindful of their own safety and the safety of others, including parties for whom they have a professional
responsibility.

To the fullest extent of the law, neither the Publisher nor the authors, contributors, or editors, assume any liability
for any injury and/or damage to persons or property as a matter of products liability, negligence or otherwise, or
from any use or operation of any methods, products, instructions, or ideas contained in the material herein.

ISBN: 978-0-323-90548-0

For Information on all Academic Press publications
visit our website at https://www.elsevier.com/books-and-journals

Publisher: Mara Conner
Editorial Project Manager: Emily Thomson
Production Project Manager: Prasanna Kalyanaraman
Cover Designer: Victoria Pearson

Typeset by MPS Limited, Chennai, India

Contents

List of contributors

Muyideen AbdulRaheem
Department of Computer Science, University of Ilorin, Ilorin, Kwara State, Nigeria

Joseph Bamidele Awotunde
Department of Computer Science, University of Ilorin, Ilorin, Kwara State, Nigeria

Paolo Barsocchi
Institute of Information Science and Technologies, National Research Council, Pisa, Italy

M.S. Bhaskar
Renewable Energy Lab, Department of communication and Networks Engineering, College of Engineering, Prince Sultan University, Riyadh, Saudi Arabia

Akash Kumar Bhoi
KIET Group of Institutions, Delhi-NCR, Ghaziabad, India

M. Abdullah Canbaz
Computer Science Department, Indiana University Kokomo, Kokomo, IN, United States

Shamit Roy Chowdhury
School of Computer Engineering, KIIT Deemed to be University, Bhubaneswar, Odisha, India

Alfonso González-Briones
Research Group on Agent-Based, Social and Interdisciplinary Applications, (GRASIA), Complutense University of Madrid, Madrid, Spain; BISITE Research Group, University of Salamanca, Salamanca, Spain; Air Institute, IoT Digital Innovation Hub, Salamanca, Spain

H.M.K.K.M.B. Herath
Faculty of Engineering Technology, The Open University of Sri Lanka, Nugegoda, Sri Lanka

Muhammed Fazal Ijaz
Department of Intelligent Mechatronics Engineering, Sejong University, Seoul, South Korea

Mojtaba Jafaritadi
Faculty of Engineering and Business, School of Information and Communications Technology, Turku University of Applied Sciences, Turku, Finland; Faculty of Technology, Department of Computing, University of Turku, Turku, Finland

Vijay Jeyakumar
Department of Biomedical Engineering, Sri Sivasubramaniya Nadar College of Engineering, Chennai, Tamil Nadu, India

Rasheed Gbenga Jimoh
Department of Computer Science, University of Ilorin, Ilorin, Kwara State, Nigeria

G.M.K.B. Karunasena
Faculty of Engineering Technology, The Open University of Sri Lanka, Nugegoda, Sri Lanka

Riku Klén
Turku PET Centre, University of Turku, Turku, Finland

Elina Kontio
Faculty of Engineering and Business, School of Information and Communications Technology, Turku University of Applied Sciences, Turku, Finland

Palani Thanaraj Krishnan
Department of Electronics and Instrumentation Engineering, St. Joseph's College of Engineering, Chennai, Tamil Nadu, India

Anh-Tu Le
Faculty of Electronics Technology, Industrial University of Ho Chi Minh City, Ho Chi Minh City, Vietnam

B.G.D.A. Madhusanka
Faculty of Engineering Technology, The Open University of Sri Lanka, Nugegoda, Sri Lanka

Ayaskanta Mishra
School of Electronics Engineering, Kalinga Institute of Industrial Technology, Deemed to be University, Bhubaneswar, Odisha, India

Munyaradzi Munochiveyi
Department of Electrical and Electronics Engineering, University of Zimbabwe, Harare, Zimbabwe

Idowu Dauda Oladipo
Department of Computer Science, University of Ilorin, Ilorin, Kwara State, Nigeria

Siba Smarak Panigrahi
Department of Computer Science and Engineering, Indian Institute of Technology Kharagpur, Kharagpur, West Bengal, India

Bhalchandra Sunil Patil
Department of Mechanical Engineering, Indian Institute of Technology Kharagpur, Kharagpur, West Bengal, India

Alex Noel Joseph Raj
Key Laboratory of Digital Signal and Image Processing of Guangdong Province, Department of Electronic Engineering, College of Engineering, Shantou University, Shantou, P.R. China

Arun Kumar Ray
School of Electronics Engineering, Kalinga Institute of Industrial Technology, Deemed to be University, Bhubaneswar, Odisha, India

Raed M. Shubair
Department of Electrical and Computer Engineering, New York University (NYU) Abu Dhabi, Abu Dhabi, United Arab Emirates

Pushpa Singh
Department of Computer Science & Information Technology, KIET Group of Institutions, Delhi-NCR, Ghaziabad, Uttar Pradesh, India

Kushika Sivaprakasam
School of Electrical Engineering, Vellore Institute of Technology, Chennai, Tamil Nadu, India

P. Sriramalakshmi
School of Electrical Engineering, Vellore Institute of Technology, Chennai, Tamil Nadu, India

Abdulhamit Subasi
Faculty of Medicine, Institute of Biomedicine, University of Turku, Turku, Finland

Prema Sundaram
Department of Biomedical Engineering, RVS Educational Trust's Group of Institutions, Dindigul, Tamil Nadu, India

Samarendra Nath Sur
Department of Electronics and Communication Engineering, Sikkim Manipal Institute of Technology, Sikkim Manipal University, Majitar, Rangpo, Sikkim, India

Anita Swain
School of Electronics Engineering, Kalinga Institute of Industrial Technology, Deemed to be University, Bhubaneswar, Odisha, India

Hiren Kumar Thakkar
Department of Computer Engineering, Marwadi University, Rajkot, Gujarat, India

Adekola Rasheed Tomori
Computer Services and Information Technology, University of Ilorin, Ilorin, Kwara State, Nigeria

Edge-IoMT-based enabled architecture for smart healthcare system

Joseph Bamidele Awotunde[1], Muhammed Fazal Ijaz[2], Akash Kumar Bhoi[3], Muyideen AbdulRaheem[1], Idowu Dauda Oladipo[1] and Paolo Barsocchi[4]

[1]*Department of Computer Science, University of Ilorin, Ilorin, Kwara State, Nigeria* [2]*Department of Intelligent Mechatronics Engineering, Sejong University, Seoul, South Korea* [3]*KIET Group of Institutions, Delhi-NCR, Ghaziabad, India* [4]*Institute of Information Science and Technologies, National Research Council, Pisa, Italy*

1.1 Introduction

The emergence of the smart healthcare system has created new opportunities in medical industries such as medical diagnosis, prediction, treatment, and clinical appointments with the doctor by the patients, thus bringing about a reconsideration of traditional methods in the healthcare system (Adeniyi, Ogundokun, & Awotunde, 2021). The implementation of telemedicine and new technological digital health will drastically decrease unnecessary clinical doctor-patient appointments and help in early disease diagnosis. Moreover, healthcare systems focused on telemedicine and smart healthcare system would allow medical services to be real-time and cost-effective creating an extraordinarily convenient time for both patients and physicians (Adeniyi et al., 2021). The healthcare system that depends on the Internet of Medical Things (IoMT) assists individuals and aids their vital everyday life activities. The affordability and user-friendliness of the usage of IoMT has begun to revolutionize healthcare services. The IoMT and its related technologies have emerged as the most preferred use cases in the healthcare industry. IoMT-based wearable technologies have encouraged widespread use of the transformation of smart healthcare systems in recent years (Adeniyi et al., 2021; Dong et al., 2020). Besides, teleoperation and remote-operated equipment are becoming a viable method for remote healthcare surgery technology management. Smart healthcare platforms can make medical procedures more time-efficient, cost-effective, and portable, making them easier to access even in the most remote areas (Ning et al., 2020).

A decentralized database that can continuously maintain update patient information provides the healthcare industry with many benefits. When various parties require access to the same information, these benefits become particularly interesting. Edge computing technology can be used to create additional value in a smart healthcare system in elderly care, chronic diseases, and medical treatment processes. The involvement of many parties in the medical system has caused serious challenges, and this huge dataset in the healthcare system has disrupted the patients' treatment.

5G IoT and Edge Computing for Smart Healthcare. DOI: https://doi.org/10.1016/B978-0-323-90548-0.00006-1

During the treatment of a patient in a situation where many parties are involved can cause huge media distractions. This can be time-consuming during information processes especially when it involves various stakeholders, and the resource-intensive authentication becomes problematic.

All the IoMT-based system requirements can never be met with the traditional cloud computing database architectures alone, because of latency transfers from the network edge to the data center for processing. A better and powerful computing model is required that can reduce the higher latency data transfers that create a dominant strategy. The cloud computing bandwidth is quickly outpaced by traffic from thousands of users. Also, the cloud servers neglect other protocols the IoMT devices use and interact only with IP. However, edge technology helps the IoMT-based devices capture data to be analyzed close to the machines that generate and function on that data. Hence, edge computing can be used to close the gap and be the bridge linking IoMT-based devices in the processing of a huge amount of data produced. The processing becomes easy with the edge computing model and handling and outlining the data from IoMT devices becomes easy and greatly improved.

Both cloud and edge computing are similar in terms of versatility and scalability of computing, storage, and networking resources on-demand supplies, and mutually built with virtual systems. Although with the emerging trend in networking in terms of demand, the two technologies have a wide barrier. The businesses and end-users are free to use cloud computing from defining certain specifics, such as storage capacity, limits on computing, and cost of network connectivity. There is still the rising problem of real-time latency-sensitive applications within nodes to meet their delay requirements in IoMT-based systems (Bonomi, Milito, Zhu, & Addepalli, 2012; Saad, 2018). The issue of security of this huge volume of data should also be the main concern for any business-minded experts because the problem hurts their reputation and they are constrained by the law to keep all data safe.

But comparing edge to cloud computing, edge computing brings computation, storage, and networking closer to the data source, reducing travel time and latency dramatically. Instead of sending data back and forth all the time, the processes take place close to the device or at the network's edge, allowing for quicker response times. Edge applications decrease the amount of data that must be transferred, as well as the traffic generated by those transfers and the distance that data must travel.

The IoMT-based cloud provides the liberty of accessing data from the service providers anytime in any part of the world, hence, exposure the IoMT-based data to security and privacy threats. To gain more accurate diagnosis results, edge computing has been widely used to decrease the burden on the medical experts, and help in decreasing the decision time of traditional methods of the diagnosis process. There are significant improvements in the treatment, prediction, screening, drug/vaccine development processes, and application of medication in healthcare sectors with continuing expansion in IoMT-based edge computing. The applications of IoMT-based edge computing have reduced human intervention in medical processes and the cost of medical applications has reduced.

5G technology with edge computing offers great advantages in supporting the IoMT for medical diagnosis, monitoring, prediction, treatment, among others efficiently (Magsi et al., 2018). Also, supports ultra-reliable low latency communication (uRLLC) with a higher data rate, and helps in the areas of connection of various IoMT-based devices. The use of 5G technology has increased the communication and transfer of data within wireless networks. The introduction of the 5G network in the IoMT-based system has benefited many in various ways. Also, various fields like education, business organizations, and medical and governmental agencies have benefited from 5G technology. The use of 5G in medical applications with reduced energy consumption has been proved by many types of research (Sodhro & Shah, 2017). Furthermore, integrating 5G and edge computing in an IoMT-based system will enhance patient examination quality, and will be useful in the area of Wireless Body Area Networks by providing a protected system in the healthcare industry (Aldaej & Tariq, 2018; Jones & Katzis, 2018). While every sector will receive enjoyment and benefit from Interne of Things (IoT)-based edge computing systems, why should IoMT-based medical systems stay behind from the benefits of edge computing? IoMT-based platforms can be enhanced and equipped with edge computing to ensure accurate diagnosis, and treatment of patients remotely. A smart healthcare system with proper motivation and proper care will contribute immensely to the medical system and overcome its obstacles.

Therefore, this chapter discusses the areas of applicability of the architecture of Edge enabled IoMT system in the healthcare system. It will also present extraordinary opportunities brought by edge-enabled IoMT system in healthcare, and their research challenges in the healthcare system are discussed. The chapter finally proposes a framework of an edge-enabled IoMT-based system for the healthcare system.

1.2 Applications of an IoMT-based system in the healthcare industry

Massive medical costs and the maintenance of big data during any disease outbreak require technical advances so that at any time and anywhere, everybody has access to healthcare services. The development of technology has allowed telehealth to provide online healthcare facilities. For patients that are permitted to travel, for villages in rural zones, and for individuals that do not have access to medical care, remote facilities are useful. The uses for telemedicine include the transmission and storage of medical images, video conference patient counseling, continuing education, and facilities in the electronic healthcare field. Sadly, the use of telemedicine technology is hindered by technical and financial costs (Jin & Chen, 2015). To this effect, studies have given cloud computing that offers, among other things, remote support capability, accessible transparent resources, efficient large internet connectivity, scalable and resources pooling, robust medical data sharing and processing, and the sharing of big data patient records.

Digital wellness innovations provide huge incentives to reshape current healthcare programs. Digital health innovations have offered improved quality of care at a more affordable cost, from the introduction of automated therapeutic annals to portable medical devices to other innovative technology. With healthcare programs, politicians are continually researching, embracing, and implementing information and communication technology (ICT) (Sust et al., 2020). This forms the way people and patients view the structures and communicate with them. The road to digital medical care (eHealth) is a systemic evolution of the conventional medical care system that incorporates numerous devices together with universal entry to automated medical annals, online tracking systems, inmate services, wearable devices, portable medical applications, data analytics, and further transformative innovations (Meskó, Drobni, Bényei, Gergely, & Győrffy, 2017; Sust et al., 2020).

Owing to various pandemics, there is an immediate need to make good use of current technology. IoMT is known to be one of the greatest innovative innovations with tremendous promise in fighting diseases and pandemic outbreaks (Oladipo, Babatunde, Awotunde, & Abdulraheem, 2021). The IoMT consists of a sparse network where the IoMT systems feel the world and transmit valuable data across the Network. IoMT-based is one of the promising technologies that will change our lives with seamless connections and vigorous integration with other technologies (Hussain, Hussain, Hassan, & Hossain, 2020; Sundwall, Munger, Tak, Walsh, & Feehan, 2020). The IoMT-based can be useful in reducing disease spread within an environment and provided various functions like tracking, and monitoring of the patient in reducing the risk and spread of diseases (Albahri et al., 2020; Saeed, Bader, Al-Naffouri, & Alouini, 2020). Fig. 1.1 displays possible applications of IoMT-based devices that can be used effectively to reduce any disease outbreak.

IoMT-based systems in healthcare are used to monitor and control the human body's vital signs and connect to healthcare facilities using communication infrastructure (Rodrigues et al., 2018). The accessibility to a quality physician is now unlimited with the introduction of telemedicine with various factors attached to them and is getting popular in remote areas (Chui, Liu, Lytras, & Zhao, 2019). For example, patients can be tracked remotely without being physically present at the hospital using devices and sensors like blood pressure, heart rate, electrocardiography, diabetes, and

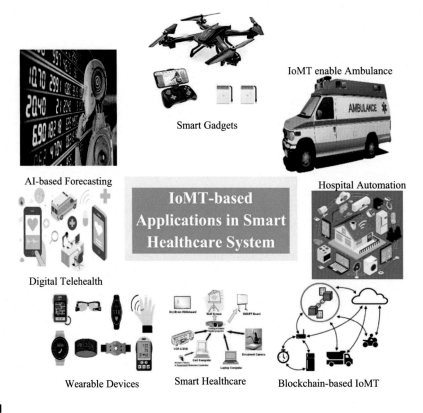

FIGURE 1.1

Potential applications of IoMT for smart healthcare system.

signs of the vital body. Examples are sensors and actuators that can be used to capture and collect data to be sent to the cloud from the patient using a local gateway. The results from processed data can be used by a medical doctor to provided and notify the patient about their status or report (Adeniyi et al., 2021).

Many studies have found that inadequate access to patient information is the explanation for most medical errors especially during infectious diseases (Sundwall, Munger, Tak, Walsh, & Feehan, 2020). The IoMT-based medical system has been regarded as a possible system to increase openness and reduce the extent of medical errors during disease diagnosis to correct health data (Chui et al., 2019; Firouzi et al., 2018). Many medical organizations have also chosen IoMT-based cloud storage to obtain and store broad patient data and maintain their electronic health records systems. Electronic health records have evolved rapidly over the last decade, providing a basis for data mining to recognize designs and styles in the big data industry in healthcare. Another common point for exchanging medical data is the interchange of electronic health records. By communicating at a common hub, these businesses facilitate healthcare sectors to transmit information rather than maintaining ties with many peer businesses (Regola & Chawla, 2013).

IoMT-based cloud systems also offer secure storage and share resources that can reduce the amount of local traffic to make organizations agile (Rubí & Gondim, 2019; Syed, Jabeen, Manimala, & Alsaeedi, 2019). By reducing the cost needed for starting up automated medical records, which is lacking in many healthcare segment facilities, this will improve the efficiency of the healthcare sector (Schweitzer, 2012). During a disease outbreak, prescriptions and diagnoses, for instance, can be shared through the cloud over different systems. Therefore, for service enhancement and higher standards, hospitals and doctors exchange patient records. The primary advantages of electronic health record cloud storage are the capacity to exchange patient records with other specialists at home and overseas, the facility to pool data in one location, and the capacity to access files anytime, anywhere. Electronic health record cloud computing enables patients to view, replicate, and transfer their secure health records (Chen, Chiang, et al., 2016). Regardless of the influences of the IoMT-based system to capture and store large health data, the prime problem is the failure of the network, protection, and privacy of patient information that users, hackers, malware, and so on are exploiting (Kumari, Tanwar, Tyagi, & Kumar, 2018b; Muhammed, Mehmood, Albeshri, & Katib, 2018).

This new emergence of these technologies is a result of their high availability, simplicity to personalize, and easy accessibility; thus enabling the providers to deliver personalized content cost-effectively on large scale easily. Also, big data analytics and IoMT are progressively gaining more attraction for the next generation of smart healthcare systems. Though the new fields evolving rapidly, they also have their shortcomings, particularly when the goal is healthcare systems with a complicated problem, difficult in energy-efficient, safe, flexible, suitable, and consistent solutions, especially when it comes to the issue of security and privacy of IoT generally. It has been projected that IoT will rise to a market scope of $300B by 2022 in healthcare covering the medical devices, systems, applications, and services sectors (Firouzi et al., 2018). IoT allows a broad range of intelligent applications and resources to solve the problems facing individuals or the healthcare sector (Medaglia & Serbanati, 2010). For instance, P to D (Patient-to-Doctor), P to M (Patient to Machine), S to M (Sensor to Mobile), M to H (Mobile to Human), D to M (Device to Machine), O to O (Object to Object), D to M (Doctor to Machine), T to R (Tag to Reader) have dynamic IoMT link capabilities. This brings people, computers, smart devices, and complex systems together intelligently to ensure a productive healthcare system (Tuli et al., 2020; Zafar, Khan, Iftekhar, & Biswas, 2020).

The IoMT has greatly contributed to the innovations in smart healthcare systems interconnected devices and medical sensors to promote knowledge-gathering, storage, communication, and sharing. The dramatic changes in traditional healthcare systems into a smarter healthcare system use various wireless technologies as a catalyst like wearable sensors, wireless sensor networks, radio frequency identification (RFID), Bluetooth, Li-Fi, and Wi-fi among others has greatly helped and change the healthcare industry (Baker, Xiang, & Atkinson, 2017; Chen, Hu, & McAdam, 2020; Fernandez & Pallis, 2014). The use of IoT has penetrated all fields in recent years in various fields like agriculture, education, transportation, and most especially in the healthcare sectors, thus paving the way towards technological transformations (Guy, 2019; Tripathi, Ahad, & Paiva, 2020). There has been tremendous growth in the healthcare system using IoMT-based devices to achieve a great level of automation. There is countersigning of the beginning of smart healthcare systems to achieve ubiquitous and holistic healthcare facilities with possible improvement where all stakeholders are interconnected using IoMT-based devices.

There is an increasing influx of people to urban areas today. Healthcare facilities are one of the most critical characteristics that have a major effect on people arriving in city centers during infectious disease outbreaks globally. Metropolises are therefore financing a digital transition to offer residents healthy environments (Marston & van Hoof, 2019). On the other hand, because of its huge number, high speed, and high variety, conventional models and methods for full conservational performance assessment are threatened by the advent of big data (Song, Fisher, Wang, & Cui, 2018). Also, because of their carbon emissions, conventional ICT systems damage the atmosphere (Petri, Kubicki, Rezgui, Guerriero, & Li, 2017). On the other hand, cloud services are a cost-effective medium for accommodating large-scale infrastructure systems have gained considerable acceptance. The use of cloud computing is, therefore, a significant phase in the green processing process that saves resources and protects the atmosphere. The use of sufficient equipment and cloud space saves the organization's resources and eliminates the costs related to cooling systems, computers, and central servers. Nevertheless, cloud computing supports renewable computing with energy savings, rendering dangerous articles less harmful (Pazowski, 2015).

By using intelligent mobile computers, IoMT-based cloud systems have inspired healthcare specialists to observe the wellbeing of patients at home remotely (Bhatia, 2020). Besides, IoT will build a network by leveraging integrated sensors to track the patient's real-time health status and control the treatment process. The IoT plays a significant role in the healthcare sector and this will continue for the next generation. Although health monitoring systems for IoT-based patients are popular, observing outdoor hospital requirements increases the IoMT's cloud computing capabilities for the handling and storing of health data (Ghanavati, Abawajy, Izadi, & Alelaiwi, 2017). Nevertheless, the sum of IoMT-based gadgets is anticipated to rise significantly in the approaching years (Al-Turjman, Nawaz, & Ulusar, 2020) and the complexity that exists in various IoMT mechanisms (system crossing point, communication protocols, data structure, system semantics) would bring interoperability and confidentiality-correlated difficulties (Edemacu, Park, Jang, & Kim, 2019). A universal healthcare framework must be robust enough to address all of these principles in this way. The incorporation of IoMT technologies in an interoperable setting and the creation of software for the collection, analysis, and extensive distribution of IoMT-based data are now becoming important.

1.3 **Application of edge computing in smart healthcare systems**

Greater efforts have been made in the area of the smart healthcare system to build and design a reliable and convenient framework for IoMT-based device systems (Pham, Mengistu, Do, & Sheng, 2018). In the biomedical industry and the rise in wearable devices, smart healthcare systems have gradually reshaped the conventional medical system (Lu et al., 2020; Wen et al., 2020). These devices and sensors are majorly used for collecting blood pressure, respiratory rate, motion function, blood glucose data, electrocardiogram (ECG), Electroencephalogram (EEG), and body temperature among others for primary medical examinations. To provide early diagnosis using this data helps in the preservation of the healthcare system and the removal of other complications during patients' treatment (Athavale & Krishnan, 2020).

IoMT-based is used in the medical system to manage doctor's advice to patients, medical tools, patients' records, disease diagnosis, and patient treatment. The application of Machine Learning (ML) algorithms with the IoMT-based system makes the smart healthcare system highly effective in the area of disease diagnosis, prediction, health monitoring system, and before human utilization (Pustokhina et al., 2020). IoMT-based systems allow telemedicine like telesurgery, telerehabilitation, and telehealth that remotely monitoring, treating, and diagnosing patients' in real-time. They use the IoMT-based model to transfer medical data to the database using IoMT-based cloud models. The models are comprised of three major components, namely the Body Sensor Network (BSN), the gateways, and the cloud server center. In recent years, the IoMT-based system supports healthcare services in real-time to distant stakeholders. The capture data using IoMT-based devices are provided to physicians and relevant stakeholders to validate and provide useful information to patients' whenever it is needed.

The IoMT-based system with edge computing lowers latency services is energy-effective and cost-effective, and provides maximum satisfaction for healthcare contributors. Most IoMT-based environments depend on a cloud platform for massive smart health systems (Janet & Raj, 2019). The model can be used to forward captured data produced from IoMT devices through the Internet to the cloud, and thereby used for diagnosis to provide useful reports using learning algorithms like ML or deep learning (DL). But the IoMT-Cloud system is inappropriate, especially where lower latency is necessary. Hence, IoMT-based systems require a faster and low latency protection technique with delay-sensitive, smart, secure, stable smart healthcare management. Edge computing is the answer to this with a prolonged type of cloud computing where IoMT-based data can be computed closer to the edge of the network where data are produced (Abdulraheem, Awotunde, Jimoh, & Oladipo, 2021).

Edge computation reduces latency, data traffic, and data distance to the network since it is running at a local processing level closed to the cloud database. Edge computing has become relevant and important since devices can recognize data instinctively, thus become useful in IoMT-based systems to reduce the latency to a lower level. Fig. 1.2 depicts the edge computing architecture, where the first part of a network uses the IoMT-based devices and sensors to collect data to be processed through a gateway using a Radio Access Network that uses edge devices to compute data aggregated by the network locally. Once the data processing has been done, the full computing operations and memory storage have been processed to the cloud.

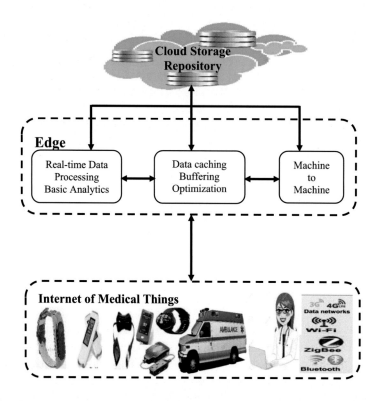

FIGURE 1.2

The architecture structure of edge computing in IoMT-based system.

The edge layer is like a junction point where enough networking, computing, and storage resources are available to manage local data collection, which can be readily obtainable and deliver fast results. Low-power system-on-chip (SoC) systems are used in most situations because they are meant to preserve the trade-off between processing performance and power consumption. Cloud servers, on the other hand, have the power to conduct advanced analytics and ML jobs to combine time series generated by a variety of heterogeneous or mixed kinds of items (Rehman, Khan, & Habib, 2020).

The IoT-based system is used to generate large medical data with the introduction of wireless technology, and customized enhance services. Such big medical data are generated from countless sources, and the cloud server is used to store, analyze, and process such data like text, multimedia, and image among others (Devarajan, Subramaniyaswamy, Vijayakumar, & Ravi, 2019). The high latency, security problems, and network traffic arise as a result of the handling of big cloud medical data. Fog computing was introduced to minimize the burden of the cloud been a new computing platform. The fog also helps in bringing the cloud service closer to the network edge and thus allows refined and secured healthcare services.

The edge of the network is an ideal place for analyzing real-time health information where data is created. The feature of edge computing placed it ahead of cloud computing like data preprocessing, local data analytics, data security and privacy, temporary storage, data trimming, distributed, decentralized storage. Both distributed edge and centralized cloud servers are needed in an IoMT-based application like health monitoring systems to efficiently perform big data analytics. The use of edge computing as an intermediary has created a better way of handling cloud databases on IoMT-based devices for real-time healthcare systems (Devarajan et al., 2019). Edge computing takes resources to the edge of the network as an extension to cloud computing. This effectively brings the benefits and power of the cloud closer to the place where the data is produced, thereby assisting and speeding up "on-the-fly solutions" for applications in a smart healthcare system. This decentralized model's main objective is to bring devices and software to the edge of the network where the data is generated. Edge computing's main aim is to reduce the amount of data that is transmitted to cloud data centers for processing and analysis. It also improves security, a key issue in the IoMT industry (Pan & McElhannon, 2017; Xu et al., 2019).

Table 1.1 compares the cloud IoMT-based and Edge IoMT-based computing using IoT-based requests. Edge computing cannot completely replace cloud computing because it is essentially an extension of the perception of cloud computing. The computing paradigm is complementary and collaborative. To process a huge amount of big data in real-time quickly, edge computing ends is very paramount, but most of the captured data is not used once. The cloud is still used for the storage of capture data, and useful in the linkage of various edge nodes, and the management of edge and virtualization resources rely solely on the cloud. The combination of both cloud and edge will bring about various IoT-based devices together to accomplish various demand situations, thereby optimizing the application benefit of both technologies.

The Cloud IoMT-based alone can no longer handle the huge amount of data generated by IoMT devices, new powerful computing models are required. The security concerns, low

Table 1.1 Comparison of IoMT-based cloud and IoMT-based fog computing.

Requirements	IoMT-based cloud computing	IoMT-based edge computing
Mark user	Internet users	Mobile users
Location of servers	Within Internet	Edge nodes
Service type	Global information	Localized information services
Geographical Distribution	Centralized	Distributed
Distance between client and server	Multiple hops	Single hop
Delay jitter	High	Low
Latency	High	Low
Data Processing	Slow	Fast
Reliability	High	Low
Type of connectivity	Leased line	Wireless
Location awareness	No	Yes
Server nodes	Few	Large
Computing cost	High	Low
N/W bandwidth	More	Low
Reply time	Minutes	Milliseconds, sub-seconds
Security	Less secure	Very secure

latency, speedy processing requirements need new powerful computing techniques to best place processing, conserve network bandwidth, and making IoMT-based systems operate in a reliable environment (Nandyala & Kim, 2016). All of these IoMT-based system requirements can never be met with traditional cloud computing architectures alone; therefore, a better and powerful computing model is required. Latency transfers data from the network edge to the data center for processing thus creates the dominant strategy. Bandwidth is quickly outpaced by traffic from thousands of users. Also, the cloud servers neglect other protocols the IoT devices use and interact only with IP. The best location for most IoMT data to be analyzed is close to the machines that generate and function on that data and this is called computing with an edge.

It is important to recognize that cloud and edge computing are two distinct, non-interchangeable technologies that cannot be used interchangeably. Time-sensitive data is processed using edge computing, while data that is not time-sensitive is processed using cloud computing. In remote areas where there is little or no access to a centralized location, edge computing is favored over cloud computing. Edge computing is the ideal option for local storage in these areas, which is equivalent to micro-network infrastructure.

Specialized and intelligent systems benefit from edge computing as well. Although these devices are similar to personal computers (PCs), they are not multifunctional computing devices. These specialized computing devices are intelligent and respond in a specific way to specific machines. Edge computing, on the other hand, suffers from this specialization especially in smart healthcare that needs fast responses. Edge computing differs from cloud computing in that it takes time to relay information to a centralized data center, which can take up to 2 s, slowing decision-making. Since signal latency can result in business losses, organizations prefer edge computing to cloud computing.

The smart healthcare system is different from most existing offloading frameworks, thus exceptionally delay-sensitive. Hence, the delay constraint in cloud servers makes it difficult to always provide satisfactory services (Dong et al., 2020). Edge computing is used to reduce transmission latency to solve this obstacle. In edge computing-enabled health monitoring systems, which can be maintained by using hybrid cloud computing, the privacy problem is established in (Pace et al., 2018). Gu, Zeng, Guo, Barnawi, and Xiang (2015) suggest a cost-efficient healthcare system with the convergence of edge computing and health monitoring. The system under review takes into account the combination of servers, the allocation of tasks for medical research, and the implementation of virtual machines.

The wearable sensor is treated with minimum power at the edge devices platform in the health monitoring system. Without decreasing the working role of the IoMT-based system, the edge devices limited energy power and, thus, reduce the application computation and energy consumption to grow edge-dependent healthcare sectors. The combination of cloud-edge computing into a healthcare monitoring system is one of the skillful strategies for integrating agile computing. The advantages of edge and computing under the application of hierarchical structure helps to extend the computation between cloud and edge devices in the analysis of the data collected using IoMT-based devices. The delay-sensitive healthcare applications have been increased using edge computing while the integration of higher storage capacity and maximum resources to compute was provided using cloud computing. The combination of cloud and edge computing enhanced the performance of IoMT-based devices in medical fields.

In various computation models, DL has been widely utilized for intelligence in numerous fields like natural language processing, object recognition, speech segregations, image classification (Ayo, Ogundokun, Awotunde, Adebiyi, & Adeniyi, 2020; Oladele, Ogundokun, Awotunde, Adebiyi, & Adeniyi, 2020). DL, due to its characteristic of self-teaching, can be used to learn input data feature hierarchically and automatically, and compression ability to highpoint the concealed patterns (Awotunde, Matiluko, & Fatai, 2014; Ayo, Awotunde, Ogundokun, Folorunso, & Adekunle, 2020). Therefore, for IoMT-based applications, the DL method has become an effective approach for learning features and classification. The DL technologies become highly thorough because of their productiveness inbuilt layers' structure. The minimum-powered edge devices are therefore not applicable in the DL system since the DL method is not capable of meeting full computational cost requirements. The key challenge involved in implementing an efficient health monitoring system for latency-consciousness is the incorporation of DL inference into edge devices that have minimal computational capabilities (Parsa, Panda, Sen, & Roy, 2017).

1.4 Challenges of using edge computing with IoMT-based system in smart healthcare system

Several technologies have been used and are comprised in smart healthcare systems, which include 5G technology, IoMT-based system, edge computing, cloud computing, medical devices and sensors, artificial intelligence, and DL. These technologies have been put together to better the performance of a smart healthcare system. In recent years, edge computing has been identified as the best computing to be applied with a system that requires lower latency and cost-efficient like IoMT-based systems. The big data 5Vs data importance had results of the huge amount of patients' data receives from the medical device such as volume, veracity, variety, and velocity. As a result, edge computing is needed to connect to receive, store, process, and communicates with IoMT-based devices. The system administrative configuration must be controlled to forestall the data fluctuation between the edge and IoMT-based cloud database. To handle various types of data like text, videos, audios, and image files, edge protocols, and data format are needed from various ways like smartphone, and a smartwatch. For regular data transfer and urgent data requests, the Smart eHealth gateway must be aware of sufficient routing.

The data collection takes place either from medical sensors or portable devices. There is a need for adequate protection for medical facilities so that patient can use their smartphone for health status updates. The use of a smart healthcare system creates possible ways of expanding the healthcare system to the whole population. The appointment time used by patients to see physicians, or waits for diagnosis outcomes can be reduced with the use of the intelligent healthcare system. This also provides direct access to real-time medical care and services. To maintain trust between patients and medical experts, the scalability of a smart healthcare system must be taking with all seriousness, and this will in turn save quality time. It is the main concern and it is not appropriate to obtain information from end-users through unauthorized entities, this, also, poses threats to the personal safety of medical data. The main problem in the introduction and deployment of the digitalized medical system is security and privacy. However, with the integration of these layers, security is needed in any layer, such as the system layer, fog layer, and cloud layer (Puthal et al., 2018).

Another problem facing fog computing is heterogeneity that refers to the various communication-capable devices. Devices like smartphones, autonomous vehicles with other IoT smart devices are at the bottom of the layers within the IoT-based system.

Cloud computing infrastructure has series of difficulties due to limited networking in a small number of datacentres services, centralized computation, security, and privacy challenges. The issues may be due to the relatively long gap between remote cloud services and edge devices. Edge cloud and edge computing tend to be a promising possibility to overcome this problem, which offers services closer to resource-poor edge IoMT-based devices and can potentially foster a new ecosystem of IoMT innovation. A variety of technological innovations, including cloud services of network functions and software-defined networking, allow for this prospect.

Sensor interoperability, system connectivity, protection, knowledge management, privacy, and device management barriers are some of the challenges of IoT and Artificial Intelligence (AI) in the smart healthcare system. Devices and sensors are used to capture data that experts used to diagnose and monitor in real-time healthcare settings, and the information obtained from heterogeneous data sources involves several problems (Lin et al., 2016). There are natural and uncontrolled problems in IoT-based devices and sensors, where such challenges are often unexplained errors in devices like smartwatches and cell phones like battery capacity, inconsistencies between unique physical attributes, frequent complexities, and environmental variations. These suggest that the uses of multimodal signal and several IoMT devices can cause different problems. To promote the general acceptance of such smart healthcare, a streamlined and simpler fusion approach should be addressed.

A major concern is the abuse of access privileges by authorized insiders either because of irresponsible, or for individual criminal reasons, and the information disclosure normally happens with healthcare facilities. Also, in exchange for illicit benefits, they can reveal confidential patient information to unauthorized persons. Health updates from celebrities and politicians can easily be leaked from a centralized cloud healthcare database to the media. This is always caused by insiders of the regulation and the records to which they can access, such as health staff who do not take care of actual patients, for instance, and retired workers who are not yet restricted from data queries. By accessing each other's protected data, a dissatisfied group can create problems for others. To infiltrate, intruders attempt to pretend to be healers. Unhealthy medical procedures have high costs, such as adverse effects on their image, fines, civil liability, and much more.

For physicians, conventional AI-based systems used in smart healthcare systems may not be suitable. Therefore, it is possible to deploy explainable AI-based systems where doctors can imagine the identification or detection of diseases. Edge-intelligent algorithms can be effectively optimized to maximize edge capital (Ghoneim, Muhammad, Amin, & Gupta, 2018; Hossain & Muhammad, 2017; Rahman, Hossain, Islam, Alrajeh, & Muhammad, 2020). Literature seldom addresses the functional utility of IoMT-activated healthcare systems. The key issue is that businesses own the most important data, and it is not available to the public. In practice, the successful implementation and use of data fusion would allow more accurate assessment and evaluation of day-to-day physical activity using low-cost monitors that can lead to simpler and better chronic disease preventive care. The next generation of wireless networks has created a tremendous opportunity for intelligent healthcare (Abdulsalam & Hossain, 2020; Alhamid, Rawashdeh, Al Osman, Hossain, & El Saddik, 2015; Vizitiu, Niţă, Puiu, Suciu, & Itu, 2020). The 5G and beyond has allowed for smart healthcare systems to be accessed more easily and quicker than ever before, thus making the edge-based and federated DL faster and simpler (Hossain, Rahman, & Muhammad, 2017; Muhammad, Alhamid, & Long, 2019).

The heterogeneity within IoT-based systems arises during data processing, data formatting, and data clearing thus creates difficulty in processing medical information. The enabling of a network to connect with various sensors is an example challenge in a smart healthcare system for the monitoring of patients. For this to take place, heterogeneity must be present when the data is transferred to another system for processing or analysis. The edge layer while going up to the next layer involves various nodes, clusters, switches, and other devices that are needed during data processing and communication facilities (Yi, Li, & Li, 2015). In the design of architecture that enables multiple monitoring of devices, heterogeneity is, therefore, an important factor to communicate with the patient using devices like heart rate, body temperature, and blood pressure sensors (Mouradian et al., 2017).

There is another research issue in fog computing due to the location of the network, their protection can create a concern. The threats that are not present in an organized cloud architecture pose a threat and arises sometimes at the edge of the network. The major threat from this problem is when an attacker transmits and changes contact between two parties, patients are in the middle of attack (Mouradian et al., 2017). There may be a compromised gateway between patient monitoring sensors and fog nodes that processes the information from the patient in the smart healthcare system. A serious problem may arise if the intruder changed the data being handled by an IoT-based system, this may create serious implications for the welfare of these patients (Awotunde, Bhoi, & Barsocchi, 2021).

There are not standard rules and regulations in the smart healthcare system for the products and services in edge computing, especially in network protocols and interfaces. There should be standardization in place to standardize the healthcare system like a dedicated agency is needed to solve this problem. This will help in data dissimilarity and helps to accomplish the real-time response. For good standardization issues like communication protocol, system interfaces should seriously be considered (Kumari, Tanwar, Tyagi, & Kumar, 2018a).

There is a broad range of message linkers through various procedures, such as Wi-Fi, similar to the application diversity issue. To be interoperable, edge computing should therefore perform the required protocol translation on various internal layers, such as network layers, message layers, and edge layer data annotation layers. Also, there should be complex regulatory systems before healthcare tools and equipment are available on the market for consumers to use. The stakeholders and end-users of e-health products should be part of the design team to provide input on their likes, dislikes, and comforts. This is going a long way to help in building user-friendly interfaces and patient-centric intelligent medical devices (Kumari et al., 2018a).

1.5 The framework for edge-IoMT-based smart healthcare system

The cloud computing technology after calculation returns the results to the terminal equipment using the capture data from the IoMT-based system by uploads medial data terminal tools to the remote cloud. But, the pressure on the cloud becomes higher if the data generated are huge, thus causing delay and energy consumption due to a huge load of data in the cloud. In this kind of satiations, cloud computing alone cannot help to provide real-time response and lower latency. The main aim of the medical IoMT-based cloud is to effectively expand the ability of cloud computing providing at the edge of the network using distributed computing resources (Lin, Song, & Jamalipour, 2019); thus, edge computing is a perfect application to meet this computing demand (Adeniyi et al., 2021; Ning, Wang, & Huang, 2018). Fig. 1.3 shown the overall framework of the proposed Edge-IoMT-based system. The framework contains four layers and key technologies. The four layers are IoMT-based devices, edge computing, cloud computing, and end-users/alert generation layers. The IoMT-based devices layer is used to collect and capture data from patients and transfer it to the edge computing through the gateway, the collected data will be processed from the edge and send the result to the cloud computing database for further use. The last layer is for the end-user that makes use of the generated results and alert generation that can be sent to the user (patients).

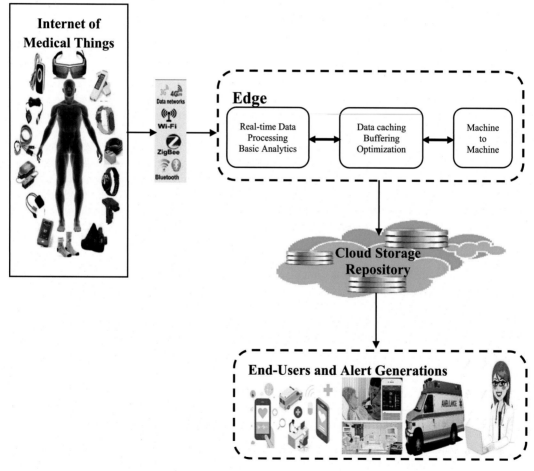

FIGURE 1.3

The proposed architecture for edge-IoMT-based smart healthcare system.

The IoMT-based layer consists of various medical-related technologies like wearable devices and medical sensors. Originally, during the transmission of health data, IoMT devices are used to collect patient health information and the interconnected devices interact with other equipment. It collects medical information such as an ECG, body temperature, heart rate, pulse oximetry, cholesterol level, and heartbeat while the IoMT-based devices are kept in the body. To evaluate the health problems of the patient, this information is sent to edge computing.

The various edge network nodes can be used as smart terminal devices in devices like laptops, smartphones among others, or have network devices gateways like routers and Wi-Fi. The nodes are distributed within the cloud database and terminal equipment in IoMT-based devices. Edge computing provided a more agile and sensitive network due to some limited hops within various nodes and the terminal equipment used in the IoMT-based system platform with network resources and cloud storage. This will prevent and reduce the delay of requests and other security problems that are important in smart healthcare systems.

1.6 Case study for the application of edge-IoMT-based systems enabled for the diagnosis of diabetes mellitus

Here, the UCI dataset for diabetes mellitus (DM) with relevant medical data is generated, and IoMT-based sensors for diagnosing patients affected with diabetes rigorously. The edge IoMT-based system was proposed to run the data close to network nodes. A ML model based on Long-Short-Term Memory (LSTM) was used for the diagnosis of DM, and the experiment was conducted using a standard UCI dataset repository. Even when dealing with extremely large amounts of data, ML models play a significant role in the decision-making process (Adeniyi et al., 2021; Ogundokun et al., 2021). Defining data types such as velocity, variety, and volume is part of the process of applying data analysis techniques to particular areas (Awotunde, Jimoh, Oladipo, & Abdulraheem, 2021). Standard data analysis models include neural network models, classification models, and clustering methods, as well as the use of efficient algorithms (Kumar, Lokesh, Varatharajan, Babu, & Parthasarathy, 2018). Data can be produced from a variety of sources with different data types, so designing methods that can handle these characteristics is essential.

The UCI database consists of a DM dataset with relevant features for the diagnosis of the disease. The dataset has the history of the patient data collected from the hospital. The capture of patient information is stored in the cloud database. The data capture layer is used for the collection of important data from the cloud storage module. The edge was introduced to secure the cloud database, create a platform that allows the capture data to be run close to the cloud node network, takes time to relay information to a centralized data center and low latency signal. The Java programming language and Amazon cloud have been used to implement the proposed system. The preprocessing, such as common and disease affected with frequency, is a major task in this study. The experiments were carried out using UCI datasets, and the proposed study was evaluated using criteria such as precision, sensitivity, and specificity.

The staging procedure of LSTM is as follows:

The LSTM's sigmoid feature recognizes data that is not needed for any process and uses it as (X_t) at time $t - 1$ and produces output (V_{t-1}) at time $t - 1$. The sigmoidal function determines which part of the output will be isolated from the previous output. The Forgetting Gate (f_t) is the name of this level. Forget about the f_t gate; the value is a vector that matches each number in the cell (C_{t-1}) and ranges from 0 to 1.

$$f_t = \sigma\left(w_f[V_{t-1}, X_t] + b_f\right) \tag{1.1}$$

where σ is the sigmoidal function, $w_f =$ weights and $b_f =$ forget gate.

Both equations have two states: the current input's ignoring and storing conditions, as well as the cell state's X_t. The sigmoidal layer and the tanh layer are the two layers. To the sigmoidal layer given, use 0 or 1 to determine if the new information needs to be changed. The tanh updates the second layer's weights and transfers the values between them (-1 to 1). The principles are chosen based on their degree of significance. Both values are changed as shown in Eq. (1.4) and the new cell state is formed.

$$m_t = \sigma\left(w_f[V_{t-1}, X_t] + b_m\right) \tag{1.2}$$

$$N_t = tanh\left(w_f[V_{t-1}, X_t] + b_n\right) \tag{1.3}$$

$$C_t = C_{t-1}f_t + N_t m_t \tag{1.4}$$

The output V_t is multiplied by the new $tanh(C_t)$ layer created in the final stage, which is based on the output of sigmoidal gates Q_t.

$$Q_t = \sigma\left(w_f[V_{t-1}, X_t] + b_q\right) \tag{1.5}$$

$$V_t = Q_t tanh(C_t) \tag{1.6}$$

The weights and biases for the output gates are w_q and b_q respectively.

1.6.1 Experimental results

The proposed system was evaluated using several instances during the experiments. Four various algorithms like K−Nearest Neighbor (KNN), Support Vector Machine (SVM), Random Forest (RF), and Naïve Bayes (NB) that have been used for DM diagnosis to compare the performance of the proposed classification model. Table 1.2 and Fig. 1.4 show the diagnosis accuracy of the proposed model and the existing methods. Various records like 2000, 4000, 6000, 8000, and 10,000 were considered for conducting five various experiments.

Table 1.2 Comparison of the accuracy of the proposed method with other machine learning.

Models	2000	4000	6000	8000	10,000
KNN	87%	90%	86%	89%	90%
SVM	73%	76%	75%	77%	80%
Random Forest	91%	92%	90%	94%	97%
Naïve Bayes	77%	80%	76%	84%	89%
Proposed model	**95%**	**97%**	**98%**	**94%**	**98%**

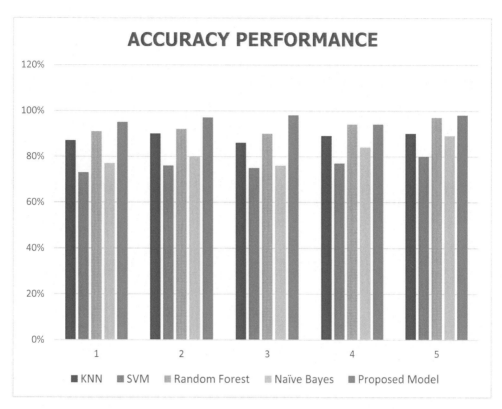

FIGURE 1.4

The accuracy performance of the proposed model.

Table 1.3 Comparison of the specificity of the proposed method with other machine learning.					
Models	**2000**	**4000**	**6000**	**8000**	**10,000**
KNN	85%	90%	87%	91%	90%
SVM	72%	75%	73%	78%	83%
Random Forest	93%	93%	92%	93%	95%
Naïve Bayes	76%	80%	77%	85%	90%
Proposed model	**94%**	**97%**	**99%**	**96%**	**99%**

From Table 1.2 and Fig. 1.4, the results of the accuracy show that the proposed LSTM performed better when compared with the existing methods with an accuracy of 98%, followed by RF with an accuracy of 97%, and the lowest of the five models is the SVM with an accuracy of 80%. This is because of the use of LSTM and the classification time constraints.

From Table 1.3 and Fig. 1.5, the results of the specificity show that the proposed LSTM performed better when compared with the existing methods with an accuracy of 99%, followed by RF

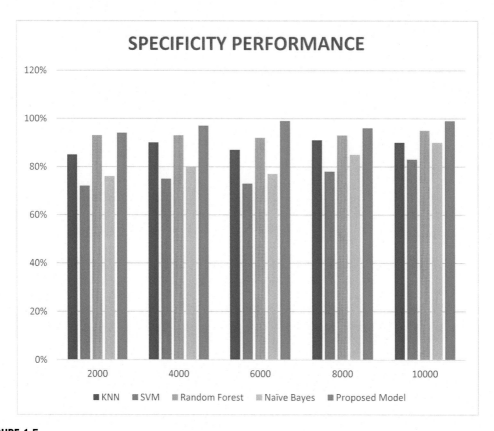

FIGURE 1.5

The specificity performance of the proposed model.

with an accuracy of 95%, and the lowest of the five models is the SVM with an accuracy of 83%. This is due to the introduction of edge computing to boost the cloud paradigm by removing time constraints during classification and reduce latency.

From Table 1.4 and Fig. 1.6, the results of the sensitivity show that the proposed LSTM performed better when compared with the existing methods with an accuracy of 96%, and the lowest of the five models is the SVM with an accuracy of 75%. This is due to the introduction of edge computing to boost the cloud paradigm by removing time constraints during classification and reduce latency.

1.7 Future prospects of edge computing for internet of medical things

The edge technology "Mobile Edge Computing (MEC)" created by European Telecommunications Standards Institute (ETSI) used 4G, 5G, and Radio Access Networks (RANs) as the main target

Table 1.4 Comparison of the sensitivity of the proposed method with other machine learning.

Models	2000	4000	6000	8000	10,000
KNN	83%	87%	84%	87%	87%
SVM	68%	72%	69%	74%	75%
Random Forest	87%	87%	85%	86%	91%
Naïve Bayes	72%	78%	73%	81%	88%
Proposed model	**89%**	**87%**	**95%**	**93%**	**96%**

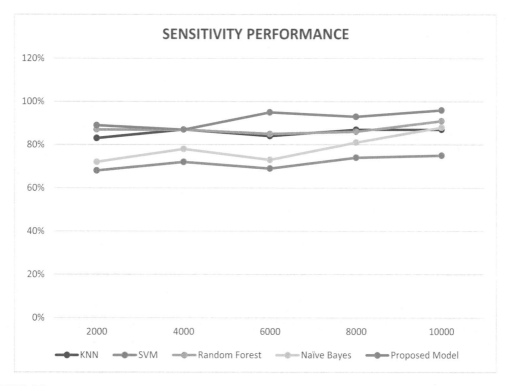

FIGURE 1.6

The sensitivity performance of the proposed model.

(Beck, Werner, Feld, & Schimper, 2014; Klas, 2015). MEC offers edge computing in the smart healthcare system at the processing resources base stations, and by recommending a collocation of storage. Initiated by Microsoft, the Micro Data Centers (MDCs) by expanding cloud data centers with small-sized versions of data centers (Avelar, 2015), hence, expand the cloud services close to the end-users. The cloudlet definition is very close to MDC as defined by The Carnegie Mellon

University (CMU), where the virtualized data center was very small, and in distributed fashion, serves users near the edge (Bilal, Khalid, Erbad, & Khan, 2018; Ren, Zhang, He, Zhang, & Li, 2019; Satyanarayanan, Bahl, Caceres, & Davies, 2009). These terms are also related to other similar objects and concepts like Nano-data centers (Valancius, Laoutaris, Massoulié, Diot, & Rodriguez, 2009).

Edge technologies can work together and cooperate, even though various edge technologies are defined independently (Borcoci, 2016). Taking into account the futuristic aspects of the IoMT-based structure and recent developments in technology collaboration, such as Heterogeneous Networks (HetNets) and interconnected Content Delivery networks (CDNs). The IoEs can be efficient and more effective using various edge technologies working together. The IDC estimated that 45% of the IoT-produced data would be processed, stored, and analyzed on the edge by the year 2019 (Klonoff, 2017; Mutlag, Abd Ghani, Arunkumar, Mohammed, & Mohd, 2019).

Edge computing is still in its infancy stages with no standardized concepts, designs, and proprieties. Edge technologies are characterized by different researchers from their viewpoints that are required for non-unvarying skills. This pattern was also established in cervical cancer (CC) before the National Institute of Science and Technology (NIST) standardization of the official concept of cloud computing in 2011 (Langmead & Nellore, 2018). In the relationship between edge technologies, IoT, and the cloud, the lack of a common description contributes to misconceptions. In the literature, where authors say that edge computing technologies can "move" or "replace" the cloud with fog or decentralize the cloud paradigm to edges, examples of such misconceptions are listed. Cloud computing is not enough for the processing and storage of IoT-based big data generated, thus these have to be a move to the edge of the network for the data processing (Naranjo et al., 2016). The innovations brought in by edge computing should be seen as a replacement for the cloud model, because this was just created as supplementary agents for cloud services, and to help in expanding cloud services for real-time applications are to meet (Sarkar & Misra, 2016). In big data analytics, the cloud is necessary for resource-intensive batch tasks and long-time processing of capture data using IoT-based devices. Similarly, various authors regard edge computing as microdata centers, and the understanding and perceiving of edge technologies are often confounding (Aazam & Huh, 2015), while others concentrate specifically on the concept of improving and equipping additional processing and storage capacities for networking components (Sarkar & Misra, 2016).

The use of edge computing in the processing of huge amounts of data that IoT devices produce would provide strategies for big data analysis at the edge of the network to provide real-time data processing. Edge computing has brought elastic on-demand for the processing of big data locally without necessarily send the capture data to the cloud, thus removed the problems of higher latency and bandwidth consumption. Big data collection, aggregation, and preprocessing can be handled by a mix of edge and cloud computing, minimizing data transport and cloud storage. For example, at the edge layer, local data can be collected and processed for an environmental surveillance system to provide timely input in emergency circumstances.

As multimedia demands more bandwidth, processing, and memory, it is a difficult task to manage such huge amounts in terms of decision making, processing, and storage (Aazam & Huh, 2016). Edge computing is intended to help reduce the total end-to-end use, delivery, efficient functioning, and processing of multimedia resources in such situations (Bilal & Erbad, 2017; Chen, Chen, et al., 2016; Chen, Zhang, et al., 2015). High costs are also incurred through multimedia distribution. CDNs, such as CloudFront, charge high prices when considering the provision of Tbps data. YouTube Live and Twitch have been reported to have reached 1 Tbps during peak hours in 2014 (Pires & Simon, 2015). The study conducted using Twitch trace analysis has been stated to have exceeded the distribution of 1.5 Tbps of video content to viewers worldwide (Chen, Jiang, et al., 2015). It must also be taken into account that any word improves Internet availability and data speeds, which means higher consumer access to data.

Also, the combination of cloud and edge computing in building a robust IoMT-based system will help in advancing a strong and fast processing system in a smart healthcare system. Also, the ultra-responsiveness and ultra-reliability need algorithms running in the device to guarantee touch applications. The edge system provided an enabling platform for running various tasks related to Tactile Internet applications in edges or cloud layers. There is a need for novel algorithms for task planning due to network traffic that may occur on the computing nodes with other variables where tasks are performed and delay sometimes will far surpass the 1 sm threshold. This is necessary to ensure that activities are carried out within the cumulative threshold and are not surpassed 1 Ms. To have optimal task scheduling, novel ML and artificial intelligence algorithms are required. To predict behavior and reactions, they may run exclusively in the edge stratum or even in a distributed manner through a cloud and/or edge strata. Edge computing helps to reduce the total network load during network connection in the IoMT-based platform, and this will help meet the latency requirement of 1 Ms. A various sophisticated algorithm, like neural network-based techniques and simple regression models, can be considered in modeling the system (Firouzi, Farahani, Barzegari, & Daneshmand, 2020; Sakr, Georganas, Zhao, & Shen, 2007).

1.8 Conclusions and future research directions

The emergence of cloud computing has touched almost all human life domains, especially smart healthcare system, which have greatly benefited from the paradigm. Cloud computing has brought a technological revolution needed by IoT-based services like high processing, storage capabilities, heterogeneity, and computation resources among others. Nevertheless, the various increase in IoMT-based devices enabled and application in the smart healthcare system in real-time response require technologies that will replace the traditional cloud paradigm with several challenges like computation, scalability, and latency. Edge computing can be applying to overcome the about problems. Edge computing technology with lower latency, high computational paradigm, and scalability will help IoMT-based system devices in real-time data collection and processing. The deployment of cloud data is far away from the network; this causes the response time delay in real-time. Moreover, the cloud may cause significant overhead on the backbone network to the user application due to the huge amount of big data sent to the cloud. Hence, the application of edge computing will bring the storage resources and computational closer to the end-user devices, thus reduce the burden on the cloud. The edge computing architecture is heterogeneous devices, and geographically distributed ubiquitously connected at the end of a network to provide collaboratively variable and flexible communication, storage services, and computation. Edge computing has various characteristics that place it in a better position when compared with cloud computing like low latency, real-time process, high level of the security model, and high response time. Therefore, this chapter presents a general review of edge computing technology in the smart healthcare system. The applicability of edge on smart healthcare systems is discussed, the challenges and the prospects of edge computing are elaborately discussed. The edge IoMT-based system computing presents better infrastructure by providing low latency, distributed processing, better security, fault tolerance, and good privacy when compared with the cloud IoMT-based infrastructure. The edge IoMT-based sufficiently produces various nodes, virtualized data centers, and edge device networks to connect the IoT-based devices to implement large storage and rich cloud computing. Edge computing offers millisecond to sub-second latency faster than real-time interaction, performs better in terms of low-latency applications, and supports multitenancy where cloud computing cannot have provided. In future work, the prospects of edge models will be extended in the smart healthcare system to provide a quick and real-time health diagnosis and monitoring for patients suffering from any diseases. Finally, edge computing is very necessary because the cloud's recorded healthcare data may be subject to different types of security risks.

References

Aazam, M., & Huh, E. N. (2015, March). Dynamic resource provisioning through fog micro datacenter. In *2015 IEEE international conference on pervasive computing and communication workshops (PerCom workshops)* (pp. 105–110). IEEE.

Aazam, M., & Huh, E. N. (2016). Fog computing: The cloud-IoT/IoE middleware paradigm. *IEEE Potentials, 35*(3), 40–44.

Abdulraheem, M., Awotunde, J. B., Jimoh, R. G., & Oladipo, I. D. (2021). An efficient lightweight cryptographic algorithm for IoT security. *Communications in Computer and Information Science, 1350,* 444–456.

Abdulsalam, Y., & Hossain, M. S. (2020). Covid-19 networking demand: An auction-based mechanism for automated selection of edge computing services. *IEEE Transactions on Network Science and Engineering.*

Adeniyi, E. A., Ogundokun, R. O., & Awotunde, J. B. (2021). *IoMT-based wearable body sensors network healthcare monitoring system. IoT in healthcare and ambient assisted living* (pp. 103–121). Singapore: Springer.

Albahri, A. S., Alwan, J. K., Taha, Z. K., Ismail, S. F., Hamid, R. A., Zaidan, A. A., ... Alsalem, M. A. (2020). IoT-based telemedicine for disease prevention and health promotion: State-of-the-art. *Journal of Network and Computer Applications, 173,* 102873.

Aldaej, A., & Tariq, U. (2018, April). IoT in 5G aeon: An inevitable fortuity of next-generation healthcare. In *2018 1st International conference on computer applications & information security (ICCAIS)* (pp. 1–4). IEEE.

Alhamid, M. F., Rawashdeh, M., Al Osman, H., Hossain, M. S., & El Saddik, A. (2015). Towards context-sensitive collaborative media recommender system. *Multimedia Tools and Applications, 74*(24), 11399–11428.

Al-Turjman, F., Nawaz, M. H., & Ulusar, U. D. (2020). Intelligence in the Internet of medical things era: A systematic review of current and future trends. *Computer Communications, 150,* 644–660.

Athavale, Y., & Krishnan, S. (2020). A telehealth system framework for assessing knee-joint conditions using vibroarthrographic signals. *Biomedical Signal Processing and Control, 55,* 101580.

Avelar, V. (2015). Practical options for deploying small server rooms and micro data centers. *Schneider Electric, 174,* White paper.

Awotunde, J. B., Bhoi, A. K., & Barsocchi, P. (2021). Hybrid cloud/fog environment for healthcare: An exploratory study, opportunities, challenges, and future prospects. *Hybrid Artificial Intelligence and IoT in Healthcare* (pp. 1−20). Singapore: Springer.

Awotunde, J. B., Jimoh, R. G., Oladipo, I. D., & Abdulraheem, M. (2021). Prediction of malaria fever using long-short-term memory and big data. *Communications in Computer and Information Science, 1350*, 41−53.

Awotunde, J. B., Matiluko, O. E., & Fatai, O. W. (2014). Medical diagnosis system using fuzzy logic. *African Journal of Computing & ICT, 7*(2), 99−106.

Ayo, F. E., Awotunde, J. B., Ogundokun, R. O., Folorunso, S. O., & Adekunle, A. O. (2020). A decision support system for multi-target disease diagnosis: A bioinformatics approach. *Heliyon, 6*(3), e03657.

Ayo, F. E., Ogundokun, R. O., Awotunde, J. B., Adebiyi, M. O., & Adeniyi, A. E. (2020, July). *Severe acne skin disease: A fuzzy-based method for diagnosis*. Lecture Notes in Computer Science (including subseries Lecture Notes in Artificial Intelligence and Lecture Notes in Bioinformatics), 12254 LNCS, pp. 320−334.

Baker, S. B., Xiang, W., & Atkinson, I. (2017). Internet of things for smart healthcare: Technologies, challenges, and opportunities. *IEEE Access, 5*, 26521−26544.

Beck, M. T., Werner, M., Feld, S., & Schimper, S. (2014, November). Mobile edge computing: A taxonomy. In *Proceedings of the sixth international conference on advances in future internet* (pp. 48−55). Citeseer.

Bhatia, M. (2020). Fog computing-inspired smart home framework for predictive veterinary healthcare. *Microprocessors and Microsystems, 78*, 103227.

Bilal, K., & Erbad, A. (2017, April). Impact of multiple video representations in live streaming: A cost, bandwidth, and QoE analysis. In *2017 IEEE international conference on cloud engineering (IC2E)* (pp. 88−94). IEEE.

Bilal, K., Khalid, O., Erbad, A., & Khan, S. U. (2018). Potentials, trends, and prospects in edge technologies: Fog, cloudlet, mobile edge, and micro data centers. *Computer Networks, 130*, 94−120.

Bonomi, F., Milito, R., Zhu, J., & Addepalli, S. (2012, August). Fog computing and its role in the internet of things. In *Proceedings of the first edition of the MCC workshop on mobile cloud computing* (pp. 13−16).

Borcoci, E. (2016, August). Fog computing, mobile edge computing, cloudlets-which one. In *SoftNet conference* (pp. 1−122).

Chen, F., Zhang, C., Wang, F., Liu, J., Wang, X., & Liu, Y. (2015). Cloud-assisted live streaming for crowd-sourced multimedia content. *IEEE Transactions on Multimedia, 17*(9), 1471−1483.

Chen, N., Chen, Y., You, Y., Ling, H., Liang, P., & Zimmermann, R. (2016, April). Dynamic urban surveillance video stream processing using fog computing. In *2016 IEEE second international conference on multimedia big data (BigMM)* (pp. 105−112). IEEE.

Chen, S. C. I., Hu, R., & McAdam, R. (2020). Smart, remote, and targeted health care facilitation through connected health: Qualitative study. *Journal of Medical Internet Research, 22*(4), e14201.

Chen, S. W., Chiang, D. L., Liu, C. H., Chen, T. S., Lai, F., Wang, H., & Wei, W. (2016). Confidentiality protection of digital health records in cloud computing. *Journal of Medical Systems, 40*(5), 124.

Chen, Z., Jiang, L., Hu, W., Ha, K., Amos, B., Pillai, P., ... Satyanarayanan, M. (2015, May). Early implementation experience with wearable cognitive assistance applications. In *Proceedings of the 2015 workshop on wearable systems and applications* (pp. 33−38).

Chui, K. T., Liu, R. W., Lytras, M. D., & Zhao, M. (2019). Big data and IoT solution for patient behaviour monitoring. *Behaviour & Information Technology, 38*(9), 940−949.

Devarajan, M., Subramaniyaswamy, V., Vijayakumar, V., & Ravi, L. (2019). Fog-assisted personalized healthcare-support system for remote patients with diabetes. *Journal of Ambient Intelligence and Humanized Computing, 10*(10), 3747−3760.

Dong, P., Ning, Z., Obaidat, M. S., Jiang, X., Guo, Y., Hu, X., ... Sadoun, B. (2020). Edge computing-based healthcare systems: Enabling decentralized health monitoring in Internet of Medical Things. *IEEE Network*.

Edemacu, K., Park, H. K., Jang, B., & Kim, J. W. (2019). Privacy provision in collaborative ehealth with attribute-based encryption: Survey, challenges, and future directions. *IEEE Access*, *7*, 89614–89636.

Fernandez, F., & Pallis, G. C. (2014, November). Opportunities and challenges of the Internet of Things for healthcare: Systems engineering perspective. In *2014 4th International conference on wireless mobile communication and healthcare-transforming healthcare through innovations in mobile and wireless technologies (MOBIHEALTH)* (pp. 263–266). IEEE.

Firouzi, F., Farahani, B., Barzegari, M., & Daneshmand, M. (2020). AI-driven data monetization: The other face of data in IoT-based smart and connected health. *IEEE Internet of Things Journal*.

Firouzi, F., Rahmani, A. M., Mankodiya, K., Badaroglu, M., Merrett, G. V., Wong, P., & Farahani, B. (2018). Internet-of-Things and big data for smarter healthcare: From device to architecture, applications, and analytics.

Ghanavati, S., Abawajy, J. H., Izadi, D., & Alelaiwi, A. A. (2017). Cloud-assisted IoT-based health status monitoring framework. *Cluster Computing*, *20*(2), 1843–1853.

Ghoneim, A., Muhammad, G., Amin, S. U., & Gupta, B. (2018). Medical image forgery detection for smart healthcare. *IEEE Communications Magazine*, *56*(4), 33–37.

Gu, L., Zeng, D., Guo, S., Barnawi, A., & Xiang, Y. (2015). Cost-efficient resource management in fog computing supported medical cyber-physical systems. *IEEE Transactions on Emerging Topics in Computing*, *5*(1), 108–119.

Guy, J. S. (2019). Digital technology, digital culture, and the metric/nonmetric distinction. *Technological Forecasting and Social Change*, *145*, 55–61.

Hossain, M. S., & Muhammad, G. (2017). Emotion-aware connected healthcare big data towards 5G. *IEEE Internet of Things Journal*, *5*(4), 2399–2406.

Hossain, M. S., Rahman, M. A., & Muhammad, G. (2017). Cyber-physical cloud-oriented multi-sensory smart home framework for elderly people: An energy efficiency perspective. *Journal of Parallel and Distributed Computing*, *103*, 11–21.

Hussain, F., Hussain, R., Hassan, S. A., & Hossain, E. (2020). Machine learning in IoT security: Current solutions and future challenges. *IEEE Communications Surveys & Tutorials*.

Janet, B., & Raj, P. (2019). *Smart city applications: The smart leverage of the internet of things (IoT) paradigm. Novel practices and trends in grid and cloud computing* (pp. 274–305). IGI Global.

Jin, Z., & Chen, Y. (2015). Telemedicine in the cloud era: Prospects and challenges. *IEEE Pervasive Computing*, *14*(1), 54–61.

Jones, R. W., & Katzis, K. (2018, April). 5G and wireless body area networks. In *2018 IEEE wireless communications and networking conference workshops (WCNCW)* (pp. 373–378). IEEE.

Klas, G. I. (2015). *Fog computing and mobile edge cloud gain momentum open fog consortium, ETSI mec, and cloudlets*. Google Scholar.

Klonoff, D. C. (2017). Fog computing and edge computing architectures for processing data from diabetes devices connected to the medical internet of things. *Journal of Diabetes Science and Technology*, *11*(4), 647–652.

Kumar, P. M., Lokesh, S., Varatharajan, R., Babu, G. C., & Parthasarathy, P. (2018). Cloud and IoT-based disease prediction and diagnosis system for healthcare using Fuzzy neural classifier. *Future Generation Computer Systems*, *86*, 527–534.

Kumari, A., Tanwar, S., Tyagi, S., & Kumar, N. (2018a). Fog computing for Healthcare 4.0 environment: Opportunities and challenges. *Computers & Electrical Engineering*, *72*, 1–13.

Kumari, A., Tanwar, S., Tyagi, S., & Kumar, N. (2018b). Verification and validation techniques for streaming big data analytics in internet of things environment. *IET Networks*, *8*(2), 92–100.

Langmead, B., & Nellore, A. (2018). Cloud computing for genomic data analysis and collaboration. *Nature Reviews Genetics*, *19*(4), 208.

Lin, K., Song, J., Luo, J., Ji, W., Hossain, M. S., & Ghoneim, A. (2016). Green video transmission in the mobile cloud networks. *IEEE Transactions on Circuits and Systems for Video Technology, 27*(1), 159−169.

Lin, P., Song, Q., & Jamalipour, A. (2019). Multidimensional cooperative caching in CoMP-integrated ultra-dense cellular networks. *IEEE Transactions on Wireless Communications, 19*(3), 1977−1989.

Lu, L., Zhang, J., Xie, Y., Gao, F., Xu, S., Wu, X., & Ye, Z. (2020). Wearable health devices in health care: Narrative systematic review. *JMIR mHealth and uHealth, 8*(11), e18907.

Magsi, H., Sodhro, A. H., Chachar, F. A., Abro, S. A. K., Sodhro, G. H., & Pirbhulal, S. (2018, March). Evolution of 5G in Internet of medical things. In *2018 International conference on computing, mathematics, and engineering technologies (iCoMET)* (pp. 1−7). IEEE.

Marston, H. R., & van Hoof, J. (2019). Who doesn't think about technology when designing urban environments for older people?" A case study approach to a proposed extension of the WHO's age-friendly cities model. *International Journal of Environmental Research and Public Health, 16*(19), 3525.

Medaglia, C. M., & Serbanati, A. (2010). *An overview of privacy and security issues on the internet of things. The internet of things* (pp. 389−395). New York: Springer.

Meskó, B., Drobni, Z., Bényei, É., Gergely, B., & Győrffy, Z. (2017). Digital health is a cultural transformation of traditional healthcare. *Mhealth, 3*.

Mouradian, C., Naboulsi, D., Yangui, S., Glitho, R. H., Morrow, M. J., & Polakos, P. A. (2017). A comprehensive survey on fog computing: State-of-the-art and research challenges. *IEEE Communications Surveys & Tutorials, 20*(1), 416−464.

Muhammad, G., Alhamid, M. F., & Long, X. (2019). Computing and processing on the edge: Smart pathology detection for connected healthcare. *IEEE Network, 33*(6), 44−49.

Muhammed, T., Mehmood, R., Albeshri, A., & Katib, I. (2018). UbeHealth: A personalized ubiquitous cloud and edge-enabled networked healthcare system for smart cities. *IEEE Access, 6*, 32258−32285.

Mutlag, A. A., Abd Ghani, M. K., Arunkumar, N. A., Mohammed, M. A., & Mohd, O. (2019). Enabling technologies for fog computing in healthcare IoT systems. *Future Generation Computer Systems, 90*, 62−78.

Nandyala, C. S., & Kim, H. K. (2016). From cloud to fog and IoT-based real-time U-healthcare monitoring for smart homes and hospitals. *International Journal of Smart Home, 10*(2), 187−196.

Naranjo, P. G. V., Shojafar, M., Vaca-Cardenas, L., Canali, C., Lancellotti, R., & Baccarelli, E. (2016, September). Big data over SmartGrid-a fog computing perspective. In *Proceedings of the 24th international conference on software, telecommunications and computer networks (SoftCOM 2016)* (pp. 22−24). Split, Croatia.

Ning, Z., Dong, P., Wang, X., Hu, X., Guo, L., Hu, B., … Kwok, R. Y. (2020). Mobile edge computing enabled 5G health monitoring for Internet of medical things: A decentralized game-theoretic approach. *IEEE Journal on Selected Areas in Communications*, 1−16.

Ning, Z., Wang, X., & Huang, J. (2018). Mobile edge computing-enabled 5G vehicular networks: Toward the integration of communication and computing. *IEEE Vehicular Technology Magazine, 14*(1), 54−61.

Ogundokun, R. O., Sadiku, P. O., Misra, S., Ogundokun, O. E., Awotunde, J. B., & Jaglan, V. (2021). Diagnosis of long sightedness using neural network and decision tree algorithms. *Journal of Physics: Conference Series, 1767*(1), 012021.

Oladele, T. O., Ogundokun, R. O., Awotunde, J. B., Adebiyi, M. O., & Adeniyi, J. K. (2020, July). *Diagmal: A malaria coactive neuro-fuzzy expert system.* Lecture Notes in Computer Science (including subseries Lecture Notes in Artificial Intelligence and Lecture Notes in Bioinformatics), 12254 LNCS, pp. 428−441.

Oladipo, I. D., Babatunde, A. O., Awotunde, J. B., & Abdulraheem, M. (2021). An improved hybridization in the diagnosis of diabetes mellitus using selected computational intelligence. *Communications in Computer and Information Science, 1350*, 272−285.

Pace, P., Aloi, G., Gravina, R., Caliciuri, G., Fortino, G., & Liotta, A. (2018). An edge-based architecture to support efficient applications for the healthcare industry 4.0. *IEEE Transactions on Industrial Informatics*, *15*(1), 481−489.

Pan, J., & McElhannon, J. (2017). Future edge cloud and edge computing for the internet of things applications. *IEEE Internet of Things Journal*, *5*(1), 439−449.

Parsa, M., Panda, P., Sen, S., & Roy, K. (2017, July). Staged inference using conditional deep learning for energy-efficient real-time smart diagnosis. In *2017 39th Annual international conference of the IEEE engineering in medicine and biology society (EMBC)* (pp. 78−81). IEEE.

Pazowski, P. (2015). Green computing: Latest practices and technologies for ICT sustainability. In *Managing intellectual capital and innovation for sustainable and inclusive society: Managing intellectual capital and innovation; Proceedings of the MakeLearn and TIIM joint international conference 2015* (pp. 1853−1860). ToKnowPress.

Petri, I., Kubicki, S., Rezgui, Y., Guerriero, A., & Li, H. (2017). Optimizing energy efficiency in operating built environment assets through building information modeling: A case study. *Energies*, *10*(8), 1167.

Pham, M., Mengistu, Y., Do, H., & Sheng, W. (2018). Delivering home healthcare through a cloud-based smart home environment (CoSHE). *Future Generation Computer Systems*, *81*, 129−140.

Pires, K., & Simon, G. (2015, March). YouTube live and twitch: A tour of user-generated live streaming systems. In *Proceedings of the 6th ACM multimedia systems conference* (pp. 225−230).

Pustokhina, I. V., Pustokhin, D. A., Gupta, D., Khanna, A., Shankar, K., & Nguyen, G. N. (2020). An effective training scheme for a deep neural network in edge computing enabled Internet of medical things (IoMT) systems. *IEEE Access*, *8*, 107112−107123.

Puthal, D., Obaidat, M. S., Nanda, P., Prasad, M., Mohanty, S. P., & Zomaya, A. Y. (2018). Secure and sustainable load balancing of edge data centers in fog computing. *IEEE Communications Magazine*, *56*(5), 60−65.

Rahman, M. A., Hossain, M. S., Islam, M. S., Alrajeh, N. A., & Muhammad, G. (2020). Secure and provenance enhanced internet of health things framework: A blockchain managed federated learning approach. *IEEE Access*, *8*, 205071−205087.

Regola, N., & Chawla, N. V. (2013). Storing and using health data in a virtual private cloud. *Journal of Medical Internet Research*, *15*(3), e63.

Rehman, H. U., Khan, A., & Habib, U. (2020). Fog computing for bioinformatics applications. *Fog Computing: Theory and Practice*, 529−546.

Ren, J., Zhang, D., He, S., Zhang, Y., & Li, T. (2019). A survey on end-edge-cloud orchestrated network computing paradigms: Transparent computing, mobile edge computing, fog computing, and cloudlet. *ACM Computing Surveys (CSUR)*, *52*(6), 1−36.

Rodrigues, J. J., Segundo, D. B. D. R., Junqueira, H. A., Sabino, M. H., Prince, R. M., Al-Muhtadi, J., & De Albuquerque, V. H. C. (2018). Enabling technologies for the internet of health things. *IEEE Access*, *6*, 13129−13141.

S Rubí, J. N., & L Gondim, P. R. (2019). IoMT platform for pervasive healthcare data aggregation, processing, and sharing based on OneM2M and OpenEHR. *Sensors*, *19*(19), 4283.

Saad, M. (2018). Fog computing and its role in the internet of things: Concept, security and privacy issues. *International Journal of Computer Applications*, *975*, 8887.

Saeed, N., Bader, A., Al-Naffouri, T. Y., & Alouini, M. S. (2020). When wireless communication faces COVID-19: Combating the pandemic and saving the economy. arXiv preprint arXiv:2005.06637.

Sakr, N., Georganas, N. D., Zhao, J., & Shen, X. (2007, July). Motion and force prediction in haptic media. In *2007 IEEE international conference on multimedia and expo* (pp. 2242−2245). IEEE.

Sarkar, S., & Misra, S. (2016). Theoretical modeling of fog computing: A green computing paradigm to support IoT applications. *IET Networks*, *5*(2), 23−29.

Satyanarayanan, M., Bahl, P., Caceres, R., & Davies, N. (2009). The case for VM-based cloudlets in mobile computing. *IEEE Pervasive Computing, 8*(4), 14−23.

Schweitzer, E. J. (2012). Reconciliation of the cloud computing model with US federal electronic health record regulations. *Journal of the American Medical Informatics Association, 19*(2), 161−165.

Sodhro, A. H., & Shah, M. A. (2017, April). Role of 5G in medical health. In *2017 International conference on innovations in electrical engineering and computational technologies (ICIEECT)* (pp. 1−5). IEEE.

Song, M. L., Fisher, R., Wang, J. L., & Cui, L. B. (2018). Environmental performance evaluation with big data: Theories and methods. *Annals of Operations Research, 270*(1−2), 459−472.

Sundwall, D. N., Munger, M. A., Tak, C. R., Walsh, M., & Feehan, M. (2020). Lifetime prevalence and correlates of patient-perceived medical errors experienced in the US ambulatory setting: A population-based study. *Health Equity, 4*(1), 430−437.

Sust, P. P., Solans, O., Fajardo, J. C., Peralta, M. M., Rodenas, P., Gabaldà, J., ... Monfa, R. R. (2020). Turning the crisis into an opportunity: Digital health strategies deployed during the COVID-19 outbreak. *JMIR Public Health and Surveillance, 6*(2), e19106.

Syed, L., Jabeen, S., Manimala, S., & Alsaeedi, A. (2019). Smart healthcare framework for ambient assisted living using IoMT and big data analytics techniques. *Future Generation Computer Systems, 101*, 136−151.

Tripathi, G., Ahad, M. A., & Paiva, S. (2020). *S2HS-A blockchain-based approach for the smart healthcare system,* . *Healthcare* (8, No. 1, p. 100391). Elsevier.

Tuli, S., Basumatary, N., Gill, S. S., Kahani, M., Arya, R. C., Wander, G. S., & Buyya, R. (2020). Healthfog: An ensemble deep learning-based smart healthcare system for automatic diagnosis of heart diseases in integrated IoT and fog computing environments. *Future Generation Computer Systems, 104*, 187−200.

Valancius, V., Laoutaris, N., Massoulié, L., Diot, C., & Rodriguez, P. (2009, December). Greening the internet with nano data centers. In *Proceedings of the 5th international conference on emerging networking experiments and technologies* (pp. 37−48).

Vizitiu, A., Niţă, C. I., Puiu, A., Suciu, C., & Itu, L. M. (2020). Applying deep neural networks over homomorphic encrypted medical data. *Computational and Mathematical Methods in Medicine, 2020*.

Wen, F., He, T., Liu, H., Chen, H. Y., Zhang, T., & Lee, C. (2020). Advances in chemical sensing technology for enabling the next-generation self-sustainable integrated wearable system in the IoT era. *Nano Energy*, 105155.

Xu, X., Liu, Q., Luo, Y., Peng, K., Zhang, X., Meng, S., & Qi, L. (2019). A computation offloading method over big data for IoT-enabled cloud-edge computing. *Future Generation Computer Systems, 95*, 522−533.

Yi, S., Li, C., & Li, Q. (2015, June). A survey of fog computing: Concepts, applications, and issues. In *Proceedings of the 2015 workshop on mobile big data* (pp. 37−42).

Zafar, S., Khan, S., Iftekhar, N., & Biswas, S. (2020). Consociate healthcare system through biometric based internet of medical things (BBIOMT) approach. *EAI Endorsed Transactions on Smart Cities, 4*(10).

Physical layer architecture of 5G enabled IoT/IoMT system

2

Anh-Tu Le[1], Munyaradzi Munochiveyi[2] and Samarendra Nath Sur[3]

*[1]Faculty of Electronics Technology, Industrial University of Ho Chi Minh City, Ho Chi Minh City, Vietnam
[2]Department of Electrical and Electronics Engineering, University of Zimbabwe, Harare, Zimbabwe [3]Department of
Electronics and Communication Engineering, Sikkim Manipal Institute of Technology, Sikkim Manipal University,
Majitar, Rangpo, Sikkim, India*

2.1 Architecture of IoT/IoMT system

The healthcare industry has changed rapidly as life expectancy has increased in step with rapid development and wealth. This has not come without its own unique challenges as more cases of chronic disease have also escalated, putting more pressure on already resource strained healthcare systems. This has seen the rise of telemedicine, as hospitals try to decongest and find innovative ways to deliver care without increasing hospitals and hospital beds. To reduce patient overload, telemedicine has become an essential innovation in the healthcare field. However, telemedicine is heterogeneous in design as it is designed to monitor a single disease at a time, for example, remote stroke monitoring and management, etc. This design feature ends up being a design flaw as it does not reduce the number of patients as diseases increase. Therefore there is need for a solution that is generalizable and scalable. Internet-of-Medical Things (IoMT) is deemed as a promising solution to better tackle this unique healthcare challenge (Gatouillat, Badr, Massot, & Sejdić, 2018).

The rapid advancements in microelectromechanical systems (MEMS) and machine-to-machine (M2M) communications has resulted in new Internet-of-Things (IoT) concepts applicable to many networking applications. IoMT for healthcare systems are among the applications that have been introduced with the emergence of IoT. IoMT is a new branch of the family of healthcare IoT devices that allow remote monitoring of chronic diseased patients. Therefore quick diagnostic results can be provided which can save patient's lives in emergencies (Ghubaish et al., 2021). According to (Ghubaish et al., 2021), IoMTs can be classified into two categories:

1. Implanted medical things (IMTs): A device that is implanted into the human body to support or replace a biological organ is defined as an IMT. Pacemakers, cochlear implants, deep brain stimulators, etc., are among some of the commonly used IMTs as shown in Fig. 2.1. Due to their location in the human body, IMTs are mostly designed to be very small and have very long battery life. Therefore energy consumption, is an important consideration, if the IMT is intended to stay inside the human body for a very long time (Ghubaish et al., 2021).
2. Wearable Internet-of-Things (WIoT): These are devices that monitor individuals biometrics as they go about their daily routine, for example, smartwatches, electrocardiogram (ECG)

5G IoT and Edge Computing for Smart Healthcare. DOI: https://doi.org/10.1016/B978-0-323-90548-0.00009-7

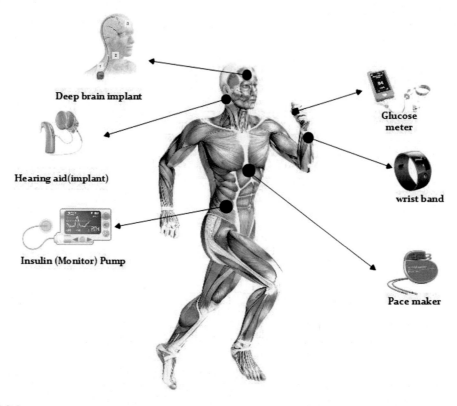

Deep brain implant

Hearing aid(implant)

Insulin (Monitor) Pump

Glucose meter

wrist band

Pace maker

FIGURE 2.1

IoMTs and their placement in and on the human body.

monitors, and blood pressure monitors as shown in Fig. 2.1. Smartwatches are one of the most popular noncritical form of monitoring for heart rate and movement. However, these devices are limited by sensor inaccuracy and poor battery life; hence, are not likely to replace IMTs from monitoring critical conditions (Ghubaish et al., 2021).

Many researchers have studied and proposed different IoT architectures such as machine-to-machine (M2M), web-of-things, autonomous and sensor based architecture. M2M architecture is the most widely used IoT architecture, and IoMT is the prevalent representation in the field of medical IoT devices (Sun, Jiang, Ren, & Guo, 2020). The current IoMT systems can be categorized into four layers, as shown in Fig. 2.2. The sensor layer starts with the collection of biometric data by IMTs sensors, then the data is transmitted over the physical link layer to a gateway host which handles some of sensor data preprocessing and simple analytics. The processed sensor data is then transmitted to the network layer via the Internet. The network layer is responsible for storage, analysis and secure access. Finally, the visualization layer is where all the data is analyzed by a physician. The patient can also visualize their health status via this layer. This is better than having to constantly carry different sets of radiology scans and blood tests from one physician to another.

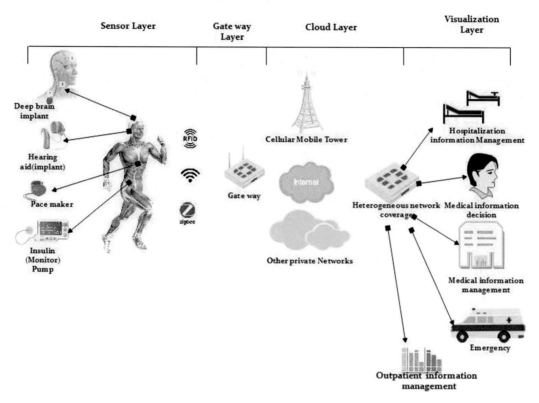

FIGURE 2.2

Overview of IoMT architecture.

Most IoMTs share the same architecture as seen in Fig. 2.2 (Ghubaish et al., 2021). In the following subsections, we zoom into the individual layers in greater detail.

2.1.1 Sensor layer

This layer is composed of wearables and small implantable sensors that collect the patient's health data. The biometric data is transmitted to the next layer via short-distance wireless technology, such as Wi-Fi, ZigBee, Bluetooth, etc. (Ghubaish et al., 2021; Sun et al., 2020).

2.1.2 Gateway layer

Due to the small form factor of IoMTs, processing and storage is not possible; therefore the data are transmitted as unprocessed to the gateway layer. This layer consists of devices that are more powerful in general than the IoMT sensors such as the patient's smartphone or smart watch, dedicated access points (APs), and microcontrollers. These devices can also serve as edge nodes in

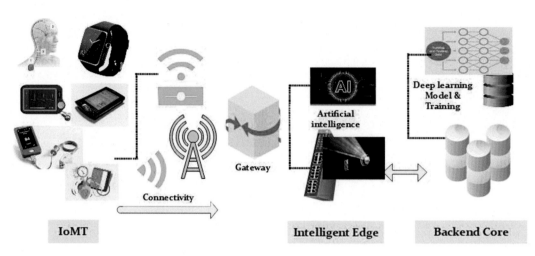

FIGURE 2.3

Edge computing integrated with AI.

edge computing. These devices can perform some computationally light preprocessing operations, such as data storage, and simple processing using machine learning (ML) or deep learning (DL) to gain insight into hidden data patterns as well as diagnose and alert critical health conditions as seen in in Fig. 2.3 (Amin & Hossain, 2021). Thereafter, the data is transferred to the cloud via the Internet for intensive processing and classification (Ghubaish et al., 2021; Sun et al., 2020).

2.1.3 Network layer

Sophisticated data storage, big data analysis, and secure access is carried out in the network/cloud layer. Advanced ML and DL are employed to analyze the streaming data from the IoMT sensors. The analysis reveals any changes in the patient's health and presents them to the health practitioner or patient for immediate action. However, cloud computing alone is not adequate in copying with the large volumes of uploaded medical data to the cloud, as this results in huge delays due to bandwidth bottlenecks. Such a situation is undesirable in healthcare monitoring, where visualization and latency must meet real-time constraints to prevent delays in addressing medical emergencies. Moreover, the data transmission leads to high energy consumption. Several works have studied solutions to expand the ability of cloud IoMT. Authors in (Limaye & Adegbija, 2018) investigate the impact of moving the computation closer to the sensor layer by introducing edge computing in the gateway layer as seen in Fig. 2.3. Furthermore, (Gatouillat et al., 2018) identifies the security of patient health data as very essential, the authors discuss possible solutions from the cyber-physical community to address this challenge. Also, (Egala, Pradhan, Badarla, & Mohanty, 2021; Garg et al., 2020) propose Blockchain technology as seen in Fig. 2.4 as a potential solution to securing patient data in the cloud. In (Ding et al., 2021), the authors propose securing medical images by using DL-based encryption and decryption techniques. In addition, (Awan et al., 2020),

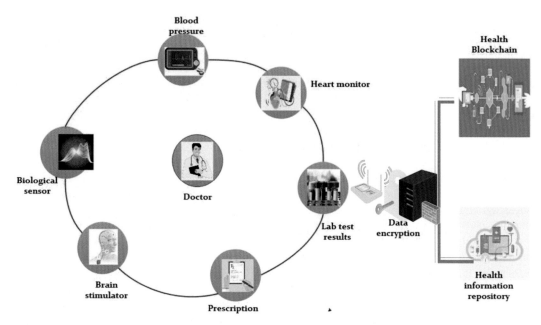

FIGURE 2.4

Blockchain secured cloud layer.

the authors propose utilizing neural networks to predict and eliminate malicious and compromised devices in the two-way communication between patient and physician.

2.1.4 Visualization layer

This layer also known as the application layer, is where the data is accessed by the physician and patients to track their health. The physician can make recommendations and upload actions to be taken by the patient based on the patient's health conditions. These actions can include drug prescriptions or referrals to other specialized physicians if the disease needs specialized care (Ghubaish et al., 2021).

2.2 Consideration of uplink healthcare IoT system relying on NOMA

2.2.1 Introduction

One of the most promising technology to enhance massive spectral efficiency is nonorthogonal multiple access (NOMA) since it can serve multiple users using similar resources at a single point of time. Compared to the traditional orthogonal multiple Access technique (OMA), which utilizes frequency, time, and code division multiple access techniques to communicate. NOMA transmits the superimposed signals utilizing similar resources as frequency, time and code by differing the respective power coefficients of each user (Baig, Ali, Asif, Khan, & Mumtaz, 2019; Li et al., 2020; Qi, Feng, Chen, & Wang, 2017; Wan, Wen, Ji, Yu, & Chen, 2018; Zheng, Wang, Wen, & Chen, 2017). NOMA utilizes the Successive Interference Cancellation (SIC) procedure to detect the original signal at the receiver since the transmitted signals are superimposed. Channels with a higher gain are less likely to be restricted by noise whereas, channels with lower gain are more likely to be restricted by noise. Studies on NOMA have reportedly shown that the reduction in transmission delay will pointedly attain massive user connectivity, and user waiting times can be reduced (Chen, Cai, Cheng, Yang, & Jin, 2017; Ding & Poor, 2016; Ding et al., 2017; Do, Le, & Lee, 2020; Do, Nguyen, Jameel, Jäntti, & Ansari, 2020; Do, Le, & Afghah, 2020; Islam, Avazov, Dobre, & Kwak, 2017; Islam, Zeng, Dobre, & Kwak, 2018; Wang, Gao, Jin, Lin, & Li, 2018; Yang, Wang, Ng, & Lee, 2017). Authors in (Islam et al., 2017) have provided a survey study on NOMA in 5G communications with the view of investigating capability, power allocation tactics, user fairness and user-pairing systems. The authors in (Islam et al., 2018) have discussed resource allocation methods for downlink NOMA concerning optimizing power allocation, low complexity, and security resource allocation. Authors in (Ding & Poor, 2016; Wang et al., 2018) have utilized massive multiple-input multiple-output (MIMO) and communication methods to replace traditional NOMA antenna which are massive in number. Whereas, the authors in (Yang et al., 2017) have considered checking the secrecy performance of NOMA when utilizing single antenna and multiple antennas, this is shown to perform better against the traditional OMA scheme. The authors in (Chen et al., 2017) discuss downlink MIMO NOMA system with the intention to study resource performance allocation, and the authors also design a NOMA scheme that can attain approximate optimal sum-rate performance in both perfect and imperfect channel state information (CSI) scenario.

The different properties of 5G communication systems require studying NOMA systems in the presence of device-to-device (D2D) communications, in which legitimate users are allowed to communicate directly without the requirement of a base station (BS) (Boccardi, Heath, Lozano, Marzetta, & Popovski, 2014; Lei, Zhong, Lin, & Shen, 2012). Authors in (Asadi, Wang, & Mancuso, 2014) have explained that D2D users can reuse the spectrum band by facing the mutual interference between the cellular and D2D links. Combining D2D with cellular systems will introduce interference to the broadcasting channels. Authors in (Do, Nguyen et al., 2020; Liuand & Erkip, 2016) have mentioned that the performance of SIC in this scenario should be studied more as the capacity-achieving schemes are not perfectly known. The authors in (Zhao, Liu, Chai, Chen, & Elkashlan, 2017) have studied the utilization of D2D in NOMA communications, where the D2D links are employed with NOMA protocol for data transmissions.

Motivated by the recent study (Zhao et al., 2017), this book chapter presents the IoT based healthcare system which contains several healthcare devices, relays and AP. We consider the analytical expression of outage probability and ergodic capacity for uplink (UL) IoT system.

2.2.2 System model

In this paper, we consider UL NOMA, which consists of an AP (A) and two IoT devices $D_i (_i \in \{1, 2\})$ to provide health care, as in Fig. 2.5. In addition, we assume the channel between D_i and A is h_i and follow Nakagami-m fading channel, and d_i is the distance between D_i and A (Liu, Ding, Elkashlan, & Poor, 2016). Moreover, the perfect channel state information (CSI) is available for signal detection.

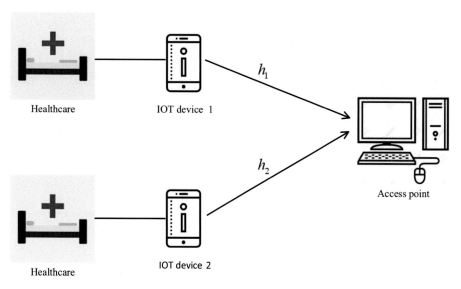

FIGURE 2.5

System model.

Here, the IoT devices D_i send the UL signal to A. The received UL signal at A is given by

$$r_A = \frac{\sqrt{P_1}h_1x_1}{\sqrt{1+d_1^\alpha}} + \frac{\sqrt{P_2}h_2x_2}{\sqrt{1+d_2^\alpha}} + n_A \tag{2.1}$$

where x_i and P_i are the signal and transmit power of D_i, respectively, α is the path-loss exponent and n_A $CN(0, N_0)$ is the Additive White Gaussian Noise (AWGN). Follow NOMA scheme (Zaidi, Hasan, & Gui, 2018a), the signal to interference and noise ratio (SINR) is detected as the signal x_1 at A is given by

$$
\begin{aligned}
\Gamma_{U\ L,x_1} &= \frac{P_1|h_1|^2(1+d_2^\alpha)}{(1+d_1^\alpha)P_2|h_2|^2 + (1+d_1^\alpha)(1+d_2^\alpha)N_0} \\
&= \frac{\eta|h_1|^2(1+d_2^\alpha)}{(1+d_1^\alpha)\eta|h_2|^2 + (1+d_1^\alpha)(1+d_2^\alpha)}
\end{aligned} \tag{2.2}
$$

where $\eta = \frac{P_1}{N_0} = \frac{P_2}{N_0}$. Next apply SIC, the SINR at A when detected the own signal x_2 is given by

$$\Gamma_{U\ L,x_2} = \frac{P_2|h_2|^2}{(1+d_2^\alpha)N_0} = \frac{\eta|h_2|^2}{1+d_2^\alpha} \tag{2.3}$$

2.2.3 Outage probability for UL NOMA

In this section, we derive the expression close-form outage probability UL of the signal x_i at A. For more insight, the asymptote of outage probability is expressed. The probability density function of

h_i is given as (Do, Le et al., 2020)

$$f_{|h_i|^2} = \left(\frac{m_i}{\lambda_i}\right)^{m_i} \frac{x^{m_i-1}}{\Gamma(m_i)} e^{-\frac{m_i}{\lambda_i}x} \tag{2.4}$$

where m is the fading severity and λ_i is the mean power.

2.2.3.1 Outage probability of x_1

The outage probability of x_1 is given by (Zaidi et al., 2018a)

$$P_{x_1} = 1 - \Pr(\Gamma_{U\ L,x_1} > \varepsilon_1) \tag{2.5}$$

where $\varepsilon_1 = 2^{R_i} - 1$ is the threshold SINR and R_i is the target rate.

Lemma 1: The close-form of x_1 can be expressed by

$$P_{x_1} = 1 - \sum_{k=0}^{m_1-1} \sum_{n=0}^{k} \binom{k}{n} \frac{(m_2+n-1)! e^{-\frac{m_1\varepsilon_1(1+d_1^\alpha)}{\lambda_1\eta}}}{k!\Gamma(m_2)}$$
$$\times \left(\frac{m_1\varepsilon_1(1+d_1^\alpha)}{\eta\lambda_1}\right)^k \left(\frac{\eta\lambda_2}{m_2(1+d_2^\alpha)}\right)^n \left(1+\frac{m_1\lambda_2\varepsilon_1(1+d_1^\alpha)}{\lambda_1 m_2(1+d_2^\alpha)}\right)^{-m_2-n} \tag{2.6}$$

Proof: With help from (2.2), (2.6) is calculated by

$$P_{x_1} = 1 - \Pr\left(\frac{\eta|h_1|^2(1+d_2^\alpha)}{(1+d_1^\alpha)\eta|h_2|^2 + (1+d_1^\alpha)(1+d_2^\alpha)} > \varepsilon_1\right)$$
$$= 1 - \Pr\left(|h_1|^2 > \frac{\varepsilon_1(1+d_1^\alpha)|h_2|^2}{(1+d_2^\alpha)} + \frac{\varepsilon_1(1+d_1^\alpha)}{\eta}\right) \tag{2.7}$$
$$= 1 - \int_0^\infty f_{|h_2|^2}(x) \int_{\frac{\varepsilon_1(1+d_1^\alpha)}{(1+d_2^\alpha)}x + \frac{\varepsilon_1(1+d_1^\alpha)}{\eta}}^\infty f_{|h_1|^2}(y)dydx$$

Substituting (2.4) into (2.7), we have

$$P_{x_1} = 1 - \left(\frac{m_1}{\lambda_1}\right)^{m_1}\left(\frac{m_2}{\lambda_2}\right)^{m_2} \frac{1}{\Gamma(m_1)\Gamma(m_2)}$$
$$\times \int_0^\infty x^{m_2-1} e^{-\frac{m_1}{\lambda_1}x} \int_{\frac{\varepsilon_1(1+d_1^\alpha)}{(1+d_2^\alpha)}x + \frac{\varepsilon_1(1+d_1^\alpha)}{\eta}}^\infty y^{m_1-1} e^{-\frac{m_1}{\lambda_1}y} dydx \tag{2.8}$$

Based on Eq. 3.351.2 in Gradshteyn & Ryzhik, 2014

$$P_{x_1} = 1 - \sum_{k=0}^{m_1-1} \frac{e^{-\frac{m_1\varepsilon_1(1+d_1^\alpha)}{\lambda_1\eta}}}{k!\Gamma(m_2)} \left(\frac{m_1\varepsilon_1(1+d_1^\alpha)}{\lambda_1(1+d_2^\alpha)}\right)^k \left(\frac{m_2}{\lambda_2}\right)^{m_2}$$
$$\times \int_0^\infty x^{m_2-1} \left(x + \frac{(1+d_2^\alpha)}{\eta}\right)^k e^{-\frac{m_2}{\lambda_2}x - \frac{m_1\varepsilon_1(1+d_1^\alpha)}{\lambda_1(1+d_2^\alpha)}x} dx \tag{2.9}$$

Then, using Eq. 1.111 and Eq. 3.351.3 in Gradshteyn & Ryzhik, 2014 it can be obtained as

$$
\begin{aligned}
P_{x_1} &= 1 - \sum_{k=0}^{m_1-1}\sum_{n=0}^{k}\binom{k}{n}\frac{e^{-\frac{m_1\varepsilon_1(1+d_1^\alpha)}{\lambda_1\eta}}}{k!\Gamma(m_2)}\left(\frac{m_1\varepsilon_1(1+d_1^\alpha)}{\lambda_1(1+d_2^\alpha)}\right)^k \\
&\quad \times \left(\frac{m_2}{\lambda_2}\right)^{m_2}\left(\frac{(1+d_2^\alpha)}{\eta}\right)^{k-n}\int_0^\infty x^{m_2+n-1}e^{-\left(\frac{m_2}{\lambda_2}+\frac{m_1\varepsilon_1(1+d_1^\alpha)}{\lambda_1(1+d_2^\alpha)}\right)x}dx \\
&= 1 - \sum_{k=0}^{m_1-1}\sum_{n=0}^{k}\binom{k}{n}\frac{(m_2+n-1)!e^{-\frac{m_1\varepsilon_1(1+d_1^\alpha)}{\lambda_1\eta}}}{k!\Gamma(m_2)}\left(\frac{m_1\varepsilon_1(1+d_1^\alpha)}{\eta\lambda_1}\right)^k \\
&\quad \times \left(\frac{\eta\lambda_2}{m_2(1+d_2^\alpha)}\right)^n\left(1+\frac{m_1\lambda_2\varepsilon_1(1+d_1^\alpha)}{\lambda_1 m_2(1+d_2^\alpha)}\right)^{-m_2-n}
\end{aligned}
\tag{2.10}
$$

This completes the proof.

2.2.3.2 Outage probability of X_2

The outage probability UL at A of x_2 is given by (Zaidi et al., 2018a)

$$
P_{x_2} = 1 - \Pr(\Gamma_{U\ L,x_2} > \varepsilon_2)
\tag{2.11}
$$

Submitting (2.3) into (2.11), the outage probability of x_2 is expressed by

$$
P_{x_2} = \sum_{k=0}^{m_2-1}\frac{1}{k!}\left(\frac{m_2\varepsilon_2(1+d_2^\alpha)}{\lambda_2\eta}\right)^k e^{-\frac{m_2\varepsilon_2(1+d_2^\alpha)}{\lambda_2\eta}}
\tag{2.12}
$$

Similarly, the close-form outage probability for x_2 is given as

$$
\begin{aligned}
P_{x_2} &= 1 - \Pr\left(|h_2|^2 > \frac{\varepsilon_2(1+d_2^\alpha)}{\eta}\right) \\
&= \frac{\int_{\varepsilon_2(1+d_2^\alpha)}^\infty f_{|h_2|^2}(x)\,dx}{\eta}
\end{aligned}
\tag{2.13}
$$

2.2.3.3 Asymptotic

In this section, we derive the asymptote of outage probability for x_i. We apply the first-order Maclaurin series expansions $e^{-x} \approx 1-x$ (Gradshteyn & Ryzhik, 2014). The asymptotic outage probability of x_1 is given as

$$
\begin{aligned}
P_{x_1}^\infty &\approx 1 - \sum_{k=0}^{m_1-1}\sum_{n=0}^{k}\binom{k}{n}\frac{(m_2+n-1)!}{k!\Gamma(m_2)}\left(1-\frac{m_1\varepsilon_1(1+d_1^\alpha)}{\lambda_1\eta}\right) \\
&\quad \left(\frac{m_1\varepsilon_1(1+d_1^\alpha)}{\eta\lambda_1}\right)^k\left(\frac{\eta\lambda_2}{m_2(1+d_2^\alpha)}\right)^n\left(1+\frac{m_1\lambda_2\varepsilon_1(1+d_1^\alpha)}{\lambda_1 m_2(1+d_2^\alpha)}\right)^{-m_2-n}
\end{aligned}
\tag{2.14}
$$

Next, the asymptotic outage probability of x_2 can be shown as

$$
P_{x_2}^\infty \approx 1 - \sum_{k=0}^{m_2-1}\frac{1}{k!}\left(\frac{m_2\varepsilon_2+(1+d_2^\alpha)}{\lambda_2\eta}\right)^k\left(1-\frac{m_2\varepsilon_2(1+d_2^\alpha)}{\lambda_2\eta}\right)
\tag{2.15}
$$

2.2.4 Ergodic capacity of UL NOMA

The Ergodic capacity of x_1 is given by (Zaidi, Hasan, & Gui, 2018b)

$$C_{x_1} = \frac{1}{\ln 2} \int_0^\infty \frac{1 - F_{\Gamma_U \ L x_1}(x)}{1 + x} dx \tag{2.16}$$

The term $F_{\Gamma_U \ L x_1}(x)$ is given by

$$F_{\Gamma_U \ L x_1}(x) = 1 - \sum_{k=0}^{m_1-1} \sum_{n=0}^{k} \binom{k}{n} \frac{(m_2 + n - 1)!}{k!\Gamma(m_2)} \left(\frac{\eta\lambda_2}{m_2(1+d_2^\alpha)}\right)^n$$
$$\times \left(\frac{m_1 x(1+d_1^\alpha)}{\eta\lambda_1}\right)^k \left(1 + \frac{m_1\lambda_2 x(1+d_1^\alpha)}{\lambda_1 m_2(1+d_2^\alpha)}\right)^{-m_2-n} e^{-\frac{m_1 x(1+d_1^\alpha)}{\lambda_1\eta}} \tag{2.17}$$

The term (2.16) is given by

$$C_{x_1} = \frac{1}{\ln 2} \sum_{k=0}^{m_1-1} \sum_{n=0}^{k} \binom{k}{n} \frac{(m_2+n-1)!}{k!\Gamma(m_2)} \left(\frac{\eta\lambda_2}{m_2(1+d_2^\alpha)}\right)^n \left(\frac{m_1(1+d_1^\alpha)}{\eta\lambda_1}\right)^k$$
$$\times \int_0^\infty \frac{x^k}{1+x} \left(1 + \frac{m_1\lambda_2(1+d_1^\alpha)x}{\lambda_1 m_2(1+d_2^\alpha)}\right)^{-m_2-n} e^{-\frac{m_1(1+d_1^\alpha)}{\lambda_1\eta}x} dx \tag{2.18}$$

Using Gaussian Chebyshev quadrature (Gradshteyn & Ryzhik, 2014) in which $\varphi_n = \cos\left(\frac{2a-1}{2I}\pi\right)$, we have

$$C_{x_1} \approx \frac{1}{\ln 2} \sum_{k=0}^{m_1-1} \sum_{n=0}^{k} \binom{k}{n} \frac{(m_2+n-1)!}{k!\Gamma(m_2)} \left(\frac{\eta\lambda_2}{m_2(1+d_2^\alpha)}\right)^n \left(\frac{m_1(1+d_1^\alpha)}{\eta\lambda_1}\right)^k$$
$$\times \sum_{a=1}^{I} \frac{\pi}{I} \sqrt{1-\varphi_n^2} \sec^2\left(\frac{(1+\varphi_n)\pi}{4}\right) \frac{\tan\left(\frac{(1+\varphi_n)\pi}{4}\right)^k}{1 + \tan\left(\frac{(1+\varphi_n)\pi}{4}\right)}$$
$$\times \left(1 + \frac{m_1\lambda_2(1+d_1^\alpha)}{\lambda_1 m_2(1+d_2^\alpha)}\tan\left(\frac{(1+\varphi_n)\pi}{4}\right)\right)^{-m_2-n} e^{-\frac{m_1(1+d_1^\alpha)}{\lambda_1\eta}\tan\left(\frac{(1+\varphi_n)\pi}{4}\right)} \tag{2.19}$$

The ergodic capacity of x_2 is given by (Zaidi et al., 2018b)

$$C_{x_2} = \frac{1}{\ln 2} \int_0^\infty \frac{1 - F_{\Gamma_U \ L x_2}(y)}{1 + y} dy \tag{2.20}$$

$F_{\Gamma_U \ L x_2}(y)$ is given as

$$F_{\Gamma_U \ L x_2}(y) = 1 - \sum_{k=0}^{m_2-1} \frac{1}{k!} \left(\frac{m_2(1+d_2^\alpha)}{\lambda_2\eta}\right)^k y^k e^{-\frac{m_2(1+d_2^\alpha)}{\lambda_2\eta}y} \tag{2.21}$$

Based on Eq. 3.353.5 in Gradshteyn & Ryzhik, 2014

$$C_{x_2} = \frac{1}{\ln 2} \sum_{k=0}^{m_2-1} \frac{1}{k!} \left(\frac{m_2(1+d_2^\alpha)}{\lambda_2\eta}\right)^k \left[(-1)^{k-1} e^{\frac{m_2(1+d_2^\alpha)}{\lambda_2\eta}} Ei\left(-\frac{m_2(1+d_2^\alpha)}{\lambda_2\eta}\right) + \sum_{c=0}^{k}(c-1)!(-1)^{k-c}\left(\frac{m_2(1+d_2^\alpha)}{\lambda_2\eta}\right)^{-c}\right] \tag{2.22}$$

2.2.5 Numerical results and discussions

In this section, we set $a = 2$, $d_1 = 5$, $d_2 = 2$, $m = m_1 = m_2 = 2$, $\lambda_1 = \lambda_2 = 1$, $R_1 = 0.5$ and $R_1 = 1$. In Fig. 2.6, we can see the outage probability can be reduced significantly at high SNR region of the

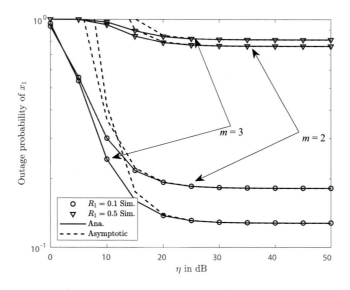

FIGURE 2.6

Outage probability of x_1 versus η with different R_1 and m.

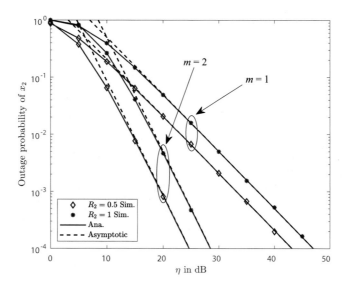

FIGURE 2.7

Outage probability of x_2 versus η with different R_2 and m.

destinations. This is the outage probability for the detection of the first destination' signal at the AP. Further, the quality of channel and required target rate led to improved outage probability. The exact curves match with Monte Carlo curves at all range of transmit SNR as expected.

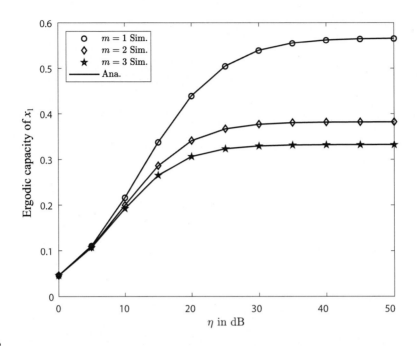

FIGURE 2.8

Ergodic capacity of x_1 versus η with different m.

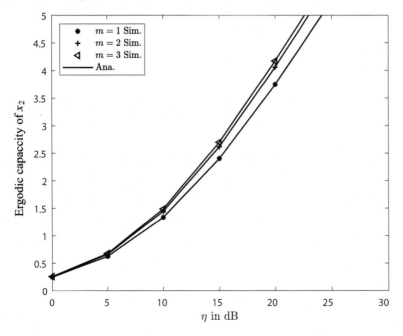

FIGURE 2.9

Ergodic capacity of x_2 versus η with different m.

In Fig. 2.7, we can see the significant reduction of outage probability for the detection of the second destination' signal at the AP. Similar performance can be observed as Fig. 2.6. The exact curves match with Monte Carlo curves at all range of transmit SNR as expected.

In Fig. 2.8, we provide system performance for ergodic capacity of the first destination's signal which is detected at the AP. The ergodic capacity can be improved at high transmit SNR. The reason is that higher transmit SNR leads to higher SINR, then corresponding ergodic capacity could be higher. The quality of channel contributes the improvement of such ergodic capacity.

We provide system performance for ergodic capacity of the second destination's signal which is detected at the AP, shown in Fig. 2.9. Similar observations can be seen for this figure, and it looks like Fig. 2.8.

2.3 Conclusions

In this work, we have studied the impact of channels on two performance metrics (outage probability and ergodic capacity) of UL health care IoT system. The AP needs to examine how it can detect the signals in UL from two destinations. We provided the closed-form expressions of two performance metrics. Further, we found that higher quality channels lead to significant improvements of these performance metrics. This book chapter provides guidelines of UL IoT with advances of the NOMA scheme.

References

Amin, S. U., & Hossain, M. S. (2021). Edge intelligence and Internet of Things in healthcare: A survey. *IEEE Access*, *9*, 45−59. Available from https://doi.org/10.1109/ACCESS.2020.3045115.

Asadi., Wang, Q., & Mancuso, V. (2014). A survey on device-to-device communication in cellular networks. *IEEE Communications Surveys and Tutorials*, *16*(4), 1801−1819, 4th Quart.

Awan, K. A., Din, I. U., Almogren, A., Almajed, H., Mohiuddin, I., & Guizani, M. (2020). Neurotrust -artificial neural network-based intelligent trust management mechanism for large-scale internet of medical things. *IEEE Internet of Things Journal*. Available from https://doi.org/10.1109/JIOT.2020.3029221.

Baig, S., Ali, U., Asif, H. M., Khan, A. A., & Mumtaz, S. (2019). Closed-form BER expression for Fourier and wavelet transform-based pulse shaped data in downlink NOMA. *IEEE Communications Letters*, *23*(4), 592−595.

Boccardi, F., Heath, R. W., Jr., Lozano, A., Marzetta, T. L., & Popovski, P. (2014). Five disruptive technology directions for 5G. *IEEE Communications Magazine*, *52*(2), 74−80.

Chen, C., Cai, W., Cheng, X., Yang, L., & Jin, Y. (2017). Low complexity beamforming and user selection schemes for 5G MIMO-NOMA systems. *IEEE Journal on Selected Areas in Communications*, *35*(12), 2708−2722.

Ding, Y., et al. (2021). DeepEDN: A deep-learning-based image encryption and decryption network for Internet of Medical Things. *IEEE Internet of Things Journal*, *8*(3), 1504−1518. Available from https://doi.org/10.1109/JIOT.2020.3012452.

Ding, Z., Lei, X., Karagiannidis, G. K., Schober, R., Yuan, J., & Bhargava, V. K. (2017). A survey on non-orthogonal multiple access for 5G networks: Research challenges and future trends. *IEEE Journal on Selected Areas in Communications*, *35*(10), 2181−2195.

Ding, Z., & Poor, H. V. (2016). Design of massive-MIMO-NOMA with limited feedback. *IEEE Signal Processing Letters, 23*(5), 629−633.

Do, D.-T., Le, A., & Lee, B. M. (2020). NOMA in cooperative underlay cognitive radio networks under imperfect SIC. *IEEE Access, 8,* 86180−86195.

Do, D.-T., Le, C.-B., & Afghah, F. (2020). Enabling full-duplex and energy harvesting in uplink and downlink of small-cell network relying on power domain based multiple access. *IEEE Access, 8,* 142772−142784.

Do, D.-T., Nguyen, M.-S. V., Jameel, F., Jäntti, R., & Ansari, I. S. (2020). Performance evaluation of relay-aided CR-NOMA for beyond 5G communications. *IEEE Access, 8,* 134838−134855.

Egala, B. S., Pradhan, A. K., Badarla, V. R., & Mohanty, S. P. (2021). Fortified-Chain: A Blockchain Based Framework for Security and Privacy Assured Internet of Medical Things with Effective Access Control. *IEEE Internet of Things Journal.* Available from https://doi.org/10.1109/JIOT.2021.3058946.

Garg, N., Wazid, M., Das, A. K., Singh, D. P., Rodrigues, J. J. P. C., & Park, Y. (2020). BAKMP-IoMT: Design of blockchain enabled authenticated key management protocol for Internet of Medical Things deployment. *IEEE Access, 8,* 95956−95977. Available from https://doi.org/10.1109/ACCESS.2020.2995917.

Gatouillat, A., Badr, Y., Massot, B., & Sejdić, E. (2018). Internet of Medical Things: A review of recent contributions dealing with cyber-physical systems in medicine. *IEEE Internet of Things Journal, 5*(5), 3810−3822. Available from https://doi.org/10.1109/JIOT.2018.2849014.

Ghubaish, A., Salman, T., Zolanvari, M., Unal, D., Al-Ali, A., & Jain, R. (2021). Recent Advances in the Internet-of-Medical-Things (IoMT) systems security. *IEEE Internet of Things Journal, 8*(11), 8707−8718. Available from https://doi.org/10.1109/JIOT.2020.3045653.

Gradshteyn, I. S., & Ryzhik, I. M. (2014). *Table of integrals, series, and products.* Academic press.

Islam, S. M. R., Avazov, N., Dobre, O. A., & Kwak, K.-S. (2017). Power-domain non-orthogonal multiple access (NOMA) in 5G systems: Potentials and challenges. *IEEE Signal Processing Letters, 19*(2), 721−742, 2nd Quart.

Islam, S. M. R., Zeng, M., Dobre, O. A., & Kwak, K.-S. (2018). Resource allocation for downlink NOMA systems: Key techniques and open issues. *IEEE Wireless Communication, 25*(2), 40−47.

Lei, L., Zhong, Z., Lin, C., & Shen, X. (2012). Operator controlled device-to-device communications in LTE-advanced networks. *IEEE Wireless Communication, 19*(3), 96−104.

Li, J., Peng, Y., Yan, Y., Jiang, X.-Q., Hai, H., & Zukerman, M. (2020). Cognitive radio network assisted by OFDM with index modulation. *IEEE Transactions on Vehicular Technology, 69*(1), 1106−1110.

Limaye, A., & Adegbija, T. (2018). HERMIT: A benchmark suite for the Internet of Medical Things. *IEEE Internet of Things Journal, 5*(5), 4212−4222. Available from https://doi.org/10.1109/JIOT.2018.2849859.

Liu, Y., Ding, Z., Elkashlan, M., & Poor, H. V. (2016). Cooperative non-orthogonal multiple access with simultaneous wireless information and power transfer. *IEEE Journal on Selected Areas in Communications, 34*(4), 938−953.

Liuand, Y., & Erkip, E. (2016). Capacity and rate regions of a class of broadcast interference channels. *IEEE Transactions on Information Theory, 62*(10), 5556−5572.

Qi, T., Feng, W., Chen, Y., & Wang, Y. (2017). When NOMA meets sparse signal processing: Asymptotic performance analysis and optimal sequence design. *IEEE Access, 5,* 18516−18525.

Sun, L., Jiang, X., Ren, H., & Guo, Y. (2020). Edge-cloud computing and Artificial Intelligence in Internet of Medical Things: Architecture, technology and application. *IEEE Access, 8,* 101079−101092. Available from https://doi.org/10.1109/ACCESS.2020.2997831.

Wan, D., Wen, M., Ji, F., Yu, H., & Chen, F. (2018). Non-orthogonal multiple access for cooperative communications: Challenges, opportunities, and trends. *IEEE Wireless Communications, 25*(2), 109−117.

Wang, B., Gao, F., Jin, S., Lin, H., & Li, G. Y. (2018). Spatial- and frequency wideband effects in millimeter-wave massive MIMO systems. *IEEE Transactions on Signal Processing, 66*(13), 3393−3406.

Yang, Q., Wang, H.-M., Ng, D. W. K., & Lee, M. H. (2017). NOMA in downlink SDMA with limited feedback: Performance analysis and optimization. *IEEE Journal on Selected Areas in Communications, 35*(10), 2281−2294.

Zaidi S.K., Hasan S.F. and Gui X., (2018a). SWIPT-aided uplink in hybrid non-orthogonal multiple access. In *Proceedings of the IEEE wireless communications and networking conference*, pp. 1−6.

Zaidi, S. K., Hasan, S. F., & Gui, X. (2018b). Evaluating the Ergodic Rate in SWIPT-Aided Hybrid NOMA. *IEEE Communications Letters, 22*(9), 1870−1873.

Zhao, J., Liu, Y., Chai, K. K., Chen, Y., & Elkashlan, M. (2017). Joint subchannel and power allocation for NOMA enhanced D2D communications. *IEEE Transactions on Communications, 65*(11), 5081−5094.

Zheng, B., Wang, X., Wen, M., & Chen, F. (2017). NOMA-based multi-pair two-way relay networks with rate splitting and group decoding. *IEEE Journal on Selected Areas in Communications, 35*(10), 2328−2341.

HetNet/M2M/D2D communication in 5G technologies

Ayaskanta Mishra[1], Anita Swain[1], Arun Kumar Ray[1] and Raed M. Shubair[2]

[1]*School of Electronics Engineering, Kalinga Institute of Industrial Technology, Deemed to be University, Bhubaneswar, Odisha, India* [2]*Department of Electrical and Computer Engineering, New York University (NYU) Abu Dhabi, Abu Dhabi, United Arab Emirates*

3.1 Introduction

Modern day digital networks are a complex mesh of interconnected devices for the transmission of data bits using various network standards, technologies, and protocols. This makes the data communication network infrastructure heterogeneous in nature. Heterogeneity is imperative in modern day packet-switched data networks. The primary challenge is to have seamless integration and interoperability of such heterogeneous network clusters. Such heterogeneity across network is predominantly contributed by the access networks (PHY-MAC layer). As per TCP/IP stack: physical and link layer specifications are different for different access networks. Physical Media Dependent Protocol (PMDP) and Physical layer Convergence Protocol (PLCP) of Layer-1 of OSI is necessary to access network specifics and differences, so an added layer of complexity in design and deployment of data network infrastructure is needed for cross-technology integrated networks. Such networks are referred to as Heterogeneous Networks or HetNets. Apart from layer-1 even the Media Access Control (MAC) is dissimilar across multiple network technologies. Thanks to IEEE 802.1 Logical Link Control (LLC) for proving a convergence to Network layer/ Internet Protocol (IP) layer for interoperability across multiple underlying Layer-1 & MAC sublayer. For example, a segment of network may be having application-specific requirements of Wireless Personal Area (WPAN) for any Wireless Sensor Network (WSN) application hence an IEEE 802.15.4 Zigbee might be a viable option for the particular cluster of networks. However, other parts of the network may have different sets of application specific network requirements for, say, a Wireless Local Area Network (WLAN) and therefore may be IEEE 802.11 Wi-Fi standard on Layer-I and MAC would be suitable for its deployment. Now the question is the "Internet" (IP-based Global data network infrastructure) is full of such use-cases where the heterogeneity is imperative and a *necessary evil* to accommodate multiple access network underlying standards. The perspective of mobile cellular networks (MCN), such as Long-Term Evolution (LTE), that is, 4th generation (4G) Mobile communication standards, shows that the network is completely packet-switched with absolution of circuit-switching from 3rd generation (3G) mobile communication standards. Hence the 5th generation mobile communication standard New-Radio (5G-NR) is also a completely packet-switched data network with more optimized Cloud based Radio Access Network (C-RAN) with other key

5G IoT and Edge Computing for Smart Healthcare. DOI: https://doi.org/10.1016/B978-0-323-90548-0.00002-4

enabling technologies like Network Functions Virtualization (NFV) and Software defined Networking (SDN). 5G-NR standard is a packet-switch data network having basic requirements to seamlessly integrate with other existing access network technologies (ANT) like IEEE 802.11 and IEEE 802.15.4 etc. Such interoperability of 5G-NR with other underlying technologies like other WPAN, WLAN, and WAN ANT are provisioned using an extra middle-layer and edge/gateway devices. The primary functionality of such gateway device is to do the required protocol level translation of the Protocol Data Unit (PDU). Another key aspect of such heterogeneous networks (HetNet) is "clustering." Clustering is the concept of dividing the network into multiple working network segments or Autonomous System (AS). These Autonomous Systems are managed independently as per their access network standards. However, there is always a requirement of having an interoperability device (Gateway device) to make any two dissimilar clusters in an op-operative communication model. The edge nodes between two dissimilar clusters are gateways in HetNets and they are the major technology enablers to seamless interoperability across multiple dissimilar network segments or clusters. Several popular HetNets Radio Access Technologies (RATs) Wi-Fi/LTE/5G-NR are used in H-CRAN of such 5G HetNets. Interoperability across HetNet clusters is achieved using gateway devices (GWs). GWs are instrumental in providing seamless integration and interoperability across dissimilar networks and protocols. The primary function of gateway node is providing a seamless integration between to dissimilar network clusters. This function is achieved through various strategies like (1) Dual stack, (2) Header Translation and (3) Tunneling. Dual stack is a strategy implemented on the Edge of two heterogonous networks. The Edge device (Gateway) is having both the connected cluster stack implemented, which enables interoperability as the gateway device can process PDUs of both the HetNet clusters. Being equipped with both the protocol stack the gateway device can process PDUs of both the HetNet clusters. Secondly, the Header translation strategy is also a viable option to be implemented on gateway devices at the Edge of two dissimilar network clusters in HetNets. Here the Gateway device would process the Header of the PDU of one side of the network and translate to the required parameters and format of the other side and vice-versa. This method would be computationally intense and put more computational overhead of the gateway device, hence gateway devices demand relatively high computational capabilities and resources with higher CPU, RAM, and faster buffer or data memory access. Lastly, we have a tunneling approach where a typical sandwiched HetNet scenario would take much benefit. For example, if there is a 5G-NR network cluster in between two IEEE 802.11 network clusters, then the 5G-NR network may encapsulate the IEEE 802.11 frames in 5G-NR Link-layer PDU. The encapsulation into the tunnel would be done when the PDU has an entry through gateway device of 5G-NR and the de-capsulation would be done when the PDU exit the 5G-NR network cluster through the existing gateway device. Apart from these three strategies for handling seamless integration of multiple HetNets, there is always a technical challenge to deal with various aspects of heterogeneity and interoperability issues of HetNets.

One such predominant application use-case using HetNets would be the "*smart healthcare system.*" The vision of Industry 4.0 includes a holistic system with smart healthcare Edge devices equipped with biomedical sensors with key features like connect, compute and communicate. These systems are referred as Internet of Healthcare Things (IoHT). For an example: An IoHT system is designed to acquire sensor data from a cluster/Wireless Body Area Network (WBAN) of low-power sensor motes which are reduced functional devices (RFDs) with limited computational resource for sensor data preprocessing and signal conditioning. For these scenario-specific requirements of Wireless Personal Area Network (WPAN) is befitting, hence Bluetooth Low Energy (BLE) or IEEE 802.15.4/Zigbee network are used in the WSN/ Body Area Network (BAN) part of a SDN-based 5G Heterogenous-Cloud-Radio Access Network (H-CRAN). However, there is a requirement of Traffic/data aggregation when numerous such sensor nodes have to send data using mMTC framework to the centralized cloud/server for data storage or analytics.

Fig. 3.1 depicts a generic HetNet in the context of 5G and other allied access network technologies.

As discussed, for any practical use-case deployment scenario of heterogeneous network; we may have multiple network clusters each managed and administrated by a separate underlying network technology (Physical and MAC layer). As illustrated in Fig. 3.1, we can have a WSN/Zigbee (IEEE 802.15.4), Wi-Fi (IEEE 802.11), satellite and Optical (SONET/SDN) network cluster alongside the 5G network. All dissimilar access network clusters are integrated together to a HetNet gateway integrator through their respective Edge interface gateway devices. These gateway devices are designed and deployed based on any one of the three methods disused prior or may be a combination of two or more methods making it a hybrid gateway as per scenario-specific requirements of

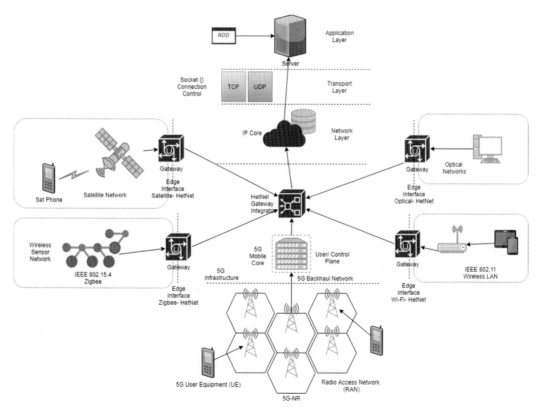

FIGURE 3.1

HetNet in the context of 5G.

the connecting clusters in HetNet. The integrator provides a seamless connectivity to the IP Core. This provides a convergence for HetNet for providing service point addressing by transport layer Socket $<ip\_address:port\_no>$ for connection control based on application-specific requirements of respective application servers. Transmission Control Protocol (TCP) for a connection-oriented hand-shaking based approach for reliable and guaranteed service framework and User Datagram Protocol (UDP) for a connection-less and best-effort service framework. The application layer of TCP/IP stack is complete software implementation based on Application Programming Interface (API) for providing service to the end-user like conventional Hypertext Transfer Protocol (HTTP), File Transfer Protocol (FTP), Telnet, Domain Name System (DNS), and even some IoT specific application protocols like Message Queue Telemetry Transport (MQTT) and Constraint Application Protocol (CoAP).

3.2 Heterogenous networks in the era of 5G

In the era of 5th generation (5G) Mobile, that is,. from 3GPP release 14 (R14) onwards has some distinct features to support massive Machine Type Communication (mMTC) framework much required for massive deployment of Machine-to-machine (M2M) communication scenario for future Internet of Things (IoT) applications. Most of the deployment using 3GPP R14 has been done by various pilot projects since 2020. One such use-case scenario of the mMTC is Smart City Bussan in South Korea. The 5G mobile standard suggest a minimum of 10Gbps data-rate using enhanced Mobile Broad-band (eMBB), enhanced Mobility support (up to 500 km/hr.) with Ultrareliable Low-Latency Communication (URLLC) infrastructure (within 1 ms End-2-End delay). These features of 5G have enabled service and business potential for many resource demanding applications like 3D Video, HD/4K Video content and streaming applications, Augmented Reality (AR), Virtual Reality (VR), Industrial Automation and Control (IAC), Self-driving Vehicles enabled smart transportation systems. Ultrahigh data-rate and low-latency smart healthcare applications, smart home, smart city, smart agriculture, smart grids, etc., mMTC supports massive deployment up to 1 million IoT-enabled smart devices (equipped with application-specific sensors and actuators) per square kilometer. It is imperative to have network heterogeneity in deployment of such IoT-based massive M2M communication as the Radio Access Network (RAN) is comprising off multiple-heterogeneous underlying technologies like IEEE 802.11 (Wi-Fi), IEEE 802.15.1 (BLE), IEEE 802.15.4 (Zigbee), Wi-MAX, LTE, 5G-NR and even Wired Network access like IEEE 802.3 Ethernet and optical networks like SONET, SDH. The challenge is seamless interoperability of Heterogeneous Networks (HetNets) in the deployment of such IoT-based mMTC using technologies like M2M and Device-to-Device (D2D). Fig. 3.2 shows the 5G M2M Heterogonous Network framework with a functional layered architectural overview. Li, Da Xu, and Zhao (2018).

3.2.1 5G mobile communication standards and enhanced features

1. *Logically independent Network:* 5G-NR is having a logically independent Network framework for application-specific requirements.

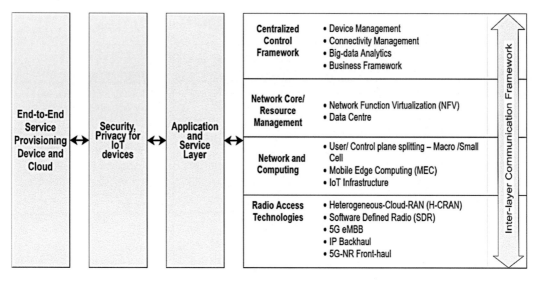

FIGURE 3.2

5G M2M Heterogonous Network framework (Li et al., 2018).

2. *Cloud RAN:* C-RAN supporting massive connections of mMTC framework with multiple standard underlying network technologies (HetNets) support.
3. *Simple core network architecture:* 5G Core Network Architecture with support for on-demand networking and computing resource allocations/ configuration using Network Function Virtualization (NFV).
4. *High data-rate:* Application support for HD/AR/VR. (at least 10Gbps on RAN)
5. *High scalability:* 5G Network with support for NFV can have fine-grained front-haul and network decomposition.
6. *Low-latency:* 5G supports interactive communication framework for various applications like AR, Video-games, 5G-IoT for low latency applications (smart healthcare).
7. *Reliable and resilient:* Improved convergence, hardened efficiency with support for interoperability amount HetNets.
8. *Security and privacy:* Improved security model for applications like digital payment.
9. *Long battery life:* Improved Energy Efficiency (EE), Low-power, Low-cost, 5G/NB-IoT devices support.
10. *High connection density:* Support for millions of M2M communications using mMTC framework.
11. *Mobility:* Enhanced mobility model and hand-off management.

Table 3.1. shows the Heterogeneous Underlying Network technologies for IoT.

Table 3.1 Heterogeneous underlying network technologies for Internet of Things.

Type of radio access networks	Peripherals communication wireless personal area network	Home and campus communication local area network (LAN)	Wide area network (WAN)
Standards	Bluetooth low Energy (BLE) IEEE 802.15.1 Zigbee (IEEE 802.15.4)	Wireless LAN (Wi-Fi) IEEE 802.11	GSM, LTE, LTE-A, 5G-NR, LoRaWAN, SIGFOX
Range	Less than 30 feet	Less than 300 feet	10–30 km
Features	Low-power Low data-rate Enhanced battery Life Low range	20–40 MHz Channel bandwidth Moderate −Power High data-rate Moderate range	Heterogeneous Bandwidth and channel allocation across multiple-technologies Long range
Applications	Wireless sensor Network (WSN) Body area network (BAN)	Smart home, monitoring, energy, security and surveillance	Industry 4.0 Smart city Smart industrial IoT Smart agriculture Smart transportation and logistics management

3.2.2 5G heterogeneous network architecture

5G network architecture support Heterogeneous Radio technologies for seamless interoperability with the help of Heterogeneous Cloud Radio Access Network (H-CRAN) framework. The functional splitting of the User plane from the control plan helps in achieving such complex network operation. Coordinated Multipoint (CoMP) transmission and reception mechanism is used for integration of User and Control plane. In 5G network architecture the features like Software Defined Network (SDN) comes handy in dealing with radio resource management in the Cloud-RAN (C-RAN). The Cooperative Signal Processing and Networking functionalities are viable by using the virtualization and cloud-based systems implemented on the C-RAN. The Network Function Virtualization (NFV) is implemented for on-demand resource management. Resources such as radio resources and computing resources, are managed using various cloud computing models: Software-as-a-Service (SaaS) and Platform-as-a-Service (PaaS). The User plan is deployed using Short-range communication (Small cells: Pico/femto) using the Remote Radio Heads (RRHs). The RRHs deals with the high-volume data-traffic for the user equipment (UE) with eMBB framework of high data-rate service. These RRHs are allocated with Base-band Units (BBUs) using SDN-based flexible front-haul on-demand. The control plane works separately using the High-Power Nodes (HPNs) relatively longer-range macrobase stations (MBSs). The primary function of these MBSs is to provide only control signals for UEs. These control signals are mostly a low data-rate signals for UE control like Cell association, Resource allocation, mobility management (hand-off), etc., hence MBSs does not have any high payload requirement. In short, the MBSs present in the control plane are low-data rate, long-range trans-receivers where the RRHs deal with high volume user data (User plane) for quite a small cell-size hence short-range communication. The splitting of Control and

User plane in 5G network architecture has enabled the H-CRAN to be deployed with multitier heterogeneous underlying radio access network (like Wi-Fi, LTE, 5G-NR, etc.) technologies. Each of the RRHs can belong to different underlying network technologies but still can be managed and configured through the SDN-based Cloud-RAN using centralized resource management using NFV. This architecture helps in achieving seamless interoperability between H-CRAN clusters using appropriate Gateway devices (GW). Software Defined RAN (SD-RAN) framework manages the on-demand resource allocation and the traffic-shaping is done based on the application-specific Quality of Service (QoS) requirements using OpenFlow and SNMP in the User/Data plane. The control plane is being handled by the Network Controller (NC) equipped with custom applications and management tools. Fig. 3.3 shows the Heterogeneous Cloud-RAN (H-CRAN)-based 5G Network Architecture and its functional components.

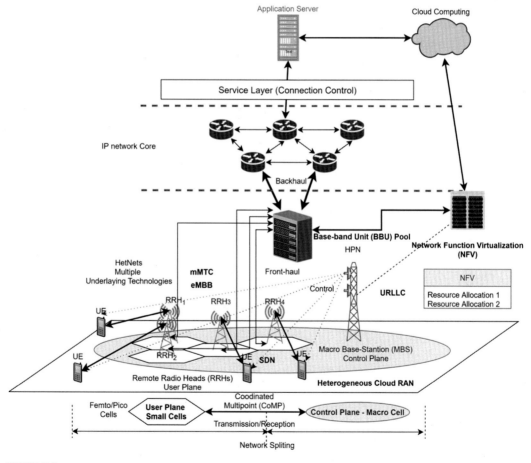

FIGURE 3.3

Heterogeneous Cloud-RAN (H-CRAN)-based 5G Network Architecture.

5G H-CRAN architecture supports multitechnology access network integration though on-demand radio resource allocation using CoMP to accommodate user-control plan splitting mechanism. The heterogenous wireless access network can take its required radio resource from RRHs with the help of MBSs (Control plane). The RRHs can take on-demand network resource from allocated BBUs using software defined radio (SDR) technology. Research in the direction of enhancing coverage and capacity of such 5G HetNets is key to the performance improvements of HetNets in the context of 5G. A stochastic geometry model based 2-tier intercellular hand-off mechanism is proposed in Ouamri, Oteşteanu, Alexandru, and Azni (2020) to increase the capacity and coverage with reduced cost of deployment. Spectral/EE in 5G is ten times and the throughput in 5G is twenty-five times than that of 4G/LTE. H-CRAN based wireless access network is the viable solution to deal with the on-demand radio resource requirements with enhanced spectral and EE. 5G network architecture supports cloud computing with the aid of "Node-C": a cloud computing capable node. The centralized large-scale co-operative processing helps in decreasing or suppressing cochannel interference. Node-C has operational capacities to allocate on-demand radio resource though BBUs to the RRHs. SDN-based H-CRAN deployment in 5G HetNets are instrumental in performance enhancement by adaptive large-scale cooperative spatial signal processing and co-operative radio resource management using techniques like NFV, self-organized front-haul, constrained resource optimization and energy harvesting. 5G H-CRAN has the following technical specifications (Peng, Li, Zhao, & Wang, 2015):

1. Ultrasmall cell HetNets (Pico/Femto for User plane) with a large number of low-cost RRHs based front-haul. Randomly connected with BBU pool.
2. RRHs are close to UE hence Low-Power Node (LPN). Low-range but high capacity (smaller cells hence a greater number of RRHs in a geographical area increase the capacity).
3. BBU pool supports resource allocation to RRHs via centralized and co-operative processing (mitigate interference) using techniques like statistical multiplexing.
4. H-CRANs are cost effective and energy efficient but at the same time complex to manage.
5. Splitting of Data and Control plan using CoMP transmission and reception. MBS are higher power /less bandwidth control plane with RRHs are low-power/high bandwidth data plane. Separative of control and data plane enables gives better network manageability over a complex SDN based HetNet architecture.
6. Co-ordinated multipoint (CoMP) architecture helps in mitigating intertier and intratier interference.
7. Improved spectral efficiency (SE) and EE with the help of co-operative signal processing and NFV on the H-CRAN with LPN/RRHs (for Data) and MBS (for Control).
8. Physical layer of H-CRAN has functionalities of co-operative radio resource management using spatial signal processing, Multiple Input Multiple Output (MIMO) antenna.
9. H-CRAN has a MAC layer using Node-C (Cloud based MAC functions).
10. SDN and NFV functions are instrumental in achieving state-of-the-art techniques like self-configuration, self-optimization, self-healing and ultradense machine type communication (mMTC).
11. Application architecture supports anytime, anywhere Gigabit data to UE with a Edge-less experience with the help of Heterogenous MBSs, ultradense RRHs, cloud computing (Node-C) for SDN based on-demand resource processing on the air-interface.

12. Node-C using cloud computing framework converges all Ancestral Communication Entities (ACEs) (i.e., MBSs, micro BSs, pico BSs, etc.). Node-C are having powerful computing resource and capabilities for large-scale co-operative signal and network processing for PHY/ MAC and upper layer functionalities. C-RAN manages RRHs and resource allocation through BBU pool. This new approach of cloud computing using Node-C has technical edge over the traditional Radio network Controller (RNC) and Base-Station Controller (BSC) based network architecture in previous generation 3GPP standards.
13. Gateway devices in H-CRAN are having two major functionalities: (1) Co-operative Multi-Radio Resource Management (CM-RRM) and (2) Media Independent Hand-over (MIH).
14. SDN-NFV based H-CRAN architecture using Node-C provides 5G HetNet the advantage of self-organizing and plug-and-play capability to dynamic ACE nodes.
15. 5G HetNets using IP network integration over SDN using Node-C for realizing the NFV. IEEE 802.11 ac/ad with mmWave 5G-NR can be an example of 5G HetNet use-case.

Fig. 3.4 shows the NFV framework used in H-CRAN for 5G HetNets for resource management and Network function virtualization. The Node-C and computationally intensive nodes cable of running multiple virtual machines on the cloud to emulate the ACEs on a live network deployment.

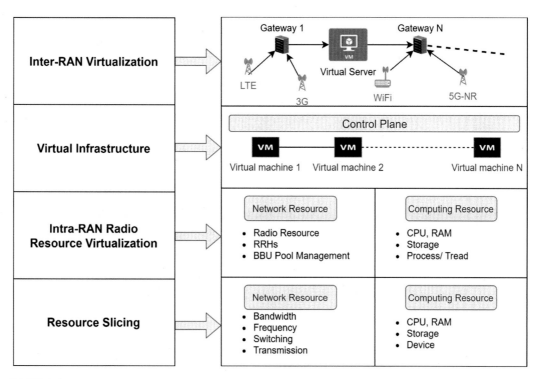

FIGURE 3.4

Network functions virtualization framework in H-CRAN for 5G HetNet (Peng et al., 2015).

NFV framework in H-CRAN for 5G HetNet is having four functional layers. For top-to-bottom approach:

1. *Inter-RAN Virtualization:* This layer has multiple gateway devices each catering a service to a heterogenous set of network segments like 3G, LTE, Wi-Fi and 5G-NR etc. Virtual servers are deployed for NFV functions required for interoperability among HetNets.
2. *Virtual Infrastructure:* Node-C-based cloud computing approach for Virtual infrastructure. Instead of having dedicated functional nodes, this layer uses virtualization of Network functions deployed over distributed Virtual machines (VMs) with the help of a generic/converging control plane.
3. *Intra-RAN Radio Resource Virtualization:* This layer deals with the Radio resource virtualization inside a RAN. Resources of a RAN are managed as two separate entities: (1) Network resource like RRHs, BBU Pool management and Radio resource management and (2) Computing resource like CPU, RAM, storage, process and threads.
4. *Resource slicing:* This is the bottom most layer mostly dealing with the real physical resource sliced into Network and Computing resources. (1) Network physical resource like bandwidth, frequency, switching and transmission. (2) Computing resource of device, hardware like CPU, RAM, storage, etc.

3.2.3 Intelligent software defined network framework of 5G HetNets

The intelligent SDN framework of H-CRAN architecture of 5G HetNets where Node-C based cloud computing-enabled nodes are deployed to harness the NFV using SDN-controllers and OpenFlow (ONF) switches with the help of co-operative multipoint communication between RRHs in data plane and MBSs in control plane (Sun, Gong, Rong, & Lu, 2015). In 5G H-CRAN architecture all of the network functions and signal processing functions are done in software over generic cloud computing platform (with the help of VMs) rather than dedicated hardware nodes. In prior 3GPP standards in 3G and 4G/LTE networks the network architecture has functional dedicated hardware nodes like MME, EPC/CN, S/P-GW connected to E-UTRAN with eNodeBs. However, the H-CRAN architecture in 5G has no dedicated functional nodes but rather software level network function virtualization over Node-Cs. Most of the network and signal processing functions are handled by software deployed over H-CRAN and this provides a huge advantage over the prior 3GPP standards as it provides are extra flexibility and programmability of resource management in both communication resource like bandwidth, frequency, channel, switching, signal processing and computing resource like CPU, RAM, storage etc. in CRAN. All Node-C-based devices are widely customizable in terms of communication and computing resource providing a high level of scalability and adaptability to change in network hence easier to deploy and manage in a very complex set of network attributes.

3.2.4 Next-Gen 5G wireless network

Next-Gen 5G-NR based wireless networks have new functionalities to accommodate IoT, Internet of Vehicles (IoV) using state-of-the-art technologies like mMTC and M2M and D2D communication. In physical layer MIMO, Smart antenna Beamforming, mmWave technologies and in upper layer C-RAN, NFV, SDN framework and CoMP based Network splitting (Data and control plane separate) are the key technology enablers for future wireless communication paradigm. The 5G-NR H-CRAN based architecture are having numerous design advantages over traditional Base Station (BS) centric networks. Small cells (Pico/femto), Low-power RRHs with enhance network capacity, high device (UE) density for mMTC are key enablers for IoT based network deployment. Better spectral and EE with intelligent antenna technology like MIMO, mmWave, beamforming creates opportunities for IoV and D2D communication. Deployment of H-CRAN based intelligent SDN based network has improved seamless integration of HetNets (Agiwal, Roy, and Saxena, 2016).

Accommodating heterogenous networks is itself a challenge for seamless integration and enhanced performance. A 3GPP-inspired HetNet model using passion cluster for improved downlink coverage is proposed in Saha, Afshang, and Dhillon (2018). The complexity of a HetNet is huge overhead for any real-world deployment scenario. A stochastic geometry-based modeling approach is proposed for mobile UEs and base Station gNBs for radio resource management in a complex HetNet deployment scenario. This helps with better resource management of HetNets and deployment planning with enhanced downlink coverage.

SDN-based C-RAN in 5G has enormous potential in handling crucial network functions of modern mobile communication. We will discuss a practical case study of one such SDN-based platform for better understanding. 5G-EmPower is a SDN platform for 5G RAN is proposed in Coronado, Khan, and Riggio (2019). 5G-EmPower is a flexible, programmable, open-source SDN platform for 5G HetNets C-RAN. 5G-EmPower is developed as an open-source platform suing APACHE 2.0 license. The platform has three main functional modules. (1) Active Network Slicing, (2) Mobility Management and (3) Load-balancing. The platform uses key 5G functionalities like mMTC, URLLC, HetNet C-RAN, NFV and SDN-OpenFlow. 5G active network slicing supports a mixed/hybrid network with small (pico/femto) cells (for data plane) and macro cells (for control plane). 5G-EmPower has three software modules: (1) Radio Access Agnostic API (Data/control plane), (2) Software agents- operate on several Radio access networks (HetNets) e.g., Wi-Fi (IEEE 802.11 ac/ad), 4G/LET and 5G-NR, and (3) SDN-controller. The EmPower platform is deployed as a complete operating System (OS) on the cloud with two interfaces: (1) North bound: Operation, Admin and Management and (2) South bound: HetNet Access Agents for Wi-Fi, LTE and 5G-NR. The North bound interface connected to the management plane having key functions like mobility management, load balancing. The EmPower OS is the control plane having functionalities like global network view, Web services, topology discovery, device management. Toward the south bound interface, a hardware abstraction layer is designed for convergence of HetNet with the help of Agent Wi-Fi, LTE and 5G-NR. The platform uses slicing resource management framework with its own proposed PDU with the help of an appropriate layer (PHY, MAC, RLC, PDCP)-specific wrapper with payload and control field structure in its proposed design.

3.2.5 Internet of Things toward 5G and heterogenous wireless networks

IoT having tremendous potential in the recent times. IoT in the era of 5G is having many key technical advantages like high data-rate, more bandwidth, UR-LLC, better QoS, low interference using beamforming, MIMO, mmWave, eMBB, mMTC, M2M, D2D.A generic service-orient architecture may hold good for all IoT applications in the context of 5G and HetNets. The 5G-IoT architecture has the following key functional layers: (Chettri & Bera, 2020)

1. *IoT sensor layer:* Physical devices, smart sensors and actuators
2. *Network layer:* LPWAN, SigFox, LoRaWAN, Zigbee, NB-IoT
3. *Communication layer:* backbone ip
4. *architecture layer:* cloud, big-data analytics
5. *Application layer:* IoT application (smart home/city/grid/transportation/agriculture, Industrial IoT, etc.)

Every layer has its own specific functions that are customizable and flexible in intralayer attributes. The modular structure of the layered architecture makes it generic with scope of massive designability, programmability and flexibility in every functional layer as per the scenario-specific requirements of IoT networks.

The IoT network uses multiple ANT based on the scenario-specific requirements of frequency, range, data-rate, channel bandwidth, signaling, and modulation, etc. Each of the IoT applications has its own specific set of design attributes and specification based on the deployment scenario. Table 3.2 below shows some of the most used wireless access network technology and their comparative overview based on the above discussed technical specifications. It is imperative to take these design specifications into consideration in deploying wireless network using 5G HetNets for any IoT application-based network. HetNet considered here in the context of IoT mostly belongs to the wireless network under the Low Power-Wireless Personal Area Network (LP-WPAN) category. Some of the most popular wireless network technologies in purview of our study here are SigFox, LoRa, Wi-Fi, Zigbee, and NB-IoT. Most of the IoT applications demand tiny constraint device communication using mMTC framework with Cyber-Physical systems (using smart sensor/actuator control applications). These sensor/actuator-based communications mostly use low-power embedded systems as the UEs hence the packet size / PDU being small in size but large in device number, hence making huge numbers of small block data traffic using massive M2M type communication.

Table 3.2 IoT network technologies and their technical specifications (Chettri & Bera, 2020).

Technology	Frequency	Range	Data rate	Channel bandwidth	Modulation	Standard
SigFox	868−915−928 MHz	20 + km	100 kbps	250−500 kHz	BPSK	ETSI
LoRa	915−928 MHz	15 km	50 kbps	100 kHz	CSS	LoRa Alliance
Zigbee	902−928 MHz / 2.4GHz	< 1 km	250 kbps	2 MHz	BPSK/ QPSK	Zigbee Alliance
Wi-Fi	2.4GHz/ 5GHz	100 m	10—150 Mbps (MIMO)	20−40 MHz	DSSC	IEEE 802.11
NB-IoT LTE (carrier inbound/ guard-band mode)	700, 800, 900 MHz	1 km (Urban) 10 km (Rural)	200 kbps	200 kHz	QPSK	3GPP

3.2.6 **5G-HetNet H-CRAN fronthaul and TWDM-PON backhaul: QoS-aware virtualization for resource management**

An optimum resource management technique using QoS-aware virtualization in H-CRAN of 5G-HetNet is proposed in Zhang, Huang, Zhou, and Chen (2020). The authors have proposed a Particle Swam Optimization (PSO)-based QoS mapping algorithm for Time Wavelength Division Multiplexing- Passive Optical Network (TWDM-PON) used in 5G-HetNet backhaul. The objective of this work is to optimize resource management by improving QoS. 5G virtualization and load-balancing mechanism based on QoS mapping algorithm in H-CRAN and backhaul is used to improve performance of 5G-HetNets. Large scale co-operative signal processing and network functions is used in the H-CRAN based front-haul comprising of RRHs and MBSs. However, the backhaul of 5G-HetNets comprised of BBU pools are the new bottleneck for resource allocation. The challenge is that the backhaul must keep up with the resource demand of H-CRAN in terms of high capacity, low latency, network availability, high energy, and cost efficiency. This potential change of improving QoS and reducing bottleneck in the backhaul is addressed using the proposed QoS-aware algorithm. The performance of TWDM-PON-based backhaul is improved by optimization of resource allocation using Network Virtualization (NV) and load balancing strategy aided with PSO based QoS mapping algorithm for better service provisioning (SP). Co-operative MultiPoint (CoMP) communication helps in reducing interference hence improving QoS. There are two more practical deployment challenges: (1) Use of multiple SPs using the same backhaul where the resource management is complex and (2) Inconsistent QoS-based priority queue (PQ) deployment between TWDM-PON (backhaul) and H-CRAN (fronthaul). To address the technical challenges the proposed NV based resource management architecture has two functional approach: (1) Wavelength allocation based on load balancing and (2) QoS mapping algorithm to mitigate QoS mismatch between TWDM-PON and H-CRAN. Fig. 3.5 shows the 5G HetNets architecture based on TWDM-PON (backhaul) and H-CRAN (fronthaul) for optimum resource management.

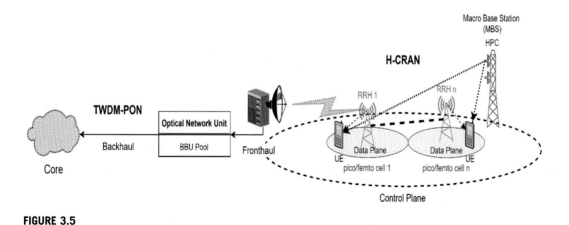

FIGURE 3.5

5G HetNets architecture based on TWDM-PON (backhaul) and H-CRAN (fronthaul) (Zhang, Huang, et al., 2020).

3.2.7 Spectrum allocation and user association in 5G HetNet mmWave communication: a coordinated framework

5G-HetNets architecture enables traffic off-loading with dense deployment of RRHs based small cells (data plane) operating on different frequency band based-on multiple radio ANT (heterogenous network) like 5G-NR, 4G/LTE, Wi-Fi, NB-IoT etc. Multitier HetNets share the microwave spectrum. Coordinated heuristic noncooperative game theory (state-of-the-art) method is proposed for user association and spectrum allocation in 5G-HetNets (Khawam, Lahoud, El Helou, Martin, & Feng, 2020). The parameters considered in the proposed methods are: mmWave density, cell load, user distribution, massive MIMO with high heterogeneous network settings. In the proposed work, a 6GHz spectrum is considered for allocation in small-micro cell deployment based on cell load, user distribution, spectrum efficiency and frequency reuse. Noncooperative game theory uses Nash-equilibrium to attain the best convergence for optimum cell density, cluster size, user density, traffic distribution; those are the global objectives of any 5G-HetNet deployment scenario.

3.2.8 Diverse service provisioning in 5G and beyond: an intelligent self-sustained radio access network slicing framework

Network slicing for concurrent SP of HetNets for QoS sensitive networks in the era of 5G and beyond is the need of the hour especially keeping in mind the heterogeneity in RAN and diverse SP requirements of HetNet without compromising with QoS. The challenge is to effectively slice the RAN with diverse QoS requirements (for multiple concurrent SP) and dynamic heterogeneity with the use of different wireless access technologies at RAN. However, a self-sustained RAN slicing framework (self-managed Network) using adaptive control strategy is proposed in Mei, Wang, and Zheng (2020). The proposed framework has intelligent features like self-organizing/optimization, self-slicing control based-on the most recent approach of Self-Learning Networks (SLN). The RAN slicing method proposed in this work is hierarchical in structure. This multitier RAN slicing has three levels: (1) Network-level for RAN and radio resource management, (2) gNodeB-level for cell management and (3) packet-scheduling-level for resource management. This approach enables more granularity in terms of resource management strategy giving more fine adaptive control over the QoS on a very diverse scenario of multiple SP requirements over 5G HetNets. The proposed transfer learning approach using AI-based algorithms helps in migrating from model-based learning to autonomic self-learning network (SLN)-based RAN slicing for QoS enhancement in diverse SP in 5G HetNets.

3.3 Device-to-Device communication in 5G HetNets

3GPP standard for 5G mobile communication provides an ad hoc communication framework for direct D2D communication between UEs without relying on the gNodeB (gNB) (does not require RRHs in data-plane and MBS in control plane). This D2D mechanism of handling local traffic (UE-to-UE traffic inside close proximity radio coverage area) helps in traffic off-loading and reduces bottlenecks in 5G RAN fronthaul as well as backhaul. The D2D mechanism particularly improves the performance of the network in mMTC type communications with massive UE density in a geographical radio coverage area. 5G architecture supports millions of UEs in a radio coverage area using mMTC framework. However, this creates a huge overhead on resource allocation in the 5G RAN infrastructure. Most of the mMTC traffic creates a challenging resource allocation requirement of huge number of small data-blocks as the UEs in this scenario would be millions of resource constraint embedded devices particularly of IoT and Machine-to-Machine (M2M) type communication. The D2D communication strategy being deployed the UEs would communicate directly with their peers hence the load on 5G RAN infrastructure would drastically reduce with improvements in overall QoS in the network. Fig. 3.6 shows the D2D communication in 5G HetNets. The close radio proximity UEs can communicate with their neighboring UEs (as shown in the figure) on-demand over a D2D ad hoc radio channel instead of depending on the 5G infrastructure. Only the remote node traffics (external traffic not local) are communicated via the gNB (base station) using existing 5G RAN infrastructure. This creates a hybrid communication architecture with ad hoc D2D communication (UE-UE) and infrastructure-based communication (UE−gNB).

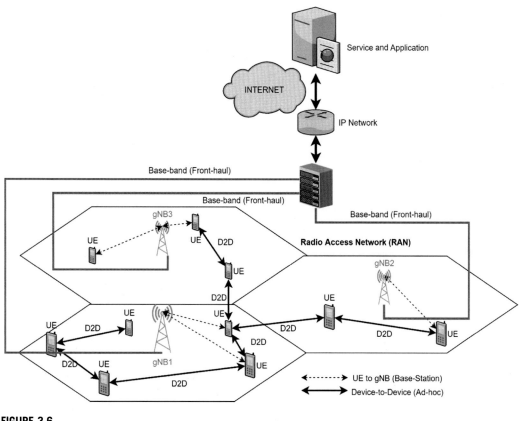

FIGURE 3.6

Device-to-device communication in 5G HetNets.

This mechanism of dealing with Local traffic (UE-to-UE traffic inside close proximity radio coverage area) by sending directly from UE to UE in an ad hoc communication without relying on an gNodeB/gNB in a 5G RAN drastically reduces the overall load on the front-haul as well as back-haul of 5G-NR and would help in reducing the bottleneck at gNB due to large number of UEs especially for mMTC. However, deploying D2D communication framework in a Cell would have its own set of challenges. The major technical challenges in D2D based ad hoc communications are as follows:

1. Multiple wireless hops.
2. High latency.
3. High complexity in resource allocation for a hybrid scenario where both ad hoc (UE-UE) as well as infrastructure (UE-gNB) communication with separate radio links takes place.
4. Computational overhead.
5. Mitigation of potential interference between two links (1. UE-UE: ad hoc D2D and 2. UE-gNB infrastructure). However, 5G-NR standard proposes beamforming technique to deal with interference in this case but the process is complex and would create a substantial computational overhead on UE for D2D communication.
6. Vulnerable to eavesdropping as several wireless broadcast links are established without centralized network control which is a huge network security challenge in D2D communication.

A D2D-enabled resource management of 5G and beyond HetNets is proposed in Irrum et al. (2021). The proposed system has a secured framework of communication with optimum resource management strategy deployed for D2D communication in 5G HetNets and beyond. The proposed secured communication framework considered both the type of communication scenario (1) Cellular users (CUs)—infrastructure-based communication and (2) D2D Users (DUs)—ad hoc communication. The authors have proposed a mixed integer nonlinear programming (MINLP)-based approach to improve secrecy rate, throughput, EE over separate channels for CUs and DUs. The proposed system achieved promising results in maximum secrecy rate with physical layer modeling using MINLP.

D2D communication technique in 5G HetNet would be instrumental in off-loading of local traffic from the 5G H-CRAN based fronthaul as well as optical backhaul by optimizing resource allocation. D2D communication would substantially improve the overall network performance as the local traffic would be handled separately over ad hoc wireless links mitigating severe bottleneck on H-CRAN with massive machine type communication environment. However, for enabling such D2D communication many challenges have to be addressed appropriately. Some of the major technical solution required for optimized deployment of D2D communication framework on 5G HetNet can be summarized as below:

1. UEs must be computationally regressive enough to handle complexity of multiple ad hoc wireless links as per the scenario-specific requirement of on-demand and dynamic local traffic in a radio coverage area. Devices should have minimum commuting resources to handle the complex signal processing required for D2D communications.
2. UEs must have smart antenna technologies and D2D communication reply mostly on complex beamforming techniques to mitigate potential interference from separate wireless links as the

communication is hybrid having ad hoc D2D wireless channel and UE-gNB (base-station) infrastructure based wireless channel.

3. The resource allocation strategy must be optimized by focusing on complex hybrid communication paradigm as well as scenario-specific requirement of massive D2D type dedicated wireless channels which might be the case for mMTC/ M2M based local traffic for over millions of device deployment scenario.

3.4 Machine-to-Machine communication in 5G HetNets

3GPP standard R14 and newer has proposed mMTC for 5G-NR architecture supporting millions of devices in the CRAN through gNodeBs. This is technically viable through mmWave communication with a huge number of channels using millions of narrow-band (NB)-IoT devices for realizing mMTC in true sense. 5G provides URLLC for low-latency (<1 Ms in RAN) to make this Machine-to-Machine (M2M) communication viable. M2M communication has native support in 5G-NR enabling millions of NB-IoT devices to communicate over CRAN. The SDN architecture provides the flexibility to support large number of software defined channels to accommodate millions of devices over RAN. Even the flexibility of data-control plane splitting and CoMP communication is key to attainment of flexibility much needed for mMTC framework in 5G-NR. Cognitive Radio technology driven Smart Objects (CRSOs) are instrumental in the deployment of M2M communication for a huge number of devices over a same radio coverage area with spectrum constrains especially when there are large number of narrow-band channels to be associated with gNB (in a Cell).

However, there are quite a few challenges with the deployment of mMTC. A few of the predominant challenges in M2M deployment are as follows:

1. Legacy systems are designed only for few hundred devices per BS and not millions.
2. Achieving low latency on a live network is a challenge especially for huge number of devices mMTC hence critical low-latency application like smart healthcare is a challenge.
3. The M2M communication has a small payload size. This means millions of small data-blocks for diverse devices is a challenge to allocate real-time network resource pool.
4. Control and channel estimation in M2M communication is a huge overhead.

3.4.1 Machine-to-Machine communication in 5G: state of the art architecture, recent advances and challenges

Unlike human-to-human communication, M2M has distinct features as listed below:

1. mMTC type large number of small data packets massive number of device deployment in a geographical radio coverage area.
2. Need for interoperability as the devices (hardware and software), networks are dissimilar and heterogenous.
3. Provision for support of HetNets as both wired and wireless scenarios of deployment may be feasible as per scenario-specific requirements of M2M application.

4. 5G integration is key to achieve enhancement in M2M deployment.
5. Guaranteed QoS-based service is required for M2M communication.
6. Exponential growth and market potential of M2M/IoT application in recent times.
7. Legacy systems like 4G and Wi-Fi based networks does not have sufficient resource to handle mMTC type communication hence 5G HetNet integration is imperative as 5G has deign advantages over older legacy systems with SDN-based C-RAN resource management flexibility, Cognitive Radio, mmWave and MIMO technologies.
8. 5G network architecture supports massive M2M (mMTC), enhances data-rate (eMBB), less End-to-End latency (URLLC), improved EE over small cells, increased battery life for M2M devices and increased network capacity to accommodate massive number of devices per cell.
9. Different Standard Development Organizations (SDOs) like 3GPP, ETSI, IEEE, W3C are active in the development of cutting-edge solutions in the IoT and M2M market space.

M2M communication in 5G has three domains of operations as described below (Mehmood, Haider, Imran, Timm-Giel, & Guizani, 2017):

1. *M2M device domain:* sensor/actuators equipped low-power embedded devices forward data to the nearest nodes and gateways (multihop ad hoc networks) are part of the device domain.
2. *M2M communication domain:* data forwarding nodes like M2M gateways (M2M-GWs), Rode Side Units (RSUs), gNBs, RRHs, MSBs, BBU pool, WAN, xDSL, Optical backhaul all can be part of communication domain.
3. *M2M server/ application domain:* M2MMiddleware, service-layer and business provisioning layers are part of server/application domain.

Technical advantages of enabling 5G technology for M2M communication are listed below:

1. *Improved core network/ backhaul:* 5G network with TWDM-PON based backhaul and NFV has improved core network resource and QoS-based performance enhancements.
2. *Improved radio access/ RAN technologies:* H-CRAN based HetNet support, SDN, D2D, data and control plane splitting and enhanced radio resource allocation.
3. *Improved trans-receivers:* technologies like mmWave, MIMO, beamforming and smart antenna have substantially improved the Trans-receivers in 5G networks.
4. *Enhanced radio interface:* 5G provides enhanced radio interface with more spectral radio resource and operational in mmWave frequency range ($> 30GHz$).
5. *Cell enhancement:* 5G network has small (pico/femto) cells in data-plane to enhance device density and network capacity and MBSs for control-plane.

3.4.2 Recent advancement in the Internet of Things related standard: oneM2M perspective

In the era of IoT mMTC based communication framework is imperative for integration of millions smart devices (Sensor/actuator control) to the 5G HetNet H-CRAN for diverse SP. IoT-based network deployment demands a single converged communication framework for seamless integration and interoperability especially when different (heterogenous) wireless access technologies are used for scenario-specific requirement of massive M2M communication. oneM2M is a unified communication framework for massive M2M type communication with 5G and heterogenous networks. IoT and mMTC-based oneM2M frameworks provide a collaborative approach for interoperability among heterogenous devices and networks. oneM2M provides an abstract semantic based unified layered architecture for seamless communication among heterogenous devices in the era of 5G5G HetNets. A semantic-enabled IoT platform would help in understanding IoT data in a standard and generic perspective for all types of dissimilar devices for achieving seamless interoperability. OneM2M interoperability specifications guarantee IoT devices of different vendors to communicate with each other over a common M2M framework. OneM2M architecture has specific three functional layers, distinct node types and their functionals attributes (Park, Kim, Joo, & Song, 2016). The three-layer architecture of oneM2M framework as given below:

1. Application entity (AE): application layer
2. Common service entity (CSE): service layer. There are twelve CSE entities proposed by oneM2M given in Table 3.3 below:
3. Network service entity (NSE): network layer

Reference points: Three reference points (interfaces) are provisioned for interlayer communication. "*Mca*" interface between AE−CSE, "*Mcn*" interface between CSE-NSE and "*Mcc*" interface between multiple CSE-CSE.

oneM2M Nodes types: There are three major types of nodes in oneM2M architecture:

1. *Application service node (ASN) or application dedicated nodes (and):* M2M end devices
2. *Middle node (MN):* M2M gateways
3. *Infrastructure node (IN):* M2M servers and platforms with support for big-data analytics and cloud-based SP

Unified service oriented M2M functional Architecture: an overview of oneM2M reference model is presented for interoperability in 5G5G HetNets for massive machine type communication. Fig. 3.7 shows the oneM2M architecture and its functional components.

OneM2M Release-1 (2014) has proposed a set of Common Service Functions (CSFs). The important CSFs are given in Table 3.4.

OneM2M Release-2 (2015) has proposed some advancement over the oneM2M-R1. The main functions of oneM2M-R2 are as follows:

1. *Platform/internetworking functions*: Supports five types of internetworking; (a) AllJoyn, (b) OIC, (c) Light-weight M2M, (d) Generic and (e) 3GPP Release-13.
2. *Home domain support*: oneM2M-R2 has support for home appliances information model.
3. *Industry domain support*: oneM2M-R2 has support for time-series data for industrial IoT applications.

Table 3.3 Common service entities proposed by oneM2M.

1. Registration	5. Data management repository	9. Communication management
2. Discovery	6. Subscription and notification	10. Network service exposure
3. Security	7. Device management	11. Location
4. Group management	8. Application and service management	12. Service charging and accounting

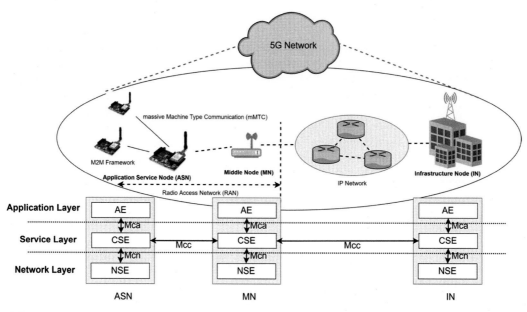

FIGURE 3.7

OneM2M architecture and its functional components.

Table 3.4 Common service functions in oneM2M R1 (2014).

Common service functions (CSFs)	Basic functionalities
Communication management and delivery handling (CMDH)	Connection, communication and scheduling
Data management and repository (DMR)	Data—storage, mediation, aggregation, formatting
Device management (DMG)	Device (ASN, MN, IN) capabilities management
Discovery (DIS)	Text-matching, information searching of attributes
Group management (GMG)	Management of group requests
Network service exposure, service execution, and triggering (NSSE)	Communication management with underlying 3GPP, 4G-LTE, 5G5G-NR Zigbee, etc.
Registration (REG)	Service registration for AEs and CSEs
Security (SEC)	Access Control. Authentication/ authorization
Subscription and notification (SUB)	Tracks resources based on subscription and then notifies

4. *Semantics*: oneM2M-R2 has specific three sets of semantic support for semantic-based interoperability; (a) oneM2M base ontology, (b) Semantic discovery and (c) Semantic description.

5. *Security*: oneM2M-R2 has two approaches for resilient network security model in place; (a) End-to-End security and (b) Dynamic authorization.

OneM2M provides service layer interoperability over multiple proprietary systems. The challenges over a legacy heterogenous systems are: (1) inflexibility, (2) lack of modularity and capabilities of service extension, and (3) lack of provision for integration of new services. The solution for all these challenges is oneM2M which is global framework and would work with all legacy and proprietary systems with a common service layer functional architecture (Swetina, Lu, Jacobs, Ennesser, & Song, 2014). CAPEX and OPEX for proprietary solutions are always more as compared to a single, common, open, and generic one like oneM2M. OneM2M provides a horizontally common platform for multiple M2M type industrial verticals. The oneM2M provides (1) architecture, (2) candidate protocols, (3) security aspects, (4) device management, and (5) abstraction technologies.

OneM2M standard has been developed by a consortium for various standardization organization / Standard Development Organizations (SDOs) and more than 270 companies across the globe. oneM2M developer consortium includes eight major SDOs: (1) Association of Radio Industries and Businesses (ARIB), Japan; (2) Telecommunication Technology Committee (TTC), Japan; (3) Alliance for Telecommunications Industry Solutions (ATIS), United States; (4) Telecommunications Industry Association (TIA), United States; (5) China Communications Standards Association (CCSA), China; (6) European Telecommunications Standards Institute (ETSI), Europe; (7) Telecommunications Technology Association (TTA), South Korea; (8) Telecommunications Standards Development Society (TSDSI), India.

oneM2M alliance has three major industry forms: (1) Open-Mobile Alliance (OMA); (2) Broadband Form (BBF); (3) Home Gateway Initiative (HGI).

There are five working groups (WGs) associated with oneM2M are: (1) WG1- Requirements; (2) WG2-Architecture; (3) Protocols; (4) WG4-Security; (5) WG5-Management, Abstraction and Semantics.

3.4.2.1 Advantages of oneM2M

1. Boost M2M economy by reducing time to market.
2. Simplify development by common APIs
3. Leverages existing world-wide network with enhancements in protocols, services and business operations.
4. Provides evolution, interoperability, and standard functions.

3.4.2.2 OneM2M protocols

1. *Service layer protocols:* RESTful, URI, APIs based protocol design for oneM2M.
2. *Encapsulation in existing protocols:* Legacy TCP/IP stack is used for oneM2M PDUs when required.
3. *Binding with underlying protocols:* OneM2M binds with LoRa, Zigbee, BLE, SigFox, NB-IoT based ANT through ADN/ASN—MN (M2M gateways).
4. *Interworking with non-M2M systems:* Interoperability with legacy application layer protocols like MQTT, CoAP, XMPP etc.

5. *Security:* Lightweight Datagram Transport Layer Security (DTLS) over CoAP in implemented in oneM2M.
6. *Working mechanism:* KNX based abstraction and semantic interoperability.

3.4.2.3 OneM2M standard platform: a unified common service-oriented communication framework

1. End-to-End platform with common service capabilities.
2. Semantic interoperability using common APIs.
3. Seamless integration over heterogenous networks and legacy systems.
4. oneM2M software framework is especially designed for IoT and M2M traffics providing a single a common convergence platform for interoperability of various types on hardware, software, storage, communication and data.
5. Supports required M2M functionalities like security (authorization and encryption), service scheduling, and notifications.
6. Provides co-operative intelligence over devices, gateways, and clouds.
7. oneM2M is a like a generic operating system (OS) compatible for sensors, devices, gateways, and clouds.
8. Common, generic, and global in standard hence provides a seamless interoperability for all M2M based nodes.

Fig. 3.8 shows oneM2M as a common service-oriented platform providing global interoperability for dissimilar devices and network entities hence making seamless integration of heterogenous systems possible for M2M communication (Gezer & Taşkın, 2016).

The OneM2M architecture has three operational layers:

1. *Connectivity layer:* IoT devices using mMTC type communication to connect to the HetNet gateways (GWs) which has traffic aggregation (TA) and seamless interoperability to the IP-based network backbone. Various dissimilar devices and network segments like NB-IoT, Zigbee, SigFox, LoRa etc. can be integrated to a common IP-based network.
2. *Service layer:* OneM2M unified common service layer provides semantic interoperability to various types of heterogenous device and data traffic to application layer interface through common Application Programming Interfaces (APIs).
3. *Application layer:* Scenario-specific applications like Smart home/cities, smart grid, Industrial IoT, smart healthcare and medical application etc. are supported by the application layer using APPs and specific functional APIs.

3.4.3 M2M traffic in 5G HetNets

M2M traffic is comprised of low data-rate, small packets PDU with correlative transmission. A two-phase traffic control (2PTC) architecture is proposed in Li, Rao, and Zhang (2016) for optimum M2M traffic management in M2M deployment scenario in 5G5G HetNets. The proposed multipath Traffic Engineering (TE) model has Virtual Service Gateway (V-SGW) to aggregate traffic from multiple machines (ADNs/ ASNs) and provide data to the sink. TA mechanism is used at the V-SGW for multiplexing large number of small sized data to a single logical data resource block to

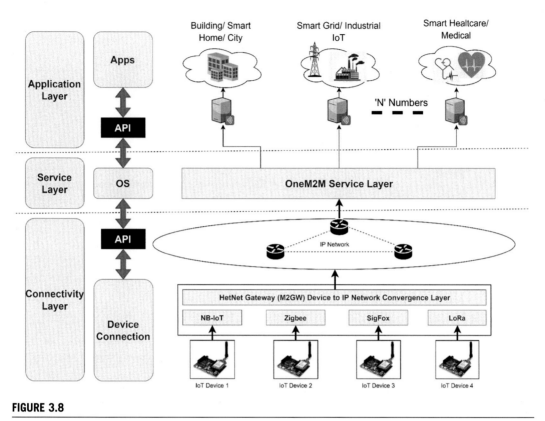

FIGURE 3.8

OneM2M communication framework.

leverage the 5G5G based H-CRAN resource block in optimized approach. Fig. 3.9 shows the two-phase traffic control (2PTC) for M2M communication.

Phase-1: Machine-to-Gateway (M2GW) traffic: Low data-rate, small packet distinct data-traffics are forwarded form machines (End-Nodes/ASNs/ADNs) through respective wireless access points via the intermediate routers to the Virtual Service Gateway (V-SGW). This phase-1 communication is called machine to gateway. *Phase-2: Gateway-to-Sink (GW2S) traffic:* TA takes place at the V-SGW and the cumulative data traffic is sent to the designation or sink node. This phase-2 communication is called gateway to sink. The proposed mixed integer programming (MIP) based 2PTC based gateway solution has following challenges in any practical deployment scenario: (1) Optimum Machine-Gateway association; (2) Minimize association cost; (3) Minimize TA/Gateway counts. The V-SGW/TA alleviates large quantity of mMTC type traffic (small packets) challenge. 5G5G enabled M2M communication framework has following technical advantages:

1. Flexible and scalable SDN based H-CRAN in 5G5G has granularity in resource management.
2. Better resource manageability over separate control plane (using MBSs -macrocells) and data plane (using RRHs—pico/femto cells) in 5G5G.

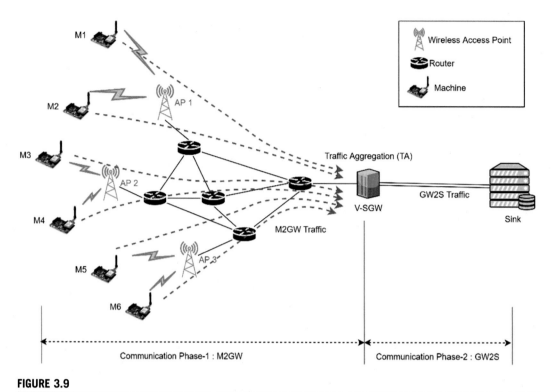

FIGURE 3.9

2PTC for M2M communication (Li et al., 2016).

3. Agile network architecture using Node-C (Cloud Nodes) and SDN controllers (SDN-C) using OpenFlow for better traffic management. Using better data-flow management and resource allocation using SDN-based framework provides improvement in required QoS parameters with resource constraints.

3.4.4 Distributed gateway selection for M2M communication cognitive 5G5G networks

M2M communication supports large number of devices (mMTC) with minimum to no human interaction. Cognitive 5G5G networks provides high spectrum efficiency (SE) and high bandwidth (spectrum resource) using mmWave technology. Multiple gateways (GWs) deployment for M2M communication is imperative to increase throughput, coverage, EE (smaller cells), and capacity. However, a single channel Carrier Sense Multiple Access (CSMA)-based MAC protocol would create challenge of interference from primary/secondary M2M devices. The solution for this challenge is a distributed gateway selection mechanism proposed in Naeem et al. (2017) using Decentralized Multiple Gateway Assignment Protocol (DGAP). A lightweight variant of the same called Low Overhead DGAP (Lo-DGAP) is used utilizing a multichannel CSMA-based MAC protocol for

M2M communication over cognitive 5G5G networks. To minimize the interference the GW device would transmit to worst primary M2M device rather than to all neighbors as per the proposed protocol. The proposed Lo-DGAP increases throughput, reduces header size (protocol overhead) and improves EE. The architecture of Lo-DGAP has a cloud computing based Central Control Unit (CCU) which facilitates the following:

1. Lo-DGAP requires less control traffic for its network operation hence increases EE substantially.
2. Lo-DGAP has a reward function mechanism to assign weights to the co-operating GW devices. This mechanism would be helpful in load-balancing across multiple GWs and hence improves throughput, capacity and spectrum efficiency.
3. Lo-DGAP is a light-weight protocol with less header size hence improves protocol efficiency.
4. Lo-GGAP required less control traffic hence improves network efficiency.

3.4.5 Algorithm for clusterization, aggregation, and prioritization of M2M devices in 5G5G HetNets

There are quite a few challenges in deployment of massive M2M communication due to very large number of devices to be connected to the C-RAN of 5G5G HetNets. The major challenges of M2M communication over a 5G-HetNet legacy system are as follows:

1. Huge number of M2M devices, and therefore a large amount of control traffic.
2. Complexity in resource management in 5G5G HetNets as centralized pre-synchronized devices are forming network cluster hence allocation of data transmission blocks to gateway is a challenge.
3. Data aggregation, classification (Group management) is a challenge due to massive number of traffic profiles.

Due to above challenges it is imperative to device a mechanism or deploy algorithm to address such challenges raised due to mMTC type large number of M2M devices and traffic. A Fuzzy Logic System (FLS) based algorithm is proposed in Klymash, Beshley, Seliuchenko, & Beshley (2017). The FLS-based proposed system has distinct attributes to solve above challenges by reducing signal load on Base Stations (BSs) by reducing control traffic. Secondly, the proposed system is promising in its results for better radio resource management with more energy efficient design hence increasing the battery life of M2M devices. The proposed cluster-based network architecture has Master and Slave M2M devices. Each master device cater service as a cluster head providing TA features on the cluster and multiple clusters communicate to a centralized M2M-Gateway (M2M-GW). The M2M-GW sends the aggregated data to the gNB of 5G5G H-CRAN in a HetNet deployment scenario. The FLS-based algorithm provides the following functions:

1. *Clusterization:* The system uses a Master-Slave communication model for M2M devices, where master device is acting as a cluster head and all the participating slave devices connected to master act as member M2M device nodes of the respective cluster.
2. *Data aggregation:* All the data from Slave nodes are sent to the master node, and at the master node, all M2M device traffics are aggregated and forwarded to M2M-GW.

3. *Classification of data and prioritization:* The proposed system has QoS Class Identifier (QCI) a metric in a scale (Li et al., 2018; Ouamri et al., 2020; Peng et al., 2015; Sun et al., 2015) using Type of Service (ToS) fields of IP packets for classification data traffic and based on the QCI priority to data traffic can be assigned.

3.5 Heterogeneity and interoperability

The IoT is described as a complicated heterogenous network where things (devices) are integrated among themselves with wired or wireless medium to exchange information. Hyung-Jun Yim et al. described the "thing" as data, events and services (Yim et al., 2017). The things have the capabilities of data sensing, data capture and storage, data processing and communicating. The features of things including location, address and control perform through several communication protocols such as ZigBee, Wi-Fi, Bluetooth, 6LoWPAN etc. However, IoT devices demand interoperable protocols for seamless integration of intelligent devices to the network. To establish a smart environment using IoT (such as smart home, smart healthcare, smart agriculture, etc.), there is a need to identify the device ID. The identification system in a single IoT platform is same where as it is challenging to identify and operate in between different devices in different IoT platforms. At this point, it is required interoperability between different heterogenous devices in different IoT platforms. Some of the IoT platforms that are in developing stages are oneM2M, Apple's HomeKit, Google's BrilloWeave, OCF's IoTrivity, Samsung's ATIK, GS1's Oliot and AllSeen Alliance's AllJoyn. For Industrial and commercial utilization of the IoT device services there is requirement of global identification and semantic information integration of IoT devices. However, the interoperability between IoT devices with other devices and with users faces some challenges such as extensive coordination, universal heterogeneity, unspecified IoT device configuration and semantic conflicts. Moreover, the aim of the requirement of interoperable protocol is to develop IoT network without modifying its system and maintaining stability and high performance.

Various literatures described about the two type of interoperability problem, that is, (1) user interoperability problem and (2) device interoperability problem. The user interoperability problem arises between the user and the device (Jardim-Goncalves, Grilo, Agostinho, Lampathaki, & Charalabidis, 2013; Panetto & Cecil, 2013). The device interoperability problem arises between two devices (Kim, Lee, Park, Moon, & Lim, 2007). Few researches focused on global middleware in heterogenous network to handle the device interoperability problem (Moon, Lee, Lee, & Son, 2005).

3.5.1 User interoperability

To resolve the issues of user interoperability, the following cases need to be solved.

1. Locating devices
2. Interconnection with devices

The problem of finding devices as per the requirement of the user is solved by using IPv6 by setting a link to the device and accepting an electronics product code (EPC) scheme for identification of that device (EPC Tag Data Standard, 2014). However, the issues of categorization of devices into existing and new catalog to avoid semantic ambiguity arise in locating a device. Considering the case of interconnection with devices after its discovery, the device needs to be in a supportive format (syntax and semantics) to the user. At this point, two problems emerge such as syntactic device interoperability and semantic device interoperability. In syntactic device interoperability, the instruction from user is executed by the device and the messages from the device are executed by the user's system. In semantic device interoperability the information from the device to the user is explained by the user and the information from the user to the device is explained by the device.

3.5.1.1 Locating the device through identification and classification

For identification of every device connected in IoT platform, EPC code standard plays an important role to accept the static data of a device. EPC code standard is developed by GS1 AISBL in uniform resource identifier (URI) format (EPC Tag Data Standard, 2014). Its unique identification system can indicate the static device object belongs to a particular product. The disadvantages of EPC are the ambiguities created in identification process which need to be taken as challenge for further research. The IP address and the URI can be taken as unique identifier. Considering the IP address, the IPv6 can be used by the device to sense, actuate and communicate (Dunkels & Vasseur, 2008). This IP address identifier can only be applied for the object in static position but for moving and roaming there must be created a unique device ID. Devices having in built web server can be accessed by URI. In Guinard and Trifa (2009), the authors stated about RESTful resources (web service API) to access the internet-based objects through URI. Different devices have different device ID; therefore users face difficulties to remember every device ID. For classification of device IDs two standards introduced in IoT, that is,. UNSPSC and ecl@ss (Xiao, Guo, Da, & Gong, 2014).

3.5.1.2 Syntactic and semantic interoperability for interconnecting devices

To handle the syntactic heterogeneity by exchanging the messages between the device and the device users without having any communication problems, few approaches have been adopted. According to various research articles, some of the technologies are introduced to remotely access the devices such as Web services (Guinard, Trifa, Karnouskos, Spiess, & Savio, 2010), open and close standard protocols (Baronti et al., 2007; Callaway et al., 2002), RESTful (Lanthaler & Gütl, 2012), and SOC (Guinard & Trifa, 2009). Along with this, few exiting heterogeneous standards, including IEEE 802.15.4 and ZigBee, are unable to satisfy the interoperability among heterogenous devices. Therefore to manage the above issues of interoperability among heterogenous devices, the middleware technologies concept has come to light though various IoT research articles (Gama, Touseau, & Donsez, 2012; Kim et al., 2007; Moon et al., 2005). Semantic interoperability is considered as important issues in device interconnection as here the device and user have to understand the exact meaning of the sent and received messages from and to them. To set up a semantic interoperability between device and user in IoT platform, ontology plays a key role to maintain a relationship between the objects (Wang, De, Toenjes, Reetz, & Moessner, 2012). However, ontology gives a benefit to the device situated in the range of same context but it has limitations for cross context devices. To control this problem, collaborating on the conceptual theory has been proposed where things are considered as signs as semiotics and when a device and a user from different context connected they first collaborate and built semantic signs (Guo, 2009). Moreover, the interoperability problems still need further research for smooth integration of device and users between different contexts in IoT environment.

3.5.2 Device interoperability

Device interoperability is needed to establish a seamless communication between different heterogenous devices in IoT in a particular defined domain or different domain. It requires the integration of different heterogenous device (smart device) that may be high end or low end. The advantages of high end IoT devices consisting of sufficient resources and high computation capabilities and the limitations of low end IoT devices such as less resource (energy and power) and lack in communication capabilities encourages to integrate between devices to provide better performance in terms of services. Some of the existing protocols and standards to support various applications, that is, for e-healthcare the medical device support ANT + standard and for the wearable device it is NFC and Bluetooth SMART. For the smart environment, the IoT-based sensor is connected through ZigBee. Along with this standard, some of the nonstandards communication medium including LoRa, SIGFOX also support sensing and communicating. The main characteristics of device interoperability are to exchange information between different heterogenous devices and protocols and to develop capabilities of integration of new devices in to the IoT environment. Comparative details are provided in Table 3.5.

3.6 Research issues and challenges

The increased requirement of mobility and connectivity of mobile devices in anytime and anywhere increases the deployments of wireless networks to fulfill the demands of wireless communication

Table 3.5 Classification of interoperability in Internet of Things.

Types of interoperability	Supporting technology developed	Research issues and findings
User interoperability (syntactic and semantic)	WSDL, RESTful, SOC, open and closed standard protocol	Locating devices and interconnecting them.
Device interoperability	ANT + standard, Wi-Fi, Bluetooth SMART, ZigBee, NFC, Wireless HART, 3G/4G, Z-Wave, LoRa and SIGFOX.	Exchange of information between heterogenous devices and integration of new device to IoT platform.
Network interoperability	Wired and wireless communication standards.	Exchanging information in different networks by solving the issues of QoS, resource management, routing, security, mobility, and addressing.
Platform interoperability	Apple's Homekit, Google's Brillo, Amazon AWS, Contiki, TinyOS, RIOT, and Open WSN.	Cross platform interoperability need further research to support smart IoT application.

services capabilities. At this point, the large volumes of data-oriented services and applications play a key role for the emerging of 5G5G communication technology. The 5G5G new radio system is going to establish the connectivity among people and other equipment, vehicles and machine through cellular network, IoT, V2V and M2M respectively (Camacho, Cárdenas, & Muñoz, 2018).

3.6.1 Resource allocation

The process of resource allocation includes the sharing of network resources in wireless communications by maximizing the transmitted data to the users. The conventional resource allocation method is inefficient to meet the demand of large volume of data-oriented applications and services. Therefore efficient radio resource management technique is needed for proper utilization of spectrum. However, various researchers have devised different radio resource techniques to optimize the system parameters such as system throughput, SE, fairness, Quality of Service (QoS) and Quality of Experience (QoE). The system throughputs can be measured by considering the total data rate passed through a device in a network, that is, bits per second (Chen, Li, Chen, Lin, & Vucetic, 2016; Liu, Chen, Yu, & Li, 2019). Few literatures proposed some of the resource allocation techniques which is going to be helpful in solving further issues. In Alnoman, Ferdouse, and Anpalagan (2017), the authors proposed a user association and bandwidth allocation technique for heterogenous network using fuzzy logic controllers. According to user demand and bandwidth availability, the controllers manage the bandwidth allocation to each user by maximizing the throughput in the network. At this point, the approach shows enhancement in data rate, resource utilization, and number of offloaded users.

In Wang, Chen, Tang, and Wu (2017), author considered an approach using user association and power allocation to resolve optimization issues. Here, the author considered two sub problems for solving the resource allocation and power allocations problem using two methods, that is, by graph theory using Hungarian algorithm and difference convex function approximation method. This technique improved the overall system throughput but have limitations in proving services for UEs in bad channel conditions. Feng, Mao, and Jiang (2018) proposed a technique of joint resource allocation considering centralized and distributed types. Here, the author integrates resource allocation with user association and frame design in a HetNet with MIMO. The result shows that the centralized technique has significant improvement in system throughput and robust toward different network settings than the distributed technique. In Yuan, Xiao, Bi, and Zhang (2017) author gave a game theory-based approach considering joint resource and power allocation framework to fit into a large coverage of network having a greater number of base stations (BSs). Some more literature states about the resource allocation with cognitive HetNet, multiband mmWave HetNet to solve the optimization and improve the system throughput (Ji, Jia, & Chen, 2019; Xu, Hu, Chen, Chai, & Li, 2017; Zhang, Wei, Wu, Meng, & Xiang, 2019).

Few studies have focused on resource allocation techniques considering another performance parameter SE. In Ye, Dai, and Li (2018), the authors proposed a model by applying the hybrid-clustering game algorithm and combining the auction method to reduce the cross-tier and cotier interference for improving the SE. But the limitations include priority and QoS. However, in Ya, Feng, Zhou, and Qin (2018), the authors considered RAT in HetNet to increase the network throughput and maintaining QoS by using multiagent reinforcement learning technique. The proposed model gave better results as compared to other scheduling methods in terms of optimization and system throughput. Researchers also proposed resource allocation techniques using game theory, binary search algorithm and macro cells to enhance the SE and guarantee QoS of the network (Manishankar, Srinithi, & Joseph, 2017; Xie, Zhang, Hu, Wu, & Papathanassiou, 2018). Along with QoS, another factor QoE (Quality of Experience) plays a key role in satisfying the user requirements, like performance factors in QoS such as delay, jitter, and throughput and packet loss. Wang, Fei, and Kuang (2016) considered a model for QoE in HetNet using macro cell and pico cell for sharing of spectrum in transmission process. However, the disadvantage of this proposed model is that it is limited to pico cell and increasing coverage area (cell size) worsens the QoE. Fairness is also an important parameter in resource allocation as there is need of equal distribution of resources according to the QoS requirement of the network. The studies in Gures, Shayea, Alhammadi, Ergen, and Mohamad (2020); Tang, Zhou and Kato (2020) focused on the fairness in the resource allocation process by taking into consideration of different methods such as controlling the sharing process in a centralized way, adjusting between priority and throughput using different utility function weights and weighted α-fairness based optimization technique to maintain a trade-off in performance of network fairness and system throughput.

3.6.2 Interference management

Interference management is considered as one of the important research issues as there may be chances of interference between transmitted signals from communicating devices present in same or closer frequency bands. There are two types of interference present in multitier HetNets such as (1) cross tier and (2) cotier. Cross tier interference is existing between the users present in different network tiers whereas cotier exist between the users in the same network tier. In Yang et al. (2018), authors proposed a join interference management and resource allocation technique with D2D based HetNet to reduce cross tier and cotier interference. When an alternate method for joint allocation problem compared with conventional DL/UL technique, it is found that the delay in overall system process increased. The literature in Xie et al. (2018) described FFR-based methods where reduction in cross tier and cotier interference were achieved along with increased throughput and higher capacity. In Huang, Zhou, Luo, Yang, and He (2017), authors proposed a model using Logarithmic and K-means clustering techniques to reduce the cross tier and cotier interference and achieve higher throughput by considering frequency reuse and fairness. But it has the limitation of increased in energy consumption. Researchers in Kaneko, Nakano, Hayashi, Kamenosono, and Sakai (2016) implemented the techniques of CSI overhearing, Normalized PF scheduling (PFS), and iterative water filling algorithms to avoid the cross-tier interference to not to have additional control overhead from MBS, and it can be applied in any type of channel quality aware scheduler. However, the cotier interference is not considered for this model. The paper (Xu, Hu, Chen, Song, & Lai, 2017) considered two techniques, that is, for MUEs, it is improved simulated annealing algorithm and for SUEs, it is maximize minimum distance algorithm to manage cross tier interference and high throughput and having issues of no service for SINR users. Lv, Liu, and Gao (2017) proposed a method using geometric programming and convex relaxation approach for managing cross tier interference and throughput to achieve QoS of MUEs but didn't considered the cell edge users.

3.6.3 Power allocation

Power allocation in 5G5G plays a key role as incoming communication need green technology for less consumption of energy. Distribution of power according to the requirement of users can reduce interference. Various literature describes the challenges and issues still exist in allocating power. In the paper (Pérez-Romero, Sánchez-González, Agustí, Lorenzo, & Glisic, 2016), the authors implement Q-learning and Soft-Max decision making and adoption of the logarithmic cooling technique to reduce power consumption. The fairness issue arises as it is not considered. However, the paper (Lohani, Hossain, & Bhargava, 2017) considered four techniques such as discrete binary PSO, dual decomposition method, dynamic programming, and greedy technique to reduce power consumption and achieve higher throughput having limitations in high computational complexity. Niknam, Nasir, Mehrpouyan, and Natarajan (2016) proposed a model using dual decomposition method, sub gradient method, sub optimal complexity greedy algorithm to improve system throughput and reduce power consumption but having issues of high implementation complexity. Authors in Munir, Hassan, Pervaiz, Ni, and Musavian (2016) considered noncooperative game and the dual Lagrangian decomposition method for improvement of EE. At this point, this proposed model having the issues of lack of interference management. Authors in Naqvi et al. (2018) implemented the technique such as Hungarian algorithm, Weighted Tchebycheff and dual decomposition method to get the advantages of real blockage effect and environmental geometry but having limitation in fairness. Another implementation considered in Wang, Zheng, Jia, and You (2017) used sum rate maximization problem and first order approximation based iterative algorithm and get fast convergence with improved throughput with fairness issues.

3.6.4 **User association**

User association is required to find the best available network for each and every user in HetNet. The user association is dependent on the request, range of base station (BS) from user and quality of the channel. Some of the current studies focus on issues of user association considering the performance matrices such as QoS, EE, power consumption and network utility. In Zhou, Huang, and Yang (2016), the author proposed a method considering the technique of nonlinear mixer −integer optimization and primal dual interior point method to improve EE, load balancing and throughput with a limitation of high-power consumption with less UEs. The author in Luo (2017) proposed a model using BCD, ADMM, MF techniques to gain low complexity having issues of interference. In the paper (Kim & Chang, 2017), the authors implement a model using the technique cost based approach, relaxation and decomposition method, distributed power update method to get fast convergence behavior on large scale network with issues of fairness.

3.6.5 **Computational complexity and multiaccess edge computing**

Computational complexity is considered as an important issue in the resource allocation process to establish a trade-off between the required processing time and the increasing hardware cost of the machine. To maintain an equilibrium state between resource allocation and power allocation, the algorithms based on game theory approaches must be solved through machine learning (ML) techniques as the future wireless communication having voluminous data demands artificial intelligence (AI) to be incorporated as RRM complexity in increasing (Elsayed & Erol-Kantarci, 2019). In Lv et al. (2017), authors used the technique of dynamic programming and divide into two phases, that is, planning and implementation to and generate resource allocation lookup table. Here, greedy algorithm is used to reduce the complexity but as the number of users increases the complexity also increases. Le, Sinh, Lin, and Tung (2018) gave an idea to study the complex traffic pattern using data mining but different types of traffic increase the processing time. To achieve the low latency and high mobility in 5G5G communication system, multiaccess edge computing (MEC) has been introduced to reduce the computational load. It computes the large volume of data and sends it to the cloud with in the RAN through UEs. The deployment of MEC with cloud RAN can be benefited from the network function virtualization (NFV) but faces an issue of network management in HetNets (Chabbouh, Rejeb, Choukair, & Agoulmine, 2016; Siddavaatam, Woungang, & Anpalagan, 2019; Zhang et al., 2015; Zhang, Gui, et al., 2020).

3.6.6 Current research in HetNet based on various technologies

In 5G5G communication system, the compute intensive applications and services demand various networks such as M2M communications, D2D communications, vehicle to vehicle (V2V) communications, IoT, WSN, mobile edge computing (MEC), cloud radio access network (CRAN) and cloud computing to integrate among themselves to satisfy the technological requirements. To support this type of integration of networks, the heterogenous network (HetNet) is evolved from the conventional homogeneous network. To enhance the quality of service of users, the HetNets support four types of cell network such as macro-cell network, micro-cell network, pico-cell network and femto-cell network. Along with different cell types, HetNets provide different communication scenarios using various technologies such as conventional HetNet, OFDMA based HetNet, NOMA based HetNet, Cooperative HetNet, H-CRAN, Multi antenna HetNet. In conventional network the communication supports all types of cell network, that is, macrocell, microcell, pico cell, and femto cell. The base station (BS) and the UE exchange the signal transmission through a single antenna without any relay. Although this network able to provide high throughput and improve QoS to the users but femto-cell interference still exists. To overcome the issues of interference between different subchannels, orthogonal frequency division multiple (OFDM) has been introduced. In Rezvani, Mokari, Java, & Jorswieck (2019); Song, Min-Jae, and Won-Chang (2020), the authors proposed two delay-based fairness techniques known as proportional fairness (PF) and min max fairness (MMF) to minimize the latency. The OFDMA based HetNets has been given preference (used orthogonal sub carriers) over the conventional HetNets as it can mitigate the multiple access interference among users in same cell (Bogale & Le, 2016; Ghosh & De, 2020).

Although there is negligible interference in OFDMA-based HetNets, there is a scarcity of orthogonal resources. To overcome this limitation, NOMA-based HetNets is proposed for managing large number of users and achieving high throughput (Moltafet, Azmi, Javan, Mokari, & Mokdad, 2019). In Nasser, Muta, Elsabrouty, and Gacanin (2019), the authors described the interference management technique and resource block allocation by proposing R-WFISTA algorithm and compared with traditional OMA HetNets and Conventional NOMA HetNets to improve the sum rate and outrage probability. To increase the coverage area, the relays are introduced in HetNets to become heterogeneous relay network (Omran, Sboui, Rong, Rutagemwa, & Kadoch, 2019). The integration of cloud computing and HetNets provide wide coverage area, improvement in SE, higher throughput and reduce energy consumption. Along with this, use of multiantenna over single in HetNets improves communication, system capacity and SE. Therefore MIMO channels are preferred to support multi users by designing appropriate transmitting beam-forming vectors for mitigation of interference and better system performance (Bogale & Le, 2016).

3.7 Smart healthcare using 5G5G Inter of Things: a case-study

Integration of network makes an open opportunity in health care sector, where the sensors and other devices attached to a patient's body can transmit vital data to doctor's or hospital attendant's mobile phone for immediate action. It helps in other applications of health care sector such as medical supply chain, medical device maintenance, etc. The concept of smart health care system using WSN-MCN convergence network is going to be a reality due to this machine to machine communication system. WSN plays an important role in data collection from patient's body and ready to transmit onwards for processing. The Mobile Cellular Network (MCN) component with 5G5G technology, receive the data from WSN directly or through its mobile terminal (Zhang, Shan, Hu, & Yang, 2012). This convergence network creates a mutual associative network to support smart health care system. Simple IoT based health care system also enable machine to machine communication possible but due to small coverage, less mobility and weak terminal, it fails to make a robust system. Therefore MCN with IoT (5G5G network) is converged with WSN to provide larger coverage area, high capacity terminal to support the proposed health care system.

3.7.1 Mobile cellular network architecture: 5th generation

In the field of mobile communication, the forthcoming 5G5G technology is going to create a revolution. The features and compatibility will make it more user friendly for every automation in aspects of human life. Because of high speed data communication in the cellular network, it is going to merge the old technologies within it (Zhang et al., 2012). 5G5G is going to facilitate broadband connectivity, gaming options, quick response, minimum latency, wide coverage, high definition video and sound quality etc.

3.7.1.1 5G5G system architecture

5G5G possesses an advanced architecture with upgraded network elements to cope with new technologies. The service providers use these advance features to provide value-based services. The main feature of 5G5G is inclusion of cognitive technology which is able to identify its geographical parameters such as temperature, weather, location, etc. By using this technology, the 5G5G terminals act as transceiver by responding to radio signals of local environment and continue to provide quality of service (QoS). The system model of 5G5G technology is totally based on IP for all wireless technologies. The 5G5G system consist of two components: (1) User Terminal (cell phone) (2) Radio access technologies which are independent and autonomous.

These radio technologies establish paths as IP links for the public domain or internet world. With the IP technology, the data routing is managed and control with respect to specific application or session established between the clients present here and the server somewhere else in the internet. For smooth routing, packets routing needs to be fixed according to the application policy.

3.7.1.2 Master core technology

5G5G master core is the key convergence point for other technologies such as WSN and existing mobile cellular network (MCN). Master core design is able to work in parallel with all IP based network modes, that is, WSN and 5G5G network modes (MCN). In this multimode 5G technology, it controls the RAN and other access network (Chabbouh et al., 2016). Due to compatibility characteristics of this technology, other new converging technologies are more efficient, powerful and less complicated. The World Combination Service Mode (WCSM) allows ultraspeed, for example, text written on a board can be seen at a remote place, displayed on real time apart from sound and video. 5G network in a slice format: eMBB, URLLC and mMTC. Theses slices allow the access of 5G between traditional and core network. The specific network functions can be involved for specific characteristics of the network.

In WSN, the routing algorithms used must be capable of location finding system so that the accurate location can be estimated for completing the self-reorganizing process. As the nodes are battery operated, power is a scarce resource in this system. Therefore energy aware MAC protocol clustering process is developed for achieving better life time. For this purpose, most of the nodes enter sleep mode when no transmission of received work is going on and adjusts the reserved power.

3.7.2 ZigBee IP

ZigBee sensor nodes are with less energy, less computing power, and low range to participate in global network. ZigBee IP is the upgraded version of ZigBee where the concept of TCP/IP protocol is implemented in WSN. The physical and MAC layer structures of ZigBee are redefined with IP based system. The physical layer and data link layer are based on original IEEE 802.11.4. IPv6, ICMP and RPL are included in network layer. ZigBee is IP enabled due to RPL which is associated with 6LowPAN that allow exchange of data packets using IPv6 over IEEE 802.11.4. Transport layer may adapt TCP or UDP for data transmission, while 5G terminals have good storage space, more bandwidth for communication, more computing power and IP enabled which supports the ZigBee network for a better data communication. The 5G MCN terminals can join the ZigBee network whenever they are in the range of ZigBee network.

3.7.3 Healthcare system architecture using wireless sensor network and mobile cellular network

In this section, we propose a smart health care system based on WSN-MCN convergence network as shown in Fig. 3.10. The proposed system has two components: (1) WSN based on ZigBee-IP, (2) MCN based on 5G technology. The forthcoming 5G cellular communication technology is equipped with smart devices which are capable of participating with existing ZigBee network and also support M2M communication (Lone et al., 2020). In this context, our proposed architecture for smart health care system works as follows:

The hospital area can be considered to be covered under the geographical region of 5G MCN.

1. Patients present on the hospital bed are equipped with smart sensors to measure biological parameters of the body. These sensors are connected with each other with IEEE 802.15.4 ZigBee-IP network and form a cluster. Data collected in the cluster locally from sensor nodes in ZigBee network are passing through the 5G terminal. However, it is not possible for a patient to carry the entire heavy weight wired instrument for everyday monitoring. Therefore light wearable devices are developing with advanced sensor technology for continuous monitoring of

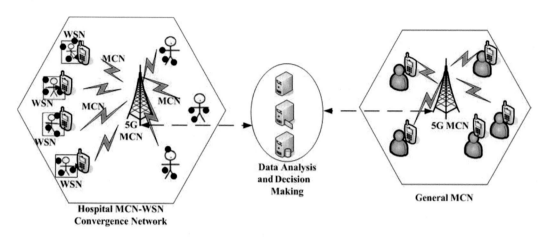

FIGURE 3.10

Health care system based on WSN-MCN network.

patient in hospital as well as in home. Table 3.1 presents different wearable devices collected through various research article and projects.

2. There are two ways of data communication exist. First, the cluster transmits all the collected data in wireless mode to the 5G cellular network which is coordinated by a 5G terminal. Second, the data from the cluster will transmit to the 5G terminal (mobile phone) present nearby, so that further transmission to the 5G cellular network is possible. This enables the dual mode communication of 5G terminals.

3. The patient data from 5G MCN are sent to the analyzing and decision-making server that can be accessed or intimated to doctors or hospital for further action.

3.7.3.1 System protocol

The protocol convergence can happen in two ways (Case-A and Case-B) can be implemented in two ways. In case-A, the 5G terminals itself participates in WSN by accepting spectrum from WSN with ZigBee IP. Therefore data from WSN can be transmitted by 5G terminal as a part of MCN. In case-B, the 5G terminal from MCN will act as a router for data transmission from WSN. The data packets of ZigBee-IP in WSN converted to payload of MCN as per the frame structure.

3.7.3.2 Data transmission by 5G terminal in ZigBee network

Initially, patient's sensor network (ZigBee) and 5G cellular network both operates in two different channels and not aware of each other's presence. First, the 5G terminal starts channel detection for the existing ZigBee network id and available channels. After detection, if it finds a busy channel, then the information is restored. The 5G terminal broadcasts its existence and working band. If a ZigBee node receives the frame to participate in the MCN, it synchronizes with the broadcasted band to transmit the data. The announcement procedure repeats if another ZigBee network find around its coverage. In the protocol convergence, WSN use the IP core of 5G MCN protocol to transmit sensory data. The 5G MCN has compatible core to recognize sensor nodes as shown in Fig. 3.11.

3.7.3.3 Data transmission through 5G terminal by ZigBee network

The 5G terminal acts as a router for passing the ZigBee data to MCN. Initially, the 5G terminal start detecting any ZigBee WSN, if detected, it broadcasts its ID and channel. If a ZigBee WSN receives this frame and found that the new location address is closer than the current 5G terminal

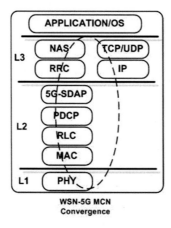

FIGURE 3.11

Protocol convergence architecture for case-A.

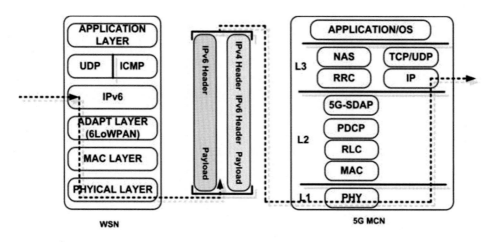

FIGURE 3.12

Protocol convergence architecture for case-B.

in which it is connected, then it updates and broadcasts its updated address. This process is followed by all 5G terminal and ZigBee units. The protocol defined for case-B is shown in Fig. 3.12. The WSN data generated from WSN-IPv6 are transmitted to 5G MCN. But when data travelled in reverse, that is, from MCN to WSN, a compatible format is generated by extracting extra header of IPv4 and 5G payloads so that it is recognized by WSN.

A survey on different types of devices in healthcare systems is presented in Table 3.6.

Table 3.6 A Survey on types of devices in healthcare system.

Types of wearable devices	Research papers and projects	Applications
Wrist-worn device	AMON (Anliker et al., 2004)	High risk cardiac and respiration
	Vivago Wrist Care (Mattila et al., 2008)	Skin temperature and movement
	Health Monitoring Watch (Mahesh et al., 2013)	Continuously measure the vital parameter such as heart rate, temperature and oxygen content in blood.
	Sense Wear Armband (Liden et al., 2002)	Heat flow
Smart phone and mobile-based device	Smart phone healthcare (Fang, Hu, Wei, Shao, & Luo, 2014)	Health monitoring
	Heart Togo (Jin, Oresko, Huang, & Cheng, 2009)	Individualized remote CVD detection
	Personal Health Monitor (Leijdekker & Gay, 2008)	Heart attack self-test for CVD patient
	Bluetooth enabled sensor with smart phone (Moro et al., 2007)	Long time monitoring
Textile based device	WEALTHY (Paradiso, Loriga, & Taccini, 2005)	Monitoring of elderly patient and chronic diseases
	BIOTEX (Coyle et al., 2010)	Personal health monitoring
	MagIC (Meriggi et al., 2010)	Daily life cardiorespiratory signal
	Smart vest (Pandian et al., 2008)	Remote health monitoring
	Smart shirt (Lee & Chung, 2009; Tada, Amano, Sato, Saito, & Inoue, 2015)	Measurement of ECG for continuous real time health monitoring
	Biosensing textile-based patch (Morris et al., 2009)	Textile based fluid handling platform for sweat monitoring
	My heart (Habetha, 2006)	Prevention and monitoring of cardio vascular diseases
Microcontroller board-based device	Proe-tex (Curone et al., 2010)	Real time monitoring in harsh environment
	Ambulatory stress monitoring (Choi, Ahme, & Gutierrez-Osuna, 2011)	Monitor a number of physiological correlates of mental stress
	LiveNet (Sung, Marci, & Pentland, 2005)	Parkinson symptom and epilepsy seizures detection

3.8 Conclusions

We have presented heterogenous network (HetNet) architecture for 5G and allied technologies like M2M and D2D in this chapter. Emerging applications like IoT and IoV demand scenario-specific deployment environment and hence a HetNet with dissimilar clusters of access network is the most viable one. However, these HetNets must be seamlessly interconnected with each other with an agile communication framework to make networks and protocols globally interoperable. Furthermore, with the development of 3GPP, a standard 5G mobile communication portfolio with a resilient architecture and advanced features like SDN, NFV, co-operative communication using H-CRAN-based fronthaul and TWDM-PON based backhaul has been realized. Many recent state-of-the-art communication frameworks like M2M and D2D are coming to aid for a more comprehensive technology-enabler for IoT and smart communication applications. Integration of HetNets, M2M, and D2D technologies with 5G network standards with future-ready features and functionalities (eMBB, URLLC, mMTC, mmWave, MIMO, Cloud-RAN, agile SDN and NFV) is the ultimate goal for today's smart communications prospects and its business potential in the near future.

References

Agiwal, M., Roy, A., & Saxena, N. (2016). Next generation 5G wireless networks: A comprehensive Survey. *IEEE Communications Surveys & Tutorials*, *18*(3), 1617−1655. Available from https://doi.org/10.1109/COMST.2016.2532458.

Alnoman A., Ferdouse L., Anpalagan A., (2017). Fuzzy-based joint user association and resource allocation in HetNets. In *Proceedings of the IEEE eighty-sixth vehicular technology conference*, VTC-Fall, pp. 1−5.

Anliker, U., Ward, J. A., Lukowicz, P., Troster, G., Dolveck, F., Baer, M., . . . Belardinelli, A. (2004). AMON: A wearable multiparameter medical monitoring and alert system. *IEEE Transactions on Information Technology in Biomedicine*, *8*(4), 415−427.

Baronti, P., Pillai, P., Chook, V. W., Chessa, S., Gotta, A., & Hu, Y. F. (2007). Wireless sensor networks: A survey on the state of the art and the 802.15. 4 and ZigBee standards. *Computer Communications*, *30*(7), 1655−1695.

Bogale, T. E., & Le, L. B. (2016). Massive MIMO and mmWave for 5G wireless HetNet: Potential benefits and challenges. *IEEE Vehicular Technology Magazine*, *11*(1), 64−75.

Callaway, E., Gorday, P., Hester, L., Gutierrez, J. A., Naeve, M., Heile, B., & Bahl, V. (2002). Home networking with IEEE 802.15. 4: A developing standard for low-rate wireless personal area networks. *IEEE Communications Magazine*, *40*(8), 70−77.

Camacho, F., Cárdenas, C., & Muñoz, D. (2018). Emerging technologies and research challenges for intelligent transportation systems: 5G, HetNets, and SDN. *International Journal on Interactive Design and Manufacturing (IJIDeM)*, *12*(1), 327−335.

Chabbouh O., Rejeb S. B., Choukair Z., Agoulmine N., (2016). A novel cloud RAN architecture for 5G HetNets and QoS evaluation. In *Proceedings of the international symposium on networks, computers and communications*, ISNCC, May 11, pp. 1−6.

Chen, Y., Li, J., Chen, W., Lin, Z., & Vucetic, B. (2016). Joint user association and resource allocation in the downlink of heterogeneous networks. *IEEE Transactions on Vehicular Technology*, *65*(7), 5701−5706.

Chettri, L., & Bera, R. (2020). A comprehensive survey on Internet of Things (IoT) toward 5G Wireless systems. *IEEE Internet of Things Journal*, *7*(1), 16−32. Available from https://doi.org/10.1109/JIOT.2019.2948888.

Choi, J., Ahme, B., & Gutierrez-Osuna, R. (2011). Development and evaluation of an ambulatory stress monitor based on wearable sensors. *IEEE Transactions on Information Technology in Biomedicine, 16*(2), 279−286.

Coronado, E., Khan, S. N., & Riggio, R. (2019). 5G-EmPOWER: A software-defined networking platform for 5G radio access networks. *IEEE Transactions on Network and Service Management, 16*(2), 715−728. Available from https://doi.org/10.1109/TNSM.2019.2908675.

Coyle, S., Lau, K. T., Moyna, N., O'Gorman, D., Diamond, D., Di Francesco, F., . . . Taccini, N. (2010). BIOTEX—Biosensing textiles for personalised healthcare management. *IEEE Transactions on Information Technology in Biomedicine, 14*(2), 364−370.

Curone, D., Lecco, E., Tognetti, A., Loriga, G., Dudnik, G., Risatti, M., . . . Magenes, G. (2010). Smart garments for emergency operators: The ProeTEX project. *IEEE Transactions on Information Technology in Biomedicine, 14*(3), 694−701.

Dunkels, A., & Vasseur, J. (2008). Ip for smart objects aliance. *Internet Protocol for Smart Objcts (IPSO) Alliance White Paper, 2*.

Elsayed, M., & Erol-Kantarci, M. (2019). AI-enabled future wireless networks: Challenges, opportunities, and open issues. *IEEE Vehicular Technology Magazine, 14*(3), 70−77.

EPC Tag Data Standard, (2014). GS1 Standard Version 1.8.

Fang D., Hu J., Wei X., Shao H., Luo Y., (2014). A smart phone healthcare monitoring system for oxygen saturation and heart rate. In *Proceedings of the international conference on cyber-enabled distributed computing and knowledge discovery*, Oct 13, pp. 245−247.

Feng, M., Mao, S., & Jiang, T. (2018). Joint frame design, resource allocation and user association for massive MIMO heterogeneous networks with wireless backhaul. *IEEE Transactions on Wireless Communications, 17*(3), 1937−1950.

Gama, K., Touseau, L., & Donsez, D. (2012). Combining heterogeneous service technologies for building an Internet of Things middleware. *Computer Communications, 35*(4), 405−417.

Gezer C., Taşkın E., (2016). An overview of oneM2M standard. In *Proceedings of the twenty-fourth signal processing and communication application conference*, SIU, pp. 1705−1708, doi: 10.1109/SIU.2016.7496087.

Ghosh S., De D., (2020). Dynamic antenna allocation in 5G MIMO HetNet using weighted majority cooperative game theory. In *Proceedings of the IEEE first international conference for convergence in engineering*, ICCE, Sep 5, pp. 21−25.

Guinard D., Trifa V., (2009). Towards the web of things: Web mashups for embedded devices. In *Proceeding of the international world wide web conference, Madrid, Spain*, Apr, pp. 8.

Guinard, D., Trifa, V., Karnouskos, S., Spiess, P., & Savio, D. (2010). Interacting with the soa-based internet of things: Discovery, query, selection, and on-demand provisioning of web services. *IEEE Transactions on Services Computing, 3*(3), 223−235.

Guo, J. (2009). Collaborative conceptualisation: towards a conceptual foundation of interoperable electronic product catalogue system design. *Enterprise Information Systems, 3*(1), 59−94.

Gures, E., Shayea, I., Alhammadi, A., Ergen, M., & Mohamad, H. (2020). A comprehensive survey on mobility management in 5G heterogeneous networks: Architectures, challenges and solutions. *IEEE Access, 8*, 195883−195913.

Habetha J., (2006). The MyHeart project-fighting cardiovascular diseases by prevention and early diagnosis. In *Proceedings of the international conference of the IEEE engineering in medicine and biology society*, Aug 31, pp. 6746−6749.

Huang, J., Zhou, P., Luo, K., Yang, Z., & He, G. (2017). Two-stage resource allocation scheme for three-tier ultra-dense network. *China Communications, 14*(10), 118−129.

Irrum, F., Ali, M., Naeem, M., Anpalagan, A., Qaisar, S., & Qamar, F. (2021). D2D-enabled resource management in secrecy-ensured 5G and beyond heterogeneous networks. *Physical Communication, 45*, 101275, ISSN 1874−4907. Available from https://doi.org/10.1016/j.phycom.2021.101275.

Jardim-Goncalves, R., Grilo, A., Agostinho, C., Lampathaki, F., & Charalabidis, Y. (2013). Systematisation of interoperability body of knowledge: The foundation for enterprise interoperability as a science. *Enterprise Information Systems*, 7(1), 7–32.

Ji, P., Jia, J., & Chen, J. (2019). Joint optimization on both routing and resource allocation for millimeter wave cellular networks. *IEEE Access*, 7, 93631–93642.

Jin Z., Oresko J., Huang S., Cheng A. C., (2009). HeartToGo: A personalized medicine technology for cardiovascular disease prevention and detection. In *Proceedings of the IEEE/NIH life science systems and applications workshop*, April 9, pp. 80–83.

Kaneko, M., Nakano, T., Hayashi, K., Kamenosono, T., & Sakai, H. (2016). Distributed resource allocation with local CSI overhearing and scheduling prediction for OFDMA heterogeneous networks. *IEEE Transactions on Vehicular Technology*, 66(2), 1186–1199.

Khawam, K., Lahoud, S., El Helou, M., Martin, S., & Feng, G. (2020). Coordinated framework for spectrum allocation and user association in 5G HetNets with mmWave. *IEEE Transactions on Mobile Computing* (01). Available from https://doi.org/10.1109/TMC.2020.3022681, 1-1.

Kim, D., Lee, C. E., Park, J. H., Moon, K., & Lim, K. (2007). Scalable message translation mechanism for the environment of heterogeneous middleware. *IEEE Transactions on Consumer Electronics*, 53(1), 108–113.

Kim, T., & Chang, J. M. (2017). QoS-aware energy-efficient association and resource scheduling for HetNets. *IEEE Transactions on Vehicular Technology*, 67(1), 650–664.

Klymash M., Beshley H., Seliuchenko M., Beshley M., (2017). Algorithm for clusterization, aggregation and prioritization of M2M devices in heterogeneous 4G/5G network. In *Proceedings of the fourth international scientific-practical conference problems of infocommunications. science and technology, PIC S&T*, pp. 182–186, doi: 10.1109/INFOCOMMST.2017.8246376.

Lanthaler M., Gütl C., (2012). On using JSON-LD to create evolvable RESTful services. In *Proceedings of the third international workshop on RESTful Design*, Apr 17, pp. 25–32.

Le L. V., Sinh D., Lin B.S., Tung L. P., (2018). Applying big data, machine learning, and SDN/NFV to 5G traffic clustering, forecasting, and management. In *Proceedings of the fourth IEEE conference on network softwarization and workshops, NetSoft*, Jun 25, pp. 168–176.

Lee, Y. D., & Chung, W. Y. (2009). Wireless sensor network based wearable smart shirt for ubiquitous health and activity monitoring. *Sensors and Actuators B: Chemical*, 140(2), 390–395.

Leijdekker P., Gay V., (2008). A self-test to detect a heart attack using a mobile phone and wearable sensors. In *Proceedings of the twenty-first IEEE international symposium on computer-based medical systems*, June 17, pp. 93–98.

Li, S., Da Xu, L., & Zhao, S. (2018). 5G Internet of Things: A survey. *Journal of Industrial Information Integration*, 10, 1–9, ISSN 2452-414X. Available from https://doi.org/10.1016/j.jii.2018.01.005.

Li, X., Rao, J. B., & Zhang, H. (2016). Engineering Machine-to-Machine traffic in 5G. *IEEE Internet of Things Journal*, 3(4), 609–618. Available from https://doi.org/10.1109/JIOT.2015.2477039.

Liden, C. B., Wolowicz, M., Stivoric, J., Teller, A., Kasabach, C., Vishnubhatla, S., ... Boehmke, S. (2002). Characterization and implications of the sensors incorporated into the SenseWear armband for energy expenditure and activity detection. *Bodymedia Inc. White Papers*, 1, 7.

Liu, R., Chen, Q., Yu, G., & Li, G. Y. (2019). Joint user association and resource allocation for multi-band millimeter-wave heterogeneous networks. *IEEE Transactions on Communications*, 67(12), 8502–8516.

Lohani, S., Hossain, E., & Bhargava, V. K. (2017). Joint resource allocation and dynamic activation of energy harvesting small cells in OFDMA HetNets. *IEEE Transactions on Wireless Communications*, 17(3), 1768–1783.

Lone, T. A., Rashid, A., Gupta, S., Gupta, S. K., Rao, D. S., Najim, M., ... Singhal, A. (2020). Securing communication by attribute-based authentication in HetNet used for medical applications. *Eurasip Journal on Wireless Communications and Networking* (1), 1–21.

Luo, X. (2017). Delay-oriented QoS-aware user association and resource allocation in heterogeneous cellular networks. *IEEE Transactions on Wireless Communications*, 16(3), 1809-1022.

Lv T., Liu C., Gao H., (2017). Novel user scheduling algorithms for carrier aggregation system in heterogeneous network. In *Proceedings of the IEEE wireless communications and networking conference*, WCNC, 2017 Mar 19, 1−6.

Mahesh K. C., Shriharsha S., Seema G. S., Smitha P. V., Radhika S., Appaji M. A., Mishra G. (2013). Wearable wireless intelligent multi-parameter health monitoring watch. In *Proceedings of the Texas Instruments India Educators' conference*, Apr 4, pp. 61−64.

Manishankar S., Srinithi C. R., Joseph D., (2017). Comprehensive study of wireless networks qos parameters and comparing their performance based on real time scenario. In *Proceedings of the international conference on innovations in information, embedded and communication systems*, ICIIECS, Mar 17, pp. 1−6.

Mattila E., Korhonen I., Merilahti J., Nummela A., Myllymaki M., Rusko H., (2008). A concept for personal wellness management based on activity monitoring. In *Proceedings of the second international conference on pervasive computing technologies for healthcare*, pp. 32−36.

Mehmood, Y., Haider, N., Imran, M., Timm-Giel, A., & Guizani, M. (2017). M2M communications in 5G: State-of-the-art architecture, recent advances, and research challenges. *IEEE Communications Magazine*, *55*(9), 194−201. Available from https://doi.org/10.1109/MCOM.2017.1600559.

Mei, J., Wang, X., & Zheng, K. (2020). An intelligent self-sustained RAN slicing framework for diverse service provisioning in 5G-beyond and 6G networks. *Intelligent and Converged Networks*, *1*(3), 281−294. Available from https://doi.org/10.23919/ICN.2020.0019.

Meriggi P., Castiglioni P., Lombardi C., Rizzo F., Mazzoleni P., Faini A., ... Parati G., (2010). Polysomnography in extreme environments: The MagIC wearable system for monitoring climbers at very-high altitude on Mt. Everest slopes. In *Proceedings of the computing in cardiology*, Sep 26, pp. 1087−1090.

Moltafet, M., Azmi, P., Javan, M. R., Mokari, N., & Mokdad, A. (2019). Optimal radio resource allocation to achieve a low BER in PD-NOMA−based heterogeneous cellular networks. *Transactions on Emerging Telecommunications Technologies*, *30*(5), e3572.

Moon, K. D., Lee, Y. H., Lee, C. E., & Son, Y. S. (2005). Design of a universal middleware bridge for device interoperability in heterogeneous home network middleware. *IEEE Transactions on Consumer Electronics*, *51*(1), 314−318.

Moro M. J., Luque J. R., Botella A. A., Cuberos E. J., Casilari E., Díaz-Estrella A., (2007). J2ME and smart phones as platform for a Bluetooth Body Area Network for Patient-telemonitoring. In *Proceedings of the twenty-ninth annual international conference of the IEEE engineering in medicine and biology society*, Aug 22, pp. 2791−2794.

Morris, D., Coyle, S., Wu, Y., Lau, K. T., Wallace, G., & Diamond, D. (2009). Bio-sensing textile based patch with integrated optical detection system for sweat monitoring. *Sensors and Actuators B: Chemical*, *139*(1), 231−236.

Munir H., Hassan S. A., Pervaiz H., Ni Q., Musavian L., (2016). Energy efficient resource allocation in 5G hybrid heterogeneous networks: A game theoretic approach. In *Proceedings of the IEEE eighth-fourth vehicular technology conference*, VTC-Fall, pp. 1−5.

Naeem, M., et al. (2017). Distributed gateway selection for M2M communication in cognitive 5G networks. *IEEE Network*, *31*(6), 94−100. Available from https://doi.org/10.1109/MNET.2017.1700017.

Naqvi, S. A., Pervaiz, H., Hassan, S. A., Musavian, L., Ni, Q., Imran, M. A., ... Tafazolli, R. (2018). Energy-aware radio resource management in D2D-enabled multi-tier HetNets. *IEEE Access*, *6*, 16610−16622.

Nasser, A., Muta, O., Elsabrouty, M., & Gacanin, H. (2019). Compressive sensing based spectrum allocation and power control for NOMA HetNets. *IEEE Access*, *7*, 98495−98506.

Niknam, S., Nasir, A. A., Mehrpouyan, H., & Natarajan, B. (2016). A multiband OFDMA heterogeneous network for millimeter wave 5G wireless applications. *IEEE Access*, *4*, 5640−5648.

Omran A., Sboui L., Rong B., Rutagemwa H., Kadoch M., (2019). Joint relay selection and load balancing using D2D communications for 5G HetNet MEC. In *Proceedings of the IEEE international conference on communications workshops, ICC Workshops*, May 20, pp. 1−5.

Ouamri, M. A., Oteşteanu, M. E., Alexandru, I., & Azni, M. (2020). Coverage, handoff and cost optimization for 5G heterogeneous network. *Physical Communication*, *39*, 101037, ISSN 1874−4907. Available from https://doi.org/10.1016/j.phycom.2020.101037.

Pandian, P. S., Mohanavelu, K., Safeer, K. P., Kotresh, T. K., Shakunthala, D. T., Gopal, P., & Padaki, V. C. (2008). Smart Vest: Wearable multi-parameter remote physiological monitoring system. *Medical Engineering & Physics*, *30*(4), 466−477.

Panetto H., Cecil J., (2013). *Information systems for enterprise integration, interoperability and networking: Theory and applications*, 1−6.

Paradiso, R., Loriga, G., & Taccini, N. (2005). A wearable health care system based on knitted integrated sensors. *IEEE transactions on Information Technology in biomedicine*, *9*(3), 337−344.

Park, H., Kim, H., Joo, H., & Song, J. (2016). Recent advancements in the Internet-of-Things related standards: A oneM2M perspective. *ICT Express*, *2*(3), 126−129, ISSN 2405−9595. Available from https://doi.org/10.1016/j.icte.2016.08.009.

Peng, M., Li, Y., Zhao, Z., & Wang, C. (2015). System architecture and key technologies for 5G heterogeneous cloud radio access networks. *IEEE Network*, *29*(2), 6−14. Available from https://doi.org/10.1109/MNET.2015.7064897.

Pérez-Romero, J., Sánchez-González, J., Agustí, R., Lorenzo, B., & Glisic, S. (2016). Power-efficient resource allocation in a heterogeneous network with cellular and D2D capabilities. *IEEE Transactions on Vehicular Technology*, *65*(11), 9272−9286.

Rezvani, S., Mokari, N., Java, M. R., & Jorswieck, E. A. (2019). Fairness and transmission-aware caching and delivery policies in OFDMA-based HetNets. *IEEE Transactions on mobile computing*, *19*(2), 331−346.

Saha, C., Afshang, M., & Dhillon, H. S. (2018). 3GPP-Inspired HetNet Model using Poisson cluster process: Sum-product functionals and downlink coverage. *IEEE Transactions on Communications*, *66*(5), 2219−2234. Available from https://doi.org/10.1109/TCOMM.2017.2782741.

Siddavaatam, R., Woungang, I., & Anpalagan, A. (2019). Joint optimisation of radio and infrastructure resources for energy-efficient massive data storage in the mobile cloud over 5G HetNet. *IET Wireless Sensor Systems*, *9*(5), 323−332.

Song H. K., Min-Jae P. A., Won-Chang K. I., (2020). Inventors; Sejong University Industry-Academy Cooperation Group, assignee, MIMO-OFDM-based cooperative communication system for interference mitigation between cells in heterogeneous network and cooperative communication method using the same, United States patent U.S. Patent No. 10,862,547. 8 Dec.

Sun, S., Gong, L., Rong, B., & Lu, K. (2015). An intelligent SDN framework for 5G heterogeneous networks. *IEEE Communications Magazine*, *53*(11), 142−147. Available from https://doi.org/10.1109/MCOM.2015.7321983.

Sung, M., Marci, C., & Pentland, A. (2005). Wearable feedback systems for rehabilitation. *Journal of neuroengineering and rehabilitation*, *2*(1), 17.

Swetina, J., Lu, G., Jacobs, P., Ennesser, F., & Song, J. (2014). Toward a standardized common M2M service layer platform: Introduction to oneM2M. *IEEE Wireless Communications*, *21*(3), 20−26. Available from https://doi.org/10.1109/MWC.2014.6845045.

Tada, Y., Amano, Y., Sato, T., Saito, S., & Inoue, M. (2015). A smart shirt made with conductive ink and conductive foam for the measurement of electrocardiogram signals with unipolar precordial leads. *Fibers*, *3*(4), 463−477.

Tang, F., Zhou, Y., & Kato, N. (2020). Deep reinforcement learning for dynamic uplink/downlink resource allocation in high mobility 5G HetNet. *IEEE Journal on Selected Areas in Communications*, *38*(12), 2773−2782.

Wang, F., Chen, W., Tang, H., & Wu, Q. (2017). Joint optimization of user association, subchannel allocation, and power allocation in multi-cell multi-association OFDMA heterogeneous networks. *IEEE Transactions on Communications*, *65*(6), 2672−2684.

Wang N., Fei Z., Kuang J., (2016). QoE-aware resource allocation for mixed traffics in heterogeneous networks based on Kuhn-Munkres algorithm. In *Proceedings of the IEEE international conference on communication systems*, ICCS, Dec 14, pp. 1−6.

Wang W., De S., Toenjes R., Reetz E., Moessner K., (2012). A comprehensive ontology for knowledge representation in the internet of things. In *Proceedings of the eleventh international conference on trust, security and privacy in computing and communications*, Jun 25, pp. 1793−1798.

Wang, X., Zheng, F., Jia, X., & You, X. (2017). Resource allocation in OFDMA heterogeneous networks for maximizing weighted sum energy efficiency. *Science China Information Sciences*, 60(6), 062304.

Xiao, G., Guo, J., Da, Xu. L., & Gong, Z. (2014). User interoperability with heterogeneous IoT devices through transformation. *IEEE Transactions on Industrial Informatics*, 17(2), 1486−1496.

Xie, B., Zhang, Z., Hu, R. Q., Wu, G., & Papathanassiou, A. (2018). Joint spectral efficiency and energy efficiency in FFR-based wireless heterogeneous networks. *IEEE Transactions on Vehicular Technology*, 67(9), 8154−8168.

Xu Y., Hu Y., Chen Q., Chai R., Li G., (2017). Distributed resource allocation for cognitive hetnets with cross-tier interference constraint. In *Proceedings of the IEEE wireless communications and networking conference (WCNC)*, Mar 19, pp. 1−6.

Xu Y., Hu Y., Chen Q., Song T., Lai R., (2017). Robust resource allocation for multi-tier cognitive heterogeneous networks. In *Proceedings of the IEEE international conference on communications*, ICC, May 21, 1−6.

Ya, M., Feng, G., Zhou, J., & Qin, S. (2018). Smart multi-RAT access based on multiagent reinforcement learning. *IEEE Transactions on Vehicular Technology*, 67(5), 4539−4551.

Yang, C., Xiao, J., Li, J., Shao, X., Anpalagan, A., Ni, Q., & Guizani, M. (2018). DISCO: Interference-aware distributed cooperation with incentive mechanism for 5G heterogeneous ultra-dense networks. *IEEE Communications Magazine*, 56(7), 198−204.

Ye, F., Dai, J., & Li, Y. (2018). Hybrid-clustering game Algorithm for Resource Allocation in Macro-Femto HetNet. *TIIS*, 12(4), 1638−1654.

Yim, H. J., Seo, D., Jung, H., Back, M. K., Kim, I., & Lee, K. C. (2017). Description and classification for facilitating interoperability of heterogeneous data/events/services in the Internet of Things. *Neurocomputing*, 256, 13−22.

Yuan, P., Xiao, Y., Bi, G., & Zhang, L. (2017). Toward cooperation by carrier aggregation in heterogeneous networks: A hierarchical game approach. *IEEE Transactions on Vehicular Technology*, 66(2), 1670−1683.

Zhang, H., Huang, C., Zhou, J., & Chen, L. (2020). QoS-aware virtualization resource management mechanism in 5G backhaul heterogeneous networks. *IEEE Access*, 8, 19479−19489. Available from https://doi.org/10.1109/ACCESS.2020.2967101.

Zhang, J., Shan, L., Hu, H., & Yang, Y. (2012). Mobile cellular networks and wireless sensor networks: Toward convergence. *IEEE Communications Magazine*, 50(3), 164−169.

Zhang, N., Cheng, N., Gamage, A. T., Zhang, K., Mark, J. W., & Shen, X. (2015). Cloud assisted HetNets toward 5G wireless networks. *IEEE Communications Magazine*, 53(6), 59−65.

Zhang, Q., Gui, L., Hou, F., Chen, J., Zhu, S., & Tian, F. (2020). Dynamic task offloading and resource allocation for mobile-edge computing in dense cloud RAN. *IEEE Internet of Things Journal*, 7(4), 3282−3299.

Zhang, W., Wei, Y., Wu, S., Meng, W., & Xiang, W. (2019). Joint beam and resource allocation in 5G mmWave small cell systems. *IEEE Transactions on Vehicular Technology*, 68(10), 10272−10277.

Zhou, T., Huang, Y., & Yang, L. (2016). Energy-efficient user association in downlink heterogeneous cellular networks. *IET Communications*, 10(13), 1553−1561.

An overview of low power hardware architecture for edge computing devices

Kushika Sivaprakasam[1], P. Sriramalakshmi[1], Pushpa Singh[2] and M.S. Bhaskar[3]

[1]*School of Electrical Engineering, Vellore Institute of Technology, Chennai, Tamil Nadu, India* [2]*Department of Computer Science & Information Technology, KIET Group of Institutions, Delhi-NCR, Ghaziabad, Uttar Pradesh, India* [3]*Renewable Energy Lab, Department of communication and Networks Engineering, College of Engineering, Prince Sultan University, Riyadh, Saudi Arabia*

4.1 Introduction

In edge computing, the computational power lies closer to the source thereby reducing latency. The idea is to optimize the edge such that data can be processed closer to its source, which is critical to many services especially in the 5G networks. This feature of edge computing devices acts as a catalyst in the process of deployment of 5G networks. An ultra-low latency network is exactly what 5G technology requires, and edge computing can provide it (Hassan et al., 2019). Not only it promotes 5G networks, but edge computing also promotes various other applications such as high quality videoconferencing, lag-free online gaming, a wide range of IoT applications (Singh & Agrawal, 2021), Virtual Reality, Augmented Reality, Big Data Analytics and Crypto currency trading.

Recently vast data are generated by IoT and other connected devices. The generated data need to be stored, maintained, managed, backed up and made accessible to users through the Internet. Cloud, fog, and edge computing infrastructures permit organizations to access gain of a various range of computing resources and data storage. Cloud, fog and edge computing looks like similar, but they are different from each other. Cloud computing viewed as storing and accessing data and programs over the Internet (third party), rather than on the local hard drive of your single computer. To access data, users need to create account with the associated cloud service providers (Indu et al., 2018). IoT and Machine Learning based applications such as smart healthcare, smart agriculture, smart transportation system and smart city deal with massive amount of data (Mahdavinejad et al., 2018).

The data processing in the cloud service increases the cost of the IoT solution. Though cloud computing remains the first choice for most of the companies, for real time applications, companies are transforming their systems to edge and fog computing. The main aim of knowing concepts about edge and fog is not to replace the cloud completely but to separate sensitive information from the generic one and hence to analyze the data economically in the safe manner. The difference between fog and cloud computing is that cloud is a centralized system whereas fog computing supports distributed environment. Fog computing is an intermediary between hardware and remote servers. It controls the information which needs to be transferred to the cloud server and to be processed locally. Hence to analyze data that leads to latency issues. This issue is resolved with fog computing, where data is processed within a fog node or IoT Gateway which is placed within the Local Area Network (LAN). The overall Industrial IoT framework consists of cloud, fog and edge computing as represented in Fig. 4.1. Topmost layer is cloud layer which can have big data processing, data warehouse and business logic. Fog computing layer acts as a middleware service and without it, the cloud communicates directly with the devices which is a costly affair. In edge computing, the data is processed and analyzed on the device or sensor itself, without being relocated anywhere. Edge computing allows processing the data remains on the device itself or on the IoT Gateway in order to reduce the network traffic. Edge computing drives the artificial intelligence (AI), and communication abilities of an application directly into devices such as EPICs (Edge Programmable Industrial Controllers), PACs (Programmable Automation Controllers), and PLCs (Programmable Logic Controllers). Edge computing keeps users' personnel data on edge devices rather than keeping it on fog or cloud, hence, decreases the risk of network data leakage and provides data security and privacy (Shi et al., 2019).

Recently, where each and every object also referred as "things" can have embedded processor and can connect billions or trillions of processors through the Internet. The conventional cloud-based model processes the data, which demands limited wireless networks. Low or Ultra Low Power (ULP) hardware architecture for edge computing devices is the only possibility that can meet these demands (Brooks & Sartori, 2017). Hence with fruitful-looking prospects, edge

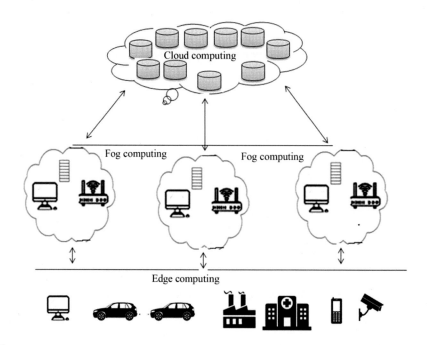

FIGURE 4.1

Industrial Internet of Things framework with cloud, fog and edge computing (Singh & Agrawal, 2021).

computing infrastructure systems need to be easy to deploy and at the same time, should be efficient. In edge computing, the computational technique falls nearer to the source and hence it reduces the latency. It can very well optimize the edge, which is critical to many services especially in 5G networks. Hence edge computing devices facilitate the process of deployment of 5G networks and a wide range of cutting-edge technologies. This paper focusses on the hardware infrastructure of edge computing devices and different ways to enhance its efficiency. For this purpose, the paper is divided into different sections to present a clear picture.

Section 4.2 explains the basic concepts of cloud, fog and edge computing infrastructure. Section 4.3 emphasizes the low power hardware architecture for edge computing devices with detailed analysis and study of different components of the hardware architecture and is divided into sections: Section 4.1 describes the objectives of hardware development in edge computing clearly defining the direction of this research. Section 4.2 gives an overview of the system architecture in an abstractive manner. Each component of the system architecture is elaborated in further sections. Section 4.3 is dedicated to the Central Processing Unit (CPU) architecture and Section 4.4 explains the input—output architecture. Section 4.5 is crucial as it deals with power consumption and management. The aim is to also touch on algorithmic optimizations which are dealt with in Section 4.6 along with the explanation of the data processing unit.

Section 4.4 consists of examples of edge gateways and devices that are based on edge computing. It draws a pragmatic picture of the theory discussed in the previous sections. Section 4.5 discusses the edge computing for intelligent healthcare applications. It elaborates the advantages of edge computing for healthcare applications, implementation challenges of edge computing in healthcare systems and applications of edge computing based smart healthcare systems. Section 4.6 projects the possible impact of edge computing, IoT and 5G on healthcare industries. The paper is concluded with Section 4.7, which consists of a summary and the future scope of research concerning the discussed topics.

4.2 Basic concepts of cloud, fog and edge computing infrastructure

Most of the power processing is carried out in the cloud. In edge computing, the real time data processing is performed on the premises. There are numerous advantages of edge computing technologies such as real time data analytics based on AI, less operating costs, enhanced application performance reduced downtime. Fog computing was introduced by Cisco systems to bring the cloud nearer to the end users. The processing is preferably performed in the LAN. It is not carried out at nodes or in gateway. The mobile edge computing (MEC) is more suitable for server-to-client applications (Torre et al., 2020) since the intelligence and management are located in the Radio Access Network (RAN).

Fog computing is performed in an abstraction layer whereas MEC and cloudlets do not need this due to its dedicated connections. Multiaccess edge computing MEC, cloudlets, micro data centers, Cloud of Things (CoT) will be the main players in the upcoming network infrastructures (http://www.lanner-america.com). Multiaccess edge computing enables the processing and computation within network which is termed as the RAN. The MEC allows operators to deal with this excessive traffic and resource demands more intelligently. It lays the foundations for future intelligent and next-generation networks. MEC provides enhanced location, augmented reality and IoT services support 5G networks. In reality, the edge and fog computing have similar objectives. A minor difference is that fog computing includes running intelligence on the end device and is more of IoT focused (http://www.lanner-america.com). An edge based IoT is shown in Fig. 4.2.

Fog computing describes a decentralized computing infrastructure. It is the extension of cloud computing (data center) to the edge of a network and helps in computational and storage applications efficiently. It is generally located between the cloud and the origin of the data. Fog nodes which are computing entities link the edge devices with the cloud. These fog nodes are capable of processing and sensing. The ultimate aim of fog computing is to reduce the massive amount of data transferred to the cloud for storage and processing. The data acquired from the sensors are processed at the edge of a network. Cloudlets are used to reduce the latency and improve the mobile applications. It helps to eradicate the latency delays usually with Wide Area Network (WAN) cloud computing. A data center in small scale which is formed in the cloud consists of trusted computing devices near to the locality of mobile user. In addition it can perform the process combined with the local network connection. The cloudlet is the major technological enabler for mobile cloud computing. It is the convergence of mobile cloud computing and cloud computing. It has the virtual mode of architecture which can be accessed by the users within their locality (García-Valls et al., 2018; Satyanarayanan et al., 2015).

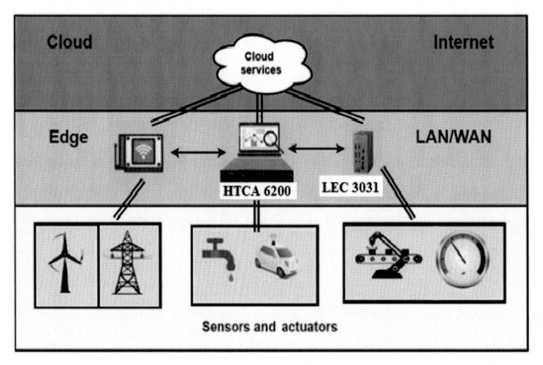

FIGURE 4.2

Edge based Internet of Things (https://cognitechx.com/what-is-edge-computing-mean-in-iot/).

Moreover, mobile users can utilize it. The cloudlets virtual machines can run the required software applications closer to the mobile devices which can be used by the mobile users. Hence the latency issues can be solved by shifting the virtual machine near to the locality of the mobile users. The mobile users need to rely on the network service providers to deploy cloudlets into the LAN network for the mobile devices to use them (Verbelen et al., 2012). Cloudlets can assist mobile devices for computations, and they cannot be used as permanent storage of data. The concept of cloudlet between cloud and mobile computing is further discussed in (Satyanarayanan et al., 2015; Verbelen et al., 2012). A mobile device is able to connect to a cloudlet and upload data to the cloudlet.

4.2.1 Role of edge computing in Internet of Things

The significant advancements in embedded systems-on-a-chip with efficient operating systems have extended the potential of the IoT. The excellent features of devices with advanced controllers not only collect and send the data to the cloud, also carry out complicated computations on the premises which are led to the concept of edge computing. It can provide the services offered by the cloud computing very nearer to the edge of the network to support various applications (Hassan et al., 2018).

The extensible edge server architecture is shown in Fig. 4.3, which exhibits that the edge computing comprises itself multiple technologies such as IoT, cloud computing and grid computing. The additional layer between connected devices and the cloud brings the end device closer (Gezer & Ruskowski, 2017).

There are various service-level objectives are needed for edge computing in the context of IoT such as minimal latency, managing network, optimized cost, energy management, resource management, data management and data types. The property of latency minimization can definitely fulfill the quality of service for smart transportation which is delay sensitive based IoT application such as unmanned vehicles and vehicle accident prevention (Anawar et al., 2018). The role of edge servers for various domains is presented in Fig. 4.4. There are few laws can be called for edge based IoT. The law of physics is applied for latency minimization which is essential for time sensitive

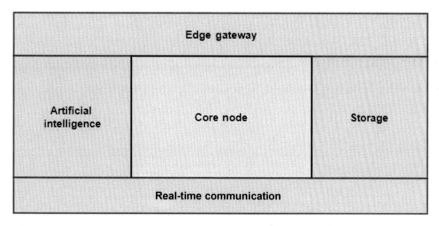

FIGURE 4.3

Extensible edge server architecture (Gezer & Ruskowski, 2017).

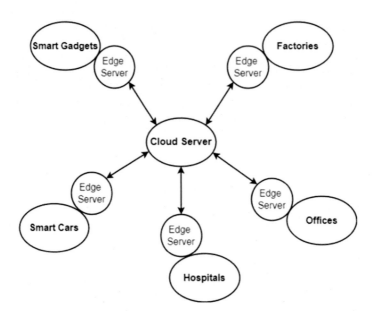

FIGURE 4.4

Role of edge servers for various domains (Gezer & Ruskowski, 2017).

based applications. The law of economics is applied where we need to minimize the cost. The law of land is involved where the data need high security and required to be processed locally (https://www.xenonstack.com/blog/edge-computing/). The edge computing based IoT must have the features of mobility, interoperability, reliability, latency and real time interactions. Data acquisition, inferential control, data analysis, decision making are the few major roles of edge computing in IoT. In addition, the technologies adopted by edge based systems need to meet certain standards (Khan et al., 2020). The software must be modified to fulfill the real time implementations (Liao et al., 2019; Rostedt, 2013). The server placed in the edge level must be flexible enough to incorporate plug-and-play features to add much more advanced functionalities through placing software and hardware blocks. High rate of data transfer is essential to transfer the large amount of data sensed from various devices, for example, large size of data acquired during surgical operations in healthcare applications to edge clouds (Yin et al., 2017).

4.2.2 Edge intelligence and 5G in Internet of Things based smart healthcare system

The primary characteristic of 5G is faster data output which helps gigantic machine communications. In addition, machine to machine communication without human interaction is possible with high reliability and low latency. The vast advancements in cloud computing technology drove a centralized approach to the system managements. Computing toward a distributed architecture is driven by IoT, mobile computing and software as a service models. The combinations of above two approaches which improve the system performance to the higher level are obtained with 5G and edge computing techniques (Hassan et al., 2019). 5G is foreknown to provide support with low latency for high interactive IoT use cases. Edge computing uses a distributed architecture model which brings the computing capabilities nearer to the premises and so it reduces latency. Edge computing unloads a gigantic amount of data from user equipment (UEs) to edge clouds (Zhang et al., 2018).

Edge computing technique attempts to integrate various types of edge devices and servers. They can perform collaboratively for the efficient processing of locally generated healthcare sensor data. In addition, edge intelligence tries to move toward smart healthcare frameworks by employing AI techniques and cognitive intelligence related to human behavior into edge architectures. Edge intelligence is applied to smart devices that are attached with sensor. These devices are available at gateways closer to the smart sensors and gateway devices which can serve as edge nodes.

Most of the healthcare systems do not consider the emergency situations of patients, and not providing any personalized resource service for users. To overcome this drawback, the Edge-Cognitive-Computing-based (ECC-based) smart healthcare system is proposed (Chen et al., 2018). This system could monitor and analyze the physical health of users using cognitive computing (Chen et al., 2019; Wan et al., 2020). The ECC-based system can adjust the computing resource allocation of the whole edge computing network according to the health-risk grade of patients.

There is various new intelligent detection techniques using deep learning and edge computing are coming up. The sensor senses the human electroencephalogram (EEG) signal and sends to a nearby edge server. The edge server introduces various preprocessing steps and assigns them to available edge devices. Consequently, the large size advanced signals are transferred to the cloud server. Edge Learning as a Service (EdgeLaaS) framework is also proposed to process health supervision data on premise itself. Edge learning nodes assist patients to choose better suggestions from the medical persons in real time whenever emergency situations occurs (Zhang et al., 2020).

The telemedicine field requires an advanced network such as 5G and 6G that offers support in real time, providing high-quality video communication without slowing down the facility network. Also the network must have standard protocols and more secured mechanisms to address security challenges not only to follow security-by-design but also security-by-operations rules (Hameed et al., 2021). Hence edge could enable more secured remote healthcare facility by assuring security standards (https://enterprise.verizon.com/resources/articles/s/future-of-healthcare-technology-5g-edge-computing/).

4.3 Low power hardware architecture for edge computing devices
4.3.1 Objectives of hardware development in edge computing

IoT systems perform a particular function in four main processes namely data collection, data processing, data transmission and actuation. Of the above-stated processes, data transmission is the most power-consuming process and the main aim of edge computing devices is to minimize the distance between the source and the destination of transmission (Saeed et al., 2018). As electronic devices and embedded systems are growing pervasively, most IoT systems are battery-powered devices. This means that the underlying hardware architecture must be power-efficient. At the same time, they must also be scalable, cheap and easy to manufacture. Therefore a clear set of objectives are defined for edge computing hardware architecture. MOSFETs which form the lowest level of the sensor nodes have been scaled down in size over the past decade to achieve a smaller chip area or more functionality in the same chip area. The smallest MOSFETs manufactured are about 5 nm in size and widely used today. However, to meet the power efficiency criteria and the cost criteria, older silicon technology processes of about 130−200 nm can be used. It is important to note that, systems performing more complex functionalities will still need smaller-sized nodes. Hence, based on the functionality of the device an optimized architecture is to be developed considering the power constraints, cost-effective and various other factors.

4.3.2 System architecture

The main component of a Microcontroller unit (MCU) is the CPU and its architecture plays an important role in determining how efficient the system is (Saeed et al., 2018). The peripherals include the input−output architecture which facilitates connections with sensors to collect data. Power consumption is the pivotal point of this analysis, hence different power management techniques are analyzed and explained. Data processing is an important function of edge computing devices and enabling MCUs to process data will in turn enable edge computing. Each of the above-mentioned components is explained in detail in the upcoming sections. The general system architecture is shown in Fig. 4.5 (Capra et al., 2019; Low Power Hardware Techniques; Overview of Computer).

4.3.3 Central processing unit architecture

Many IoT devices demand high computational power, and this can be achieved using multicore CPU architecture. Multicore processors are CPUs which have several independent processors on a single chip. In multicore processors, the workload is distributed between several cores which allows parallel computing and lower power consumption which is depicted in Fig. 4.6 (Overview of Computer; Low Power Hardware Techniques; Ermiş & Çatay., 2017). These multicores are made heterogeneous—in which different cores are assigned to different functionalities and will be

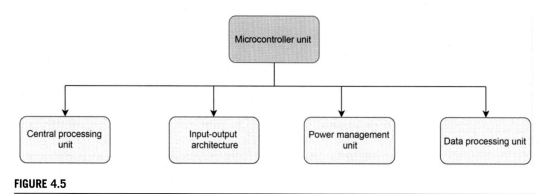

FIGURE 4.5

System architecture (Zhang et al., 2020).

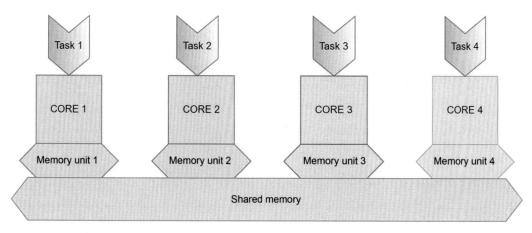

FIGURE 4.6

Central processing unit architecture (Ermiş & Çatay., 2017).

structured according to the function they perform. For instance, one large core will be used for the high computational function while a smaller core in the same chip will be used for peripheral requests. This technique increases the power efficiency because each of the cores handles different tasks and can be turned off separately when that particular task is not required to be performed. Hence, multicore CPU architecture facilitates high performance with lower power consumption (Overview of Computer; Low Power Hardware Techniques; Ermiş & Çatay., 2017).

The power consumption of CMOS microcontrollers is of two types: dynamic power and static power. The power consumed when the microcontroller is performing its programmed tasks is called dynamic power whereas the power consumed when not running code and that occurs only by applying a voltage to the device is called static power. In battery-operated systems, static power is of greater significance as most of them will spend significant portions of the application lifetime in sleep mode. in a multicore processor, dynamic voltage and frequency scaling (DVFS) can be performed on each core, enhancing power savings by decoupling their working points. An optimal configuration can be achieved by operating each core independently, as well as shutting down them selectively.

Larger process technologies(130−200 nm), have considerably lower leakage current but at the cost of higher dynamic power. This means that when larger process technologies are used, the static power decreases while the dynamic power increases. Therefore larger process technologies may be a good solution for battery-operated devices that do not require high computational power, since static power dominates the dynamic power. However, in cases which require large computational power, dynamic power dominates. When trying to optimize both dynamic and static power, a critical trade-off becomes apparent. As a result of this trade-off, it can be difficult to determine which is more important to reduce power consumption for a system. To select the correct MCU, which will minimize power for a particular system, it is important to know whether dynamic or static power has a more significant impact on the system, and this can be done by power budgeting.

Different CPU architectures will consume different amounts of energy in a single clock cycle. Hence, it is important to take the Instruction Set Architecture (ISA) into consideration. Cycles per instruction, clocking schemes and the instructions have a major impact on the performance of the device, which directly affects the power consumed in one cycle, and therefore the overall power consumption. Single core and multicore architecture are depicted in Fig. 4.7.

4.3.4 Input—output architecture

Peripherals which constitute the input—output architecture of a microcontroller play an important role in reducing power consumption. There are two factors which should be considered when comparing different peripherals- the current and the time taken to send a given amount of data. A trade-off is achieved by multiplying both these quantities, which results in the total charge consumed. This in turn determines the power consumption. This trade-off is important to achieve because it could be deceiving to consider the factors individually. For instance, the universal asynchronous receiver-transmitter (UART) has a smaller current than the serial peripheral interface (SPI) and may seem like a low power module. However, it takes more time than the SPI to transfer data. While the SPI may consume more power during the transaction, after the transaction is completed, the device can go to sleep and eliminate the power consumption of the CPU which is the highest source of power consumption.

IoT nodes need to be specifically altered with respect to the application they perform and need to be optimized with low power and high-performance characteristics. To achieve these optimal standards, an alliance has been formed which works on several multimedia protocols. This alliance

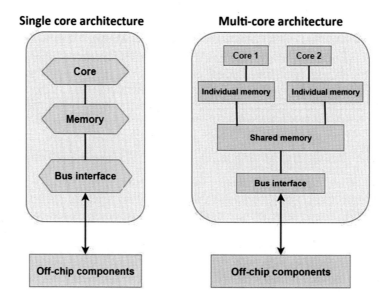

FIGURE 4.7

Single core and multicore architecture (Johnson & Dinyo, 2015).

is called the MIPI (mobile industry processor interface) alliance. The aim is to realize optimized devices with low power, high bandwidth and low electromagnetic interferences. The low power MIPI design—I3C interface has been optimized for a high clock speed, has high power efficiency, and is much faster than its previous model—I2C interface (MIPI Alliance Specifications Overview).

4.3.5 Power consumption

The lifetime of any embedded system primarily depends on the lifetime of the battery; hence optimizing the power dissipation is crucial to battery-powered systems (Shin et al., 2010). As the MCU serves as the brain of the system, it consumes the most power and also has control over the system power consumption. Since edge computing moves the processing power from the cloud to a point closer to the user, the main goal is to achieve a balance between high processing power and low power consumption. One of the widely used techniques to achieve this goal is turning off the parts of the circuit that are not performing any task. Clock gating and power gating are popular methods to perform this technique. In clock gating, logical OR and AND gates usually stop the clock from the part of the circuit that is not required in the current task, which in turn results in a significant reduction of the dynamic power consumption. However, a static leakage power persists but it can be reduced by combining clock gating and dynamic voltage scaling. In power gating, the supply voltage is disconnected from the circuit using MOS transistors, which reduces both the dynamic power and the static power thereby serving as a better solution than clock gating. The downside to this technique is that it is complex to design due to insertion of current switches which have to be settled by a combination of extra circuitry and customized tools and methodologies. Nevertheless, the technique should be selected based on the constraints of the application designed. Other basic MCU-level considerations can be made to achieve a low power consuming circuit. The fundamental rules to design low power MCUs are to increase impedance in current paths and reduce impedance in high-speed switching paths. Hence, leakage currents and operating duty cycles are reduced.

Another approach is to make the MCUs operates at the source voltage closer to the threshold value of the transistors. Conventionally, the MCUs work in regions more than the threshold level of transistors. Scaling the voltage so that the transistors can operate in regions near their threshold can significantly reduce the power dissipation. This could give rise to performance degradation issues which need to be compensated by a technique known as PVT (process, voltage, and temperature) compensation. Special circuits such as Canary circuits and Razor flip-flops are used for PVT compensation by controlling the voltage supply.

4.3.6 Data processing and algorithmic optimization

Parallel computing plays a key role in processing data at the edge. In parallel computing, different tasks are broken down into smaller independent tasks and are executed simultaneously by multiple processors. Different processors communicate with each other and the memory unit via shared memory. The results of each independent smaller task are combined upon completion as part of an overall algorithm. Instruction-level parallelism and multistage instruction pipelines are a part of all modern processors and enable the execution of more than one instruction per clock cycle. Task parallelism can be achieved by using multicore processors to execute different threads (or processes) on the same or different data. It emphasizes on the distributed nature of the processing (concurrent processing of threads). By running many threads at once, these applications can tolerate the high amounts of peripheral data and memory system latency. For instance, if one thread is delayed waiting for memory access, other threads can do useful work without being interrupted by the delayed task.

Increased bandwidth is another important feature of edge computing devices. While collecting real time data from the sensors, the information contains some redundant data, which need not to be processed. Transmitting this data to the cloud can consume an unnecessary amount of bandwidth. Edge computing prevents this unnecessary usage of bandwidth with the help of an algorithmic optimization technique known as compressive sensing. In compressive sensing, the data collected from the sensors is transmitted to a local processing unit that has got an ample amount of space and computing capacity. Only the important information and features that are required are filtered out and transmitted to the cloud which in turn significantly reduces the network bandwidth.

4.4 Examples of edge computing devices

One of the main applications of edge computing is human machine interface (HMI) products which serve as the ideal choice for IoT edge applications in factory, marine and building automation. EXOR's eX Series 700 HMI products are IoT edge devices that perform powerful controlling and networking operations. The design is very user-friendly with a capacitive touchscreen and high-resolution display. The eX715 is powered by a rechargeable Lithium battery and an ARM Cortex-A9 quad-core 800 MHz CPU. The CPU consists of four cache-coherent cores. The CPU architecture supports four main modes namely run mode, standby mode, dormant mode and shutdown mode. The run mode is the normal mode of operation, where all of the functions of the processor are available. During the standby mode, clock pulses are not gated but logic is powered up. Dormant mode enables the processor to be powered down while leaving the caches powered up and maintaining their state. Shutdown mode shuts down the entire device. These different modes play a major role in optimizing the power dissipation of the device. This device supports serial communication through SPI and contains three Ethernet ports. Ethernet can be a better option when compared to Wi-Fi as it is a faster and more reliable option with lesser interference.

Another example of edge computing devices is edge gateways which accelerate IoT at the edge by providing fast and responsive data solutions. Dell Edge Gateway 5100 is one such edge gateway solution. It runs on a dual-core processor. It supports SATA (Serial Advanced Technology Attachment) and UART communication standards. The CPU adopts the Intel Hyper-Threading Technology (Intel HT Technology) which provides dual processing threads per physical core and ensures parallel computing. Intel Virtualization Technology (VT-x) is another technology that is incorporated which supports task parallelism by isolating computing activities into separate partitions (https://www.intel.in/content/www/in/en/gaming/resources/hyper-threading.html). The Enhanced Intel SpeedStep Technology is also supported which is an advanced method of enabling high performance and meeting the power-conservation constraints of the system. Based on the processor load and the Enhanced Intel SpeedStep, the traditional Intel SpeedStep technology transits both voltage and frequency between high and low levels.

4.5 Edge computing for intelligent healthcare applications

Edge computing is a remarkable technology that allows processing at the end of the device edge. It hypothesizes that computation should always take place near a data source (Intelligent & Cloud Computing, 2021). Edge computing models are things that act as data consumers but can also act as data producers. Edge computing includes data transfer, data storage, processing, and delivery from the cloud to the user. Edge-of-Things (EoT)-based healthcare services are providing patient-care amenities related to autonomic and persuasive healthcare in which EoT broker plays the role as a middleman between the Healthcare Service Consumers (HSC) and Computing Service Providers (CSP). The data acquired by the sensor from a patient's Body Area Networks (BAN) are very sensitive. They should be stored and analyzed in secured environment. EoT based patient monitoring systems and edge computing based real time analysis is more secured (Golam Rabiul Alam et al., 2019). Edge computing as a service (ECaaS) and Cloud Computing as a Service (CCaaS) providers are more economical, so that the broker can deliver smart healthcare services to consumers with optimized cost. Edge computing plays an important role in the smart health care system (Gupta & Khamparia, 2020). It is very difficult for the traditional healthcare systems to maintain the scalability to meet the rising number of patients and to provide the best healthcare solutions. With the evolution of IoT and edge based systems, traditional healthcare systems into smart health systems. Mobile healthcare is also termed as tele-healthcare used to provide healthcare solutions to patients at the remote area. Edge computing is used to enhance the managing system and quality level of the healthcare system. In addition, it assists the system for better utilization of its resources and needs to reduce its cost (Oueida et al., 2018).

The consumable and nonconsumable resources of the healthcare system are very well monitored with the help of edge computing based network. The human resource and employing this resource in an effective way is very essential and which can be implemented with the help of edge computing effectively. It is the need of the hour to identify certain novel approaches to monitor the healthcare system economically and effectively (Kim, 2018). The wireless body area networks can be improved for the deployment of healthcare applications (Otoum et al., 2015). The massive amount of data generated in this field must be stored and analyzed very carefully and handled cautiously (Aloqaily et al., 2016). The features of the IoT and fog computing are highlighted and illustrated in (Gia et al., 2015). The difficulties in adopting novel technologies and the advantages of shifting from traditional healthcare system to cloud, fog and edge based smart healthcare system are discussed in (Kuo, 2011).

4.5.1 Edge computing for healthcare applications

A huge amount of data is generated by healthcare applications and there is a dire need for a reliable and secure system for smart health applications. As the healthcare sector deals with critical applications, the migration from the conventional system to a smart system has to be planned very meticulously. Edge computing accelerates this migration as it forms a system in which the data is manipulated at its source, hence is secure and faster than regular servers. Smart healthcare applications are broadly classified into two primary classes: patient-related class and process-related class. The patient-related class applications include wearable devices and embedded sensors that collect health data. The process-related class applications include various healthcare policies that can be automated or improved with technology. Some examples are scheduling of resources, service quality, and utilization factor of resources. This chapter demonstrates an example of an application from one of the healthcare departments and the importance of edge computing in enhancing the department, has been highlighted. Emergency departments are the most time-critical systems of the healthcare department and require foremost planning and systemization. One of the main problems of these departments is the congestion and overcrowding due to the presence of a large number of patients at a given time. This in turn leads to a complex workflow and resource management. Hence edge computing can be used to optimize the existing models and can be used for managing the processes at emergency departments. These models used to represent distributed system. The main idea of edge computing is to spread the computation tasks to different layers of the network. A part of the work is distributed to the edge rather than computing all the information in the cloud.

In this model, the scheduling algorithm is stored in the cloud while every healthcare resource will have an edge node that communicates the fulfillment of an assigned task. Then, the scheduling algorithm from the cloud reassigns another task to the edge node. Hence the task allocations and information regarding completed tasks are sent from the edge nodes to the cloud and the cloud stores the database and workflow software. Every healthcare resource has a smart device, such as a mobile phone, tablet, smart-watch that behaves as an edge node. After the task is completed, the node notifies the cloud that the resource is available and ready to be reassigned. Thus scheduling occurs in the cloud while the assignment is distributed to the edge (Oueida et al., 2018). A systematic cycle is formed that has a significant impact on the management of resources in emergency departments.

4.5.2 Advantages of edge computing for healthcare applications

The advantages of edge based healthcare system are illustrated in Fig. 4.8. The adaptation of edge computing for healthcare system finds wide application in healthcare systems. Even after the vast development and innovations in healthcare systems, there are numerous challenges in providing high quality healthcare to rural areas. Due to the lack of Internet connectivity and access, medical providers find difficulty in providing the quality healthcare to the people who are far away from health centers. All the above-mentioned issues can be alleviated with the help of Medical Internet of Things (MIoT) and edge computing applications. It is quite possible to generate, collect, store and analyze very critical data without having consistent network connectivity. It is very simple and fast to diagnose the patients with smart medical devices and the data can be easily sent to the cloud servers or to the central servers with the network. In addition, the massive data collected by medical devices such as blood glucose monitor, wearable sensors and healthcare apps make everything feasible for the healthcare professionals to diagnose the patients especially with chronic issues (Suresh & Paiva, 2021). Thus edge computing has the potential to expand the quality healthcare systems to the remote areas. Generally, the massive data collected which may be structured or unstructured from the patients are stored in the cloud server. These data must be analyzed in a way that can be used to obtain the inference by using some effective analytical tool. By that time the health condition would have been changed. Edge computing applications can be used effectively to

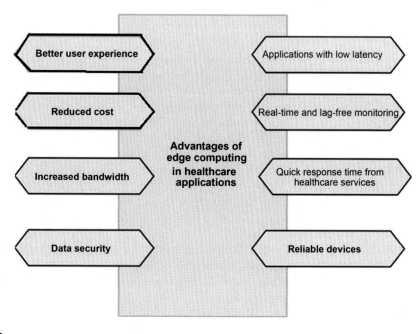

FIGURE 4.8

Advantages of edge based healthcare systems (Naveen & Manjunath, 2019).

overcome the aforementioned issues. The critical analysis can be performed on the edge devices located on the edge of the network (http://www.vxchange.com).

Real time analytics on the premises can easily predict the anomalies and healthcare professionals can take immediate actions during medical emergencies. The data which are not critical can be stored in the cloud server and data collected over the period can be segregated and analyzed by adopting certain machine learning algorithm in the remote data centers. There are numerous benefits when edge solution is employed properly in healthcare systems. This includes data proximity, integration of IoT data and health care provider system, medical care with lowest cost, patient record keeping, operating rooms leveraging advanced robotics and video equipment, patient monitoring leveraging medical devices such as insulin pumps, smart lenses and pacemakers, wearables and connected apps that track various health metrics, such as heart rate, count steps and hydration, facility utilization where sensors and data analytics help make the most efficient use of clinical facilities. Smart devices facilitate people to check in for appointments (https://blog.apc.com/2017/12/12/edge-computing-iot-healthcare/). Smart medical devices are the key edge computing use cases that completely transform the healthcare industry's customer experience. Edge computing companies can be associated with healthcare professionals. Many hospitals provide interactive educational contents to patients. Edge data centers can provide this content and make it available to the patients with reduced latency. IoT healthcare supply chain innovations offer an opportunity to gain operational efficiencies on the margins and represent one of the more compelling edge computing use cases.

4.5.3 Implementation challenges of edge computing in healthcare systems

Adequate bandwidth is the foremost challenge that is fundamental for the implementation of 5G wireless networks. Security aspects of medical devices and interoperability are some other challenges. Cost is an essential factor in procuring sufficient bandwidth. Distributed edge computing architecture is not so expensive since the data is stored on premise and so data need not be transmitted to a long distance where the cloud server is located. But it is not effective to store large amounts of data. Edge computing can provide better safety since the data is not transmitted too far as in cloud computing (Varghese et al., 2016).

The challenges of edge based healthcare system are presented in Fig. 4.9, and there is less chance for system wide malfunctioning of the edge architecture. But edge computing is not reliable for storing large amounts of data. Selecting the appropriate edge computing tool for various healthcare scenarios is one of the most important challenges. Due to privacy concerns, it is difficult to deploy open-source software immediately into the healthcare system. Medical professionals should be encouraged to embrace innovation. Healthcare specialists need to be prepared to approve new sensors that can provide information in real time. Edge computing in healthcare is domain-specific and requires support from healthcare organizations. The approval from the medical experts is the major challenge. The medical experts have to give creative ideas and helps in designing edge based use cases for real time smart healthcare system.

4.5.4 Applications of edge computing based healthcare system

By 2025, 75% of data is expected to be generated at the edge and several edge computing healthcare use cases will be implemented such as point-of-care management and monitoring, chronic disease management, remote assisted living, wellness and preventive care, drug management and hospital operations and logistics. With an increased number of connected devices in the healthcare ecosystem, some of the major problems faced by traditional digital healthcare systems are latency

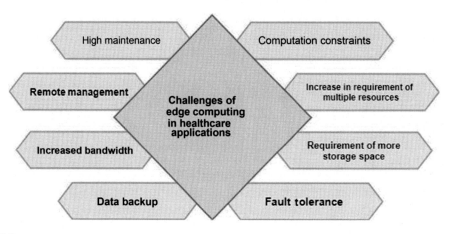

FIGURE 4.9

Challenges of edge based healthcare system (Naveen & Manjunath, 2019).

and data security. Edge computing plays a role in resolving these issues as the data is processed closer to where it is generated. The distributed nature of computation of edge computing systems also enables simpler and quicker data management. In smart healthcare systems, there is a continuous flow of critical data and an efficient and practical management policy is required to be devised.

Tele-health tools such as video check-ups and appointments with medical professionals will require seamless data transfer which can be enabled using edge computing. With enhanced connectivity and saved bandwidth, plenty of healthcare applications can be practically implemented and improvised (http://www.cbinsights.com/research/internet-of-medical-things-5g-edge-computing-changing-healthcare/; https://9to5mac.com/2020/03/08/apple-watch-blood-oxygen-saturation/). Remote diagnosis is one of the primary use cases of edge computing in smart healthcare applications. Increased bandwidth and data security are critical for remote diagnosis. Portable devices and wearables will be able to work faster and are more sensitive with the application of edge computing as the nodes are decentralized. With the rise in robotic surgery technology, there is a dire need for seamless data transmission for the reliable operation of such systems. Edge computing facilitates tele-health and continuous lag-free real time patient monitoring making it a much more reliable and efficient system. Timing is the most crucial aspect of emergency wards and intensive care units in hospitals. Managing the operations of such time-critical units will require highly reliable and secure network gateways which can be provided by edge computing gateways.

An enormous amount of real time data is continuously generated in healthcare applications and managing this data is an ordeal. Since the processes of data storage and data processing are decentralized and performed closer to the source of the data, data management is much easier and quicker. Each data source may have its own data processing unit, where a reduced amount of data is quickly processed and anomalies are detected immediately when generated. This mechanism can be used in emergency and time-critical situations. With an overall faster response in the healthcare ecosystem, the data lifecycle management is simplified to a great extent thereby reducing costs and human effort and increasing efficiency. Processing and storing medical data in the centralized edge network of the hospital can overcome the privacy challenges of smart healthcare systems. The data is not stored in a third-party cloud and hence is not vulnerable to data privacy and security threats. Processing data at the edge can also ensure the protection of sensitive data transmitted between patient and healthcare provider. A combination of edge computing and cognitive computing (computing algorithms for simulation of human thought processes) can be used to personalize the user experience by altering the resources according to the health data collected from each user.

4.5.5 Patient data security in edge computing

Maintaining the patient's records is a key issue in the medical world. Hence the experts in the medical field are looking for technological solutions. Edge computing assists in speed up the process in transferring the data without delay. So it helps to make decisions quickly. Edge-enabled tools are helpful in data management. In addition, the APIs created by companies enable medical professionals to know the patient's current condition and to know the patient's history and take preventive measures in real time. Healthcare providers can securely store patient data in accordance with the Health Insurance Portability and Accountability Act (HIPAA) (https://www.ncbi.nlm.nih.gov/books/NBK500019/). All data can be stored securely in a location called mini centers. Companies can work in conjunction with cyber security and focus on IoT security features and external networks (Jurcut et al., 2020). MIoT technology and edge-integrated AI can be expanded to make use of these resources. 5G and edge computing help surgeons to use diving technology for training and planning. A kidney injury monitoring program was developed by Potrero Medical. Forecast analytics analyzes patient data in real time with the help of edge computing and alerts medical professionals automatically before any issues are significant (https://potreromed.com/).

4.6 Impact of edge computing, Internet of Things and 5G on smart healthcare systems

The technologies associated with of IoT are enormous. The connected devices count is expected to increase to 3.2 billion by 2023 (https://www.cisco.com/c/en/us/solutions/collateral/executive-perspectives/annual-internet-report/white-paper-c11-741490.html). One of the most important factors for this massive growth is the development of 5G networks. The upcoming 5G mobile communication technology is great news for the Internet of Tings world. Since the connected devices finds the improved performance and reliability due to these advancements in the mobile communication technology. The 5G technology is going to be 10 times faster than current LTE networks. It enables the IoT devices to communicate and share data faster. With 5G, there is considerable increase in data transfer speeds. Nearly all IoT devices will benefit from greater speeds including those with healthcare and industry applications. For any IoT application, reliable and stable network conditions are very essential and especially for the applications that depend on real time updates. Reliability and performance of connected devices increases with 5G network. The 5G network communication which has the high-speed connectivity, reduced latency, and greater coverage will be the key for the IoT world (Liu & Jiang, 2016).

With 5G, smart healthcare systems can handle telemedicine appointments which can also support real time video to get transferred. The files and videos used by the medical experts will be of massive data set and it is very difficult to review process at the medical professional end. With the help of edge computing, the doctors and associated members can collaborate easily. It can reduce pain and anxiety for terminally ill patients in hospice by providing calming, distracting content. But these data consume more time to get transferred if the network bandwidth is low or sometimes may not be sent successfully. The patients need to wait longer for treatment and more patients are not attended. Moreover due to the delay, the condition of the patient might get changed. Both 5G and edge computing technologies are required to be combined to achieve ultra-low latency for use cases like healthcare, remote tele-surgery and autonomous drones. Additionally, wearables are predicted to decrease hospital costs by 16% over the next five years (https://www.healthcareexecutive.in/blog/wearable-technology). By adding 5G network which is super-fast to existing architectures increase the speed and reliable transport of massive data files of medical imagery, which can improve both access to care and the quality of care. Most of the key healthcare systems started to use AI to diagnose the patients and provide the prescription to the concern patient depend upon the specific treatment plan (Singh et al., 2020; Singh & Singh, 2020).

4.7 Conclusion and future scope of research

This chapter proposes an overview of the low power hardware architecture for edge computing devices and also presents a few examples of such devices toward smart healthcare systems. With the proliferation of IoT devices, large amounts of data are created. It is crucial to save bandwidth by distributing the processing load and data to points closer to the source. Moving toward 5G, low latency is the need of the hour and is critical for real time systems. With so much data being sent to the cloud and the digitalization era, the security and privacy of user data is a huge problem. Edge Computing can be a reliable solution for all the above problems associated with smart healthcare system. However, the pitfall is that they require more storage space and demand more investment and maintenance costs. Nevertheless, trade-offs can be made for cost-effective and scalable solutions. It is important to take into consideration that an IoT system consists of both hardware and software parts and optimizing both components is important for a fully-optimized low power system. This chapter emphasizes on the hardware architecture and the different techniques used to reduce the power consumption, advantages of edge computing for smart healthcare, challenges in implementation and applications of edge computing in smart healthcare system. The optimizations of software side such as code optimization, conditional code execution etc. are not the focus and can be an interesting topic for future research. Moreover, this paper presents only an overview of the power management techniques, an in-depth analytical investigation can be performed in the future.

References

Alam, M. G. R., Munir, M. S., Uddin, M. Z., Alam, M. S., Dang, T. N., & Hong, C. S. (2019). Edge-of-things computing framework for cost-effective provisioning of healthcare data. *Journal of Parallel and Distributed Computing, 123*, 54–60.

Aloqaily, M., Kantarci, B., & Mouftah, H. T. (2016). Multiagent/multiobjective interaction game system for service provisioning in vehicular cloud. *IEEE Access, 4*, 3153–3168.

Anawar, M. R., Wang, S., Azam Zia, M., Jadoon, A. K., Akram, U., & Raza, S. (2018). Fog computing: An overview of big IoT data analytics. *Wireless Communications and Mobile Computing, 2018*.

Brooks, D., & Sartori, J. (2017). Ultra-low-power processors. *IEEE Micro, 37*(6), 16–19.

Capra, M., Peloso, R., Masera, G., Ruo Roch, M., & Martina, M. (2019). Edge computing: A survey on the hardware requirements in the Internet of Things World. *Future Internet, 11*(4), 100.

Chen, M., Li, W., Fortino, G., Hao, Y., Hu, L., & Humar, I. (2019). A dynamic service migration mechanism in edge cognitive computing. *ACM Transactions on Internet Technology (TOIT), 19*(2), 1–15.

Chen, M., Li, W., Hao, Y., Qian, Y., & Humar, I. (2018). Edge cognitive computing based smart healthcare system. *Future Generation Computer Systems, 86*, 403–411.

Ermiş, G., & Çatay, B. (2017). Accelerating local search algorithms for the travelling salesman problem through the effective use of GPU. *Transportation research procedia, 22*, 409–418.

García-Valls, M., Dubey, A., & Botti, V. (2018). Introducing the new paradigm of social dispersed computing: Applications, technologies and challenges. *Journal of Systems Architecture, 91*, 83–102.

Gezer, V., & Ruskowski, M. An Extensible Real-Time Capable Server Architecture for Edge Computing.

Gia, T. N., Jiang, M., Rahmani, A. M., Westerlund, T., Liljeberg, P., & Tenhunen, H. (2015, October). Fog computing in healthcare internet of things: A case study on ecg feature extraction. In *2015 IEEE international conference on computer and information technology; ubiquitous computing and communications; dependable, autonomic and secure computing; pervasive intelligence and computing* (pp. 356–363). IEEE.

Gupta, D., & Khamparia, A. (Eds.). (2020). Fog, Edge, and Pervasive Computing in Intelligent IoT Driven Applications. John Wiley & Sons.

Hameed, K., Bajwa, I. S., Sarwar, N., Anwar, W., Mushtaq, Z., & Rashid, T. (2021). Integration of 5G and Block-Chain Technologies in Smart Telemedicine Using IoT. *Journal of Healthcare Engineering*, 2021.

Hassan, N., Gillani, S., Ahmed, E., Yaqoob, I., & Imran, M. (2018). The role of edge computing in internet of things. *IEEE communications magazine*, *56*(11), 110−115.

Hassan, N., Yau, K. L. A., & Wu, C. (2019). Edge computing in 5G: A review. *IEEE Access*, *7*, 127276−127289.

Indu, I., Anand, P. R., & Bhaskar, V. (2018). Identity and access management in cloud environment: Mechanisms and challenges. *Engineering science and technology, an international journal*, *21*(4), 574−588.

Intelligent and Cloud Computing, Springer Science and Business Media LLC, (2021).

Johnson, O., & Omosehinmi, D. (2015). Comparative Analysis of Single-Core and Multi-Core Systems. *International Journal of Computer Science & Information Technology (IJCSIT)*, *7*(6), 117−130.

Jurcut, A. D., Ranaweera, P., & Xu, L. (2020). Introduction to IoT security. *IoT security: Advances in authentication*, 27−64.

Khan, L. U., Yaqoob, I., Tran, N. H., Kazmi, S. A., Dang, T. N., & Hong, C. S. (2020). Edge-computing-enabled smart cities: A comprehensive survey. *IEEE Internet of Things. Journal*, *7*(10), 10200−10232.

Kim, J. (2018). The effect of patient participation through physician's resources on experience and wellbeing. *Sustainability*, *10*(6), 2102.

Kuo, M. H. (2011). Opportunities and challenges of cloud computing to improve health care services. *Journal of medical Internet research*, *13*(3), e67.

Liao, C. C., Chen, T. S., & Wu, A. Y. (2019). Real-time multi-user detection engine design for IoT applications via modified sparsity adaptive matching pursuit. *IEEE Transactions on Circuits and Systems I: Regular, Papers, 66*(8), 2987−3000.

Liu, G., & Jiang, D. (2016). 5G: Vision and requirements for mobile communication system towards year 2020. *Chinese Journal of Engineering*, *2016*(2016), 8.

Low Power Hardware Techniques: <http://ww1.microchip.com/downloads/en/appnotes/01416a.pdf>.

Mahdavinejad, M. S., Rezvan, M., Barekatain, M., Adibi, P., Barnaghi, P., & Sheth, A. P. (2018). Machine learning for Internet of Things data analysis: A survey. *Digital Communications and Networks*, *4*(3), 161−175.

MIPI Alliance Specifications Overview: https://www.mipi.org/specifications.

Naveen, S., & Kounte, M. R. (2019). Key technologies and challenges in IoT edge computing. In *2019 Third international conference on I-SMAC (IoT in social, mobile, analytics and cloud)(I-SMAC)* (pp. 61−65). IEEE.

Otoum, S., Ahmed, M., & Mouftah, H. T. (2015). Sensor Medium Access Control (SMAC)-based epilepsy patients monitoring system. In *2015 IEEE 28th Canadian conference on electrical and computer engineering (CCECE)* (pp. 1109−1114). IEEE.

Oueida, S., Kotb, Y., Aloqaily, M., Jararweh, Y., & Baker, T. (2018). An edge computing based smart healthcare framework for resource management. *Sensors*, *18*(12), 4307.

Overview of Computer Architecture: https://www.cise.ufl.edu/Bmssz/CompOrg/CDAintro.html.

Rostedt, S. (2013). Intro to Real-Time Linux for Embedded Developers. *Linux Foundation Blog*.

Saeed, F., Gazem, N., Mohammed, F., & Busalim, A. (Eds.). (2018). Recent Trends in Data Science and Soft Computing: *Proceedings of the 3rd International Conference of Reliable Information and Communication Technology (IRICT 2018)* (Vol. 843). Springer.

Satyanarayanan, M., Simoens, P., Xiao, Y., Pillai, P., Chen, Z., Ha, K., & Amos, B. (2015). Edge analytics in the internet of things. *IEEE Pervasive Computing*, *14*(2), 24−31.

Shi, W., Pallis, G., & Xu, Z. (2019). Edge computing [scanning the issue]. *Proceedings of the IEEE, 107*(8), 1474−1481.

Shin, Y., Seomun, J., Choi, K. M., & Sakurai, T. (2010). Power gating: Circuits, design methodologies, and best practice for standard-cell VLSI designs. *ACM Transactions on Design Automation of Electronic Systems (TODAES), 15*(4), 1−37.

Singh, P., & Agrawal, R. (2021). An Overloading State Computation and Load Sharing Mechanism in Fog Computing. *Journal of Information Technology Research (JITR), 14*(4), 94−106.

Singh, P., & Singh, N. (2020). Blockchain With IoT and AI: A Review of Agriculture and Healthcare. *International Journal of Applied Evolutionary Computation (IJAEC), 11*(4), 13−27.

Singh, P., Singh, N., Singh, K. K., & Singh, A. (2021). *Diagnosing of disease using machine learning. In. Machine Learning and the Internet of Medical Things in Healthcare* (pp. 89−111). Academic Press.

Suresh, A., & Paiva, S. (Eds.) (2021). Deep learning and edge computing solutions for high performance computing. *Springer International Publishing.*

Torre, R., Doan, T., & Salah, H. (2020). *Mobile edge cloud. In. Computing in Communication Networks* (pp. 77−91). Academic Press.

Varghese, B., Wang, N., Barbhuiya, S., Kilpatrick, P. & Nikolopoulos, D. S. (2016). Challenges and opportunities in edge computing. In *2016 IEEE International Conference on Smart Cloud (SmartCloud)* (pp. 20−26) IEEE.

Verbelen T., P. Simoens, F. Turck, B. Dhoedt. (2012). Cloudlets: bringing the cloud to the mobile user. In *Proceedings of the third ACM workshop on mobile cloud computing and services.* (pp. 29−36).

Wan, S., Gu, Z., & Ni, Q. (2020). Cognitive computing and wireless communications on the edge for healthcare service robots. *Computer Communications, 149*, 99−106.

Yin, H., Zhang, X., Liu, H. H., Luo, Y., Tian, C., Zhao, S., & Li, F. (2016). Edge provisioning with flexible server placement. *IEEE Transactions on Parallel and Distributed Systems, 28*(4), 1031−1045.

Zhang, J., Xia, W., Yan, F., & Shen, L. (2018). Joint computation offloading and resource allocation optimization in heterogeneous networks with mobile edge computing. *IEEE Access, 6*, 19324−19337.

Zhang, Y., Chen, G., Du, H., Yuan, X., Kadoch, M., & Cheriet, M. (2020). Real-time remote health monitoring system driven by 5G MEC-IoT. *Electronics, 9*(11), 1753.

Convergent network architecture of 5G and MEC

5

Ayaskanta Mishra[1], Anita Swain[1], Arun Kumar Ray[1] and Raed M. Shubair[2]

[1]*School of Electronics Engineering, Kalinga Institute of Industrial Technology, Deemed to be University, Bhubaneswar, Odisha, India* [2]*Department of Electrical and Computer Engineering, New York University (NYU) Abu Dhabi, Abu Dhabi, United Arab Emirates*

5.1 Introduction

5G IoT applications, such as smart healthcare, smart home, smart city, smart farming, smart retail, etc., can be enabled by multi-access edge computing (MEC). However our focus is on "*Smart healthcare*" application due to its significant impact on human life. Such 5G-enabled, IoT-based smart healthcare systems need scenario specific designs attributes especially in the context of MEC. Convergence of 5G with MEC is mission critical (MC) in design and deployment of such 5G-enabled IoT-based smart healthcare systems those deal with critical healthcare data. As the smart healthcare market matures, it needs to implement massive sensor based machine to machine (M2M) communication devices for medical service such as data transmission between bio-medical sensors network data acquisition systems, telemedicine based framework for communication of huge amount of critical medical data with medical practitioner, patient and hospital management system and even a resilient, ultra-high bandwidth, ultra-low latency for remote robotic surgeries. These remote surgeries may require real-time live high definition (HD) or ultra-high definition (UHD)/4K video feed and ultra-low latency actuator control of surgery robotic system. To support this type of priority service, the Tactile 5G-based data network infrastructure is instrumental for high throughput, ultra-reliability, and low latency communication. Nevertheless, all these objectives have to be achieved with a cost effective solution. An ultra-resilient accurate convergent network infrastructure is the need of the hour for realizing the envisioned *digital health framework* considering the scenario specific design requirements of future smart healthcare systems. Researchers and scientific community is constantly improvising the current communication standards and technologies to address the shot-falls in existing communication framework towards more complex and dynamic need of digital healthcare domain for practically realizable smart healthcare infrastructure of future. Wireless network and sensing technologies for digital healthcare system are rapidly growing. The objective is small system with high performance accuracy; prolong battery life and low cost to make the healthcare smarter. The mobile terminal of 5G can act as both sensor nodes for patient to sense and collect the medical data and also as gateway to transmit the information to required destination (Medical practitioner, hospital data server). Therefore, by comparing various communication standards, Bluetooth low energy (BLE) is found to be more suitable due to its low cost and easy availability in almost every mobile terminal. Along with it, to increase the smartness of the system, Edge AI, Embedded AI, AI PaaS, and Explainable AI will play a

5G IoT and Edge Computing for Smart Healthcare. DOI: https://doi.org/10.1016/B978-0-323-90548-0.00003-6

key role to assist and manage different resources with rapid decision making capabilities. The authors also discuss about the research directions in this area.

Fifth-generation mobile communication was initially suggested by International Telecommunication Union's IMT-2020 standard (International Telecommunication Union, 2021). IMT-2020 suggests a peak download speed of 20 Gbps and upload speed of 10 Gbps with some additional requirements. The 3rd generation partnership project (3GPP) with their release 15 onwards proposed the 5G NR (new radio) standard together with long term evolution (LTE) for achieving the IMT-2020 standards technology requirements. 3GPP release 14 was the last specifications for 4th generation (4G) LTE networks. Fifth-generation (5G) mobile communication as standardized by 3GPP release 15 (2019) onwards to provide enhanced mobile broadband (eMBB), ultra-reliable low latency communications (URLLC) and massive machine type communications (mMTC). All these three application areas of 5G has three distinct flavors of 5G communication: (1) high data-rate up to 20 Gbps (per device), (2) resilient ultra-low latency network for MC applications and (3) massive M2M communication framework to support billions of Internet of Things (IoT) enabled smart devices. 3GPP release 15 (2019) (3GPP the mobile broadband standard, 2021) standard suggested 5G phase-1, NR, mMTC/IoT, vehicle-to-everything (V2X), MC interworking with legacy system, Network slicing, API exposer—third party access to 5G services, service based architecture, further improvements to LTE and mobile communication system for railway (FRMCS). 3GPP release 16 (2020) (3GPP the mobile broadband standard, 2021) standard suggests key features like 5G phase-2 with support for V2X phase-3, industrial IoT, URLLC, NR based access in Unlicensed spectrum (NR-U), increased 5G efficiency with interference mitigation, SON, enhanced multiple-input multiple-output, location, positioning, power efficiency and mobility enhancement, integrated access and backhaul, enhanced common API framework for 3GPP northbound APIs (eCAPIF), satellite access in 5G and FRMCS (phase-2). Apart from providing 20 Gbps in downlink and 10 Gbps in uplink the 5G provides ultra-low latency for MC applications with support for billions of IoT and M2M device communication.

European Telecommunications Standards Institute (ETSI) has proposed a mobile/MEC standard (Mobile edge computing—A key technology towards 5G, 2021) for 5G. In their recommendations, ETSI has proposed a decentralized and distributed fog computing (FC) framework for 5G networks. The 5G network support application programming interface (API) based on some software centric approach like software defined network (SDN) and network functions virtualization (NFV). The distributed computing approach brings the cloud towards the edge devices by bring decentralized storage and computing platforms towards radio access network (RAN). In the context of 5G, the FC platforms are strategically placed near the gNodeB (RAN-access points/trans-receivers). Adapting the decentralized Edge-cloud (distributed computing) approach would be helpful in achieving various technical goals like: (1) Computational resource and load-balancing, (2) avoiding bottleneck at a centralized cloud, (3) minimizing latency for low-latency and mission critical applications like image processing, deep learning and augmented/virtual reality (AR/VR) applications. In recent times, there is substantial development in ultra large scale integration (ULSI) based system-on-chip (SoC). These micro-chips possess various design attributes like low-power design, enhanced computing capabilities and hence are instrumental in designing edge computing platforms near to the user device in RAN. The distributed FC approach is enabled using these enhanced SoCs realizing the MEC design concept.

In the era of 5G mobile communication, heterogeneous technologies integration and their convergence is a major challenge especially in the paradigm of mobile/MEC approach. In this chapter, we are presenting different technical key-aspects related to network and data communication and computation framework and architecture pertaining to integration and convergence aspects of 5G and MEC. As per the 3GPP and ETSI standards LTE and 5G technologies and beyond is the era for complete packet switch data network which is a complete internet protocol (IP) based network. Even the subscriber's voice calls are processed over voice over internet protocol (VoIP) over LTE and 5G making the technologies voice over LTE (VoLTE) for 4G and voice over new radio (VoNR) or voice over 5G (Vo5G) for 5G networks. As the architecture is based on packet switched data networks all the user data is quantized as distinct packets using the TCP/IP protocol layered architecture. All the fundamental networking techniques would come into play in these networks allowing it to be seamlessly integrated to the Internet (global packet switched network). When we talk about Internet, it is the conglomerate of multiple technologies, protocols and standards making it heterogeneous network (HetNets) architecture. In the context of heterogeneity comes the concept of interoperability and convergent network architecture. The Convergent network architecture has to be adapted for such HetNets to make all functional elements interoperable. Recently numerous research works propose different strategies providing interoperability in the context of HetNets in the era of 5G and MEC (Gur et al., 2020; Chih-Lin et al., 2020). Most of the research works in these directions propose some clustering based network architecture supported by gateway devices based network design approach. The gateway or edge routers are deployed in such HetNets. The gateway is a software implementation on the edge of two dissimilar network clusters. For any example IEEE 802.11 Wireless fidelity (Wi-Fi) network cluster with 5G network cluster. As per 3GPP release 16−17(draft), support integration with unlicensed ISM band (2.4 and 5 GHz) spectrum using IEEE 802.11 wireless LAN access network. The framework and convergent network architecture for such HetNet cluster are pretty much crucial in Next-Gen Wireless network where some predominant wireless network technologies would be working in collaboration with each other. In this case, a 5G-NR network cluster with Wi-Fi network (IEEE 802.11 standard) cluster. On the RAN both the technologies support multiple-input-multiple-output (MIMO) technology for enhanced data-rate by spatial channel multiplexing concept. These similarities in functional attributes of network can be harnessed using interoperable network framework using edge gateway devices. The gateway devices are meticulously designed to translate the required protocol stack implementations by means of techniques like header translation, dual stack, tunneling approaches for Protocol Data Unit based data payload delivery in HetNets. This convergent architecture makes a seamless data communication framework for dissimilar and heterogamous networks, hence providing a complete interoperability support to the access networks for payload delivery. One more use case for such convergent network architecture would be the integration of 5G network with Wireless Sensor Networks (WSN) for massive-M2M communication model for future IoT-based application network deployment scenario. The WSN access technologies like Zigbee (IEEE 802.15.4), Long Range Wide-Area Network (LoRaWAN), IEEE 802.15.1 BLE are integrated with 5G-NR networks using clustering and gateway devices to provide seamless data transfer from billions of smart IoT devices and sensor motes and even can control and drive actuator based application by providing the distinct flavors of 5G network attributes and advanced MEC based infrastructural support. Convergent network architecture for Next-Generation Wireless (NG-W) networks in the era of IoT, M2M and 5G would be useful for many mission-critical applications like Smart healthcare, smart-city, AR/VR based smart life-style, smart home automation, assisted living, smart agriculture, vehicular networks, logistic tracking and management network and many more.

Internet of healthcare things (IoHT) and telemedicine applications in future would be getting major boost with these convergent network technologies with the help of 5G and MEC integration with IoT and massive M2M communication framework. Needless to say, with HetNet technologies coming closer for cooperative network deployment and seamless data integration would be enabling future smart medical devices to sense, connect and compute in the paradigm of bio-medical sensor data acquisition, data-storage, data-processing, data-extraction and data-visualization for human inferable data for MC medical applications. Further, 5G and MEC technologies in conjunction with IoT, M2M and WSN networks would be instrumental in remote low-latency medical applications like remote surgery and actuator controls for medical instruments and obviously doing real-time data-acquisition from bio-medical sensors for critical health applications using IoT.

Artificial intelligence (AI) techniques like deep learning, neural networks along with machine learning (ML) techniques are quite popular in recent times for solving many engineering problems, especially where a large data-set is involved. With advancement in enhanced Microprocessors and AI accelerators it is now technically feasible to implement AI, deep learning and ML models on real-time and a large amount of data can be processed using these intelligent algorithms with very-less processing delay. We are witnessing a huge boom in last 5 years in AI and ML use cases because of this advanced Microprocessor Units (MPUs) and Graphics Processing Units (GPUs) with help of ULSI and modern fabrication techniques. The major industry players in these segments like Samsung, NVIDIA, AMD have achieved 7 nm chips successfully. Samsung and Taiwan Semiconductor Manufacturing Company (TSMC) are ambitious for 5 nm fabrication technique for their newer chips. Hence the recent advancements in chip industry would provide better computational platform and hence has a huge potential to support AI and ML software using newer MPUs, GPUs and accelerators. These enhanced computational platforms would be the technology enabler for future 5G and MEC based network architecture having capabilities to support more computationally regressive AI and ML application framework on real-time or with very minimal computational latency. Future looks very bright with some other Nonconventional computing approach like Quantum-Computing. Companies like ColdQuanta, IBM, Honeywell and Google and many more have already developed and implemented working models of Quantum Computers and in coming future more and more advancement are expected in this field to support more computational regressive applications like advanced AI and computational intelligence models. All these technologies will be instrumental for future data communication networks like 5G and beyond with the help of MEC approach with enhanced application support.

In Section 5.1 of the chapter, we present introduction to convergent network architecture of 5G and MEC with a technical overview. Rest of the chapter is organized as follows: Section 5.2 focuses on technical overview on 5G Network with MEC. In Section 5.3, we discuss on convergent network architecture for 5G with MEC. Section 5.4 deals with current research in 5G with MEC. We present challenges and issues in implementation of 5G with MEC in Section 5.5 and finally Section 5.6 concludes the chapter.

5.2 Technical overview on 5G network with MEC

5G network infrastructure provides ultra-low latency high bandwidth communication. The URLLC framework using technologies like millimeter wave, massive-MIMO (m-MIMO) based 5G NR with additional enhancements in technologies like SDN and NFV.

5G NR provides improved RAN with the help of m-MIMO antenna system. m-MIMO technology helps in improving the data-rate with spatial multiplexing.

Concept of reduced range is to support more network capacity with the help of technologies like pico/femto/micro cell. The cell size in 5G is small; hence the transmission power gNodeB (Base Station (BS)) is less in each cell. This small size of cell makes more number of cells in a geographic location. Increased number of cells helps in increasing the network capacity to accommodate more users and devices. This would be ideal for IoT and M2M deployment scenario.

Very high frequency operation (~ 50 GHz) in the range of millimeter wave (mm-wave) communication is an essential technological aspect to increase the channel bandwidth hence increasing the effective data-rate of communication. As very high-frequency is used in 5G even it's inevitable to reduce the transmission power to make the power spectral density in a technically feasible and practicable range and also to limit the permissible specific absorption rate threshold for safe use for living beings mostly in an urban deployment scenario where the population is exposed to so many 5G antennas.

5G silent features

- Date-rate: up to 20 Gbps (per device)
- Latency: <1 ms in RAN
- Mobility: up to 500 km/h
- Localization: 1 m (accuracy)
- Service continuity: trains, spare and dense areas
- Scale: 20 million user devices (UD) and more than trillion IoT/M2M devices with high reliability

5G key technological aspects

- Virtualization/programmability of network services
- Emerging technology support for SDN, NFV, MEC and FC
- Migration network/cloud

5.2.1 5G with multi-access edge computing (MEC): a technology enabler

Advancements in Embedded Systems and Energy efficient MPU based SoC have opened up newer avenues in computing approach in today's era. Unlike the centralized cloud based architecture, now it is very much technically feasible to have computation at the edge devices. The Edge is referred to the mobile terminals /smart phones and the Base-Station gNodeB (5G terminology) in the RAN. The classical approach of Cloud computing has an issue of bottle-neck at the centralize server because all the computation and data processing is done at the single node. However in MEC approach the computation has been distributed using a task scheduler technique. This method would be helpful in reducing the computational load on the centralized cloud infrastructure.

The 5G standard proposed in 3GPP Revision (3GPP the mobile broadband standard, 2021) and the MEC framework for 5G networks proposed in ETSI recommends few key technical design aspects of the 5G-NR based RAN and deployment of edge computing platform for incorporating the MEC into 5G network. Some of the predominant techniques suggested by latest 5G and MEC standards can be summarized below:

- 5G-NR based RAN with enhanced computational resources on the Edge of the network which enables MEC features

- Ultra-low latency (also referred to as "Zero Latency") in the RAN for supporting real-time applications on the edge
- Software Defined Radio (SDR) in RAN and SDN in backbone network infrastructure
- NFV for creating a generic communication infrastructure and virtualizing the network functions on it using SDR and SDN techniques
- High data-rate (typical 20 Gbps per device) on the RAN to support real-time data intensive application framework. This can support rich application content like UHD (4K) video streams and Smart Vehicular applications and also applications like AR, VR, AI, deep learning and ML can be implemented on the RAN using techniques like MEC
- A hybrid application framework which works on distributed computing approach. A Task Scheduler for splitting the applications between the Cores centralized Cloud and Edge devices. This would be instrumental in providing a optimum on-demand computing approach and would be helpful in reducing load from the centralized cloud server and hence be effective in mitigating the bottleneck on the centralized cloud server. This is the key feature towards MEC in the era of 5G network technology and beyond.

Fig. 5.1 shows the *MEC* framework in 5G networks. The 5G-MEC framework as suggested by 3GPP and ETSI standards support major communication and computation functionalities through some functional elements. We have presented the same with the functional division of the system through control and data plane.

Control plane: The control plane would deal with the control functionalities of 5G and MEC network and RAN aspects.

- *Resource configuration* of RAN with NFV and SDR/SDN approach
- *Edge-cloud task manager* and scheduler to spilt the application in edge computing (distributed computing) and centralized cloud computing. The Task manager would be able to take decision based on the application. Some part of the application should be executed on the Edge-device and some data intensive part of the application should be executed on the centralized cloud
- *Control* over different functional blocks of the 5G and MEC system for communication and computation framework support
- *Virtual Machine* support for virtualizations of functions and operation over NFV and SDN technique.

Data plane: The data plane would support the transportation of data (packet switched data network) from node to node delivery model using TCP/IP stack implementation in 5G.

- *Next generation (NG) core data network* for switching, forwarding and routing of data packets through the network
- *Centralized data center and cloud server* to provide the central cloud computing and storage functionalities.

Some of the highlights of MEC approach in 5G technology can be summarized as below:

- Reduce load from the centralized cloud infrastructure
- Bringing cloud near to the User device (into the RAN) to reduce the latency (theoretically zero latency) to support applications like AR/VR/Image Processing applications on the go on mobile devices

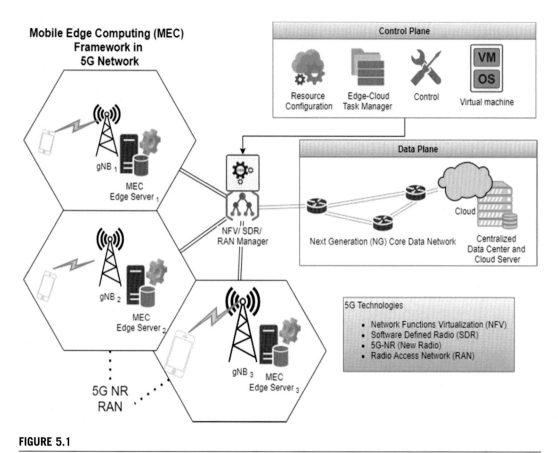

FIGURE 5.1

Multi-access edge computing (MEC) framework in 5G network.

- Computing, storage and networking resources are integrated with gNodeB (base-Station)
- Compute intensive and latency sensitive applications framework can be deployed on the edge devices and RAN
- Computational platform and storage resources like Platform-as-a-Service (PaaS), Software-as-a-Service (SaaS) Infrastructure-as-a-Service (IaaS) using the hybrid edge-cloud computing approach on the network edge with close proximity to the user devices.

5.2.2 Application splitting in MEC

Application splitting is the strategy used in 5G and MEC based communication and computation framework. The Fig. 5.2 shows the framework for application splitting for MEC based architecture in 5G. The task scheduler/manager would take decision based on the application. A part of the application which has low-latency requirement and real-time executional component should be executed on the edge computing infrastructure of 5G RAN and the part of the application which is

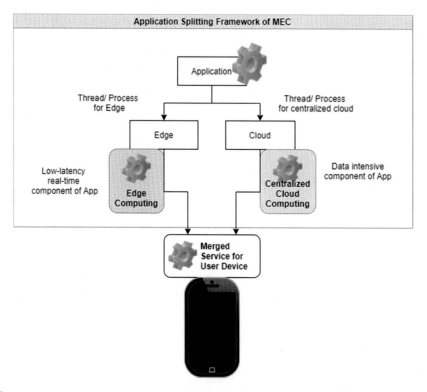

FIGURE 5.2

Application splitting framework using MEC in 5G network.

data intensive and need access to the Big-data of centralized cloud server would be executed centrally. Finally at the user device it application can be integrated using distributed computing approach. This approach is efficient in managing an application in the paradigm of MEC but at the same time it is one of the major technical challenge to distribute subtasks (Task Manager) and coordinate millions and billions of tasks and should be addressed by service provider and technology and computational framework developers.

Key-technical aspects of MEC

- Commercial off-the-shelf (COTS) application servers are integrated to gNodeB (base-Stations) of 5G networks
- Tailored services/Apps for an enhanced user experience (UX)

Functionalities of MEC Server

- Computation and storage on the Edge
- Connectivity and communication in RAN
- Access user traffic
- RAN functions using SDR and NFV

- Network information and management
- Query to response in less than <100 ms latency
- Real-time traffic analytics and Machine Intelligence
- Real-time processing of data on the Edge in the RAN using gNodeB
- Virtualization: MEC Virtualization Manager

Functionalities of Centralized/hybrid Cloud Server with collaborative computing approach with MEC

- Offline-batch processing tasks
- High latency tasks
- Data intensive tasks (big-data and data mining applications)
- Larger (Data Centers) and centralized cloud computing approach
- High computational power

5.2.3 Layered service oriented architecture for 5G MEC

Fig. 5.3 depicts the layered architecture for 5G and MEC. The 5G communication infrastructure provides platform low-latency and high data-rate for Communicate. Additionally compute and sense/actuate (control) billions of smart devices are feasible using technologies like IoT and M2M. The Fog/Distributed computing approach provide flexibility in computing. The layering is abstract concept suggested by differently modular and functional blocks. The service oriented layering architecture discussed here is in abstract conceptual level (not any standard yet). This service oriented modular architecture is suggested here with the vision of 5G and MEC integration aspects suggested by ETSI and 3GPP release 16 onwards.

Functions of modular service oriented architecture

Device access management: All the User devices are connected to 5G RAN. All the data and control for management and content delivery of User devices in RAN is handled by the MEC device access and management layer (MDAML).

Process scheduler—virtualization: Process Scheduler—Virtualization Layer (PSVL) is for Centralized Cloud and MEC based hybrid computing platform. MDAML is connected to the PSVL through an interlayer communication (Logical connection) with resource management framework. The PSVL would be allocating and managing the hybrid computational platform integration aspects. The some application modules which require a decentralized approach of computing (Edge Computing/Fog) would be assigned MEC framework by PSVL and application modules which would require a centralized and data intensive cloud-based computational resource would be assigned the centralized cloud infrastructure by the PSVL. Point to be noted the application framework support for both Edge Computing/Fog and Cloud computing is done on the application virtualization layer and splitting concept.

Edge computing framework: The distributed MEC servers at gNodeB in RAN are part of the edge computing framework. Low-latency and less computationally regressive applications would be handled by the MEC infrastructure of RAN in 5G-NR. Application like AR/VR with access layer computational requirement would be processed by the MEC computational framework.

Cloud computing framework: The centralized cloud based computational framework would be processing all data intensive applications where extensive big-data support and more computationally

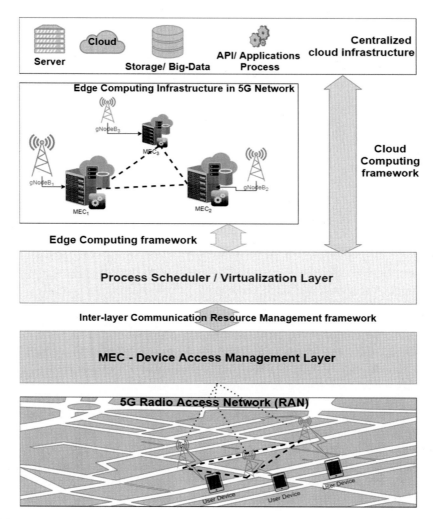

FIGURE 5.3

Layered service oriented architecture for 5G MEC.

regressive algorithms are needed to be implemented. The centralized data servers with extensive storage requirements are key aspects of cloud computing based architecture. The Cloud computing based Application programming Interfaces (APIs) are enabled for the processes to run with high resource requirement of data, RAM and CPU. Application like data analytics, ML, deep learning, data mining, multimedia with audio and video content streaming based applications would be more suitable for this centralized cloud infrastructure based cloud computing platform.

Advanced application with high demand for computation with IoT and 5G technologies will lead to unprecedented traffic requirement. This encourages the shifting of centralized cloud computing to edge computing. The new technology such as MEC will enable a paradigm shift in the way we envisage cloud computing as a part of RAN. The important characteristic of MEC is to perform the task of computing, controlling and storage at the edge of the network. MEC servers are deployed to execute delay sensitive and context aware applications in the close proximity of end users. The MEC offers large bandwidth, low latency, highly efficient network operation and service delivery to the end user. Although IoT is an integral part of 5G but the devices have limited storage and processing capabilities. Therefore, to fulfill the requirement of new compute-intense applications of 5G IoT, it intuitive to converge with MEC for smooth processing of large data traffic before sending to the cloud (Zhang, Leng, He, Maharjan, & Zhang, 2018). The successful realization of convergence of MEC with 5G requires research and development from academic and industries communities.

The standardization of MEC was provided by European Telecommunication Standard Institute (ETSI) and Industry Specification Group (ISG). MEC is also recognized by European 5G Infrastructure Public Private Partnership as a primary emerging technology for 5G networks. ETSI defined MEC as "Multi-access edge computing provides an IT service environment and cloud computing capabilities at the edge of the mobile network (RAN) and in close proximity to mobile subscribers." Deployment of MEC at the BS will enable the end user to access the cloud services with enhanced computation and less system failure. The characteristics of MEC include the following:

- On premises- MEC can work isolated from other networks by having access to only local resources which makes it less vulnerable.
- Proximity- The deployment of MEC in the nearest location helps to analyze big data and give benefit to the computation demanded applications such as AR/VR, analytics, etc.
- Low Latency- less than 1 ms.
- High Bandwidth- Download data-rate of 20 Gbps for each device.
- Location Awareness- MEC can be aware about the location of the devices through the information received from the edge distributed devices within the local access network.
- Network Context Information- The context information and real-time radio network benefit the service experience and utilization of network resources to handle the voluminous data traffic by optimizing network operation.
- Mobility Support-Enhanced mobility support up to 500 km/h.
- Security and Privacy- Advanced Encryption and authentication framework support.

Mobile cloud computing (MCC), cloudlet and FC were in demand for computation purposes before MEC. However, FC presents a layer consisting of devices such as the M2M gateway and wireless routers known as fog computing nodes (FCNs) between the cloud and the end user. These devices compute and store the data locally before sending to the cloud. The FCNs behaves as heterogeneous combination of nodes and support every device including non-IP based technologies to fit into every protocol layer and it is hidden from the end devices due to creation of Fog abstraction layer. MCC is the combination of cloud computing and mobile computing and creates a virtualized environment in the cloud consisting of resources such as compute, storage and communication which can be accessed by the end users remotely. Here the resources are placed inside the cloud.

The drawbacks of MCC attract the researchers to propose the edge cloud concept. In Cloudlets, the devices have similar capabilities like data center but in a lower scale by distributing computations to the cloudlet devices. A Cloudlet is called as "data center in a box" handling a virtual machine providing resources for end devices and users over WLAN network. Cloudlets consist of three layers such as (1) component layer (2) node layer (3) Cloudlet layer. One of the use-case implementation of Cloudlet Architecture for cognitive assistance applications involving the primary virtual machine along with other virtual machines provides cognitive functionalities in the Cloudlet to serve the request. However, the Cloudlet has the limitation to access only in Wi-Fi which cover small range of area and unable to provide ubiquitous computing.

Table 5.1 shows a comparison of different computing technologies in the context of MEC, MCC, Cloudlet and FC. We have presented all these different approach of computing with respect to their advantages, limitations and challenges with key parameters like origin, context awareness and latency. The detailed comparison shows the technical advantages of both centralized computing approach and the distributed computing. Each of the computational models has their own trade-off in dealing with the computational requirements. The cloud computing approach would be more suitable for data intensive centralized computing architecture. However the MEC, Fog or distributed computing approach would be more preferable for applications with low latency requirements. The comparison can be summarized as the best use-case implementation as a hybrid computational model with application splitting approach with both edge as well as cloud computing modules in place in the real-world deployment in the context of 5G and MEC.

To solve the above mentioned limitations/challenges of the similar technology, MEC is proposed. MEC provides the platform to converge the cloud computing capabilities within the RAN nearer to the mobile users (Sodhro, Luo, Sangaiah, & Baik, 2019). The early definition of MEC is to offload the computation intensive task to the BSs from the end devices. Later on, in 2014 the MEC ISG and ETSI started the standardization of MEC to encourage the growth of edge cloud computing in mobile networks. MEC provide an open environment to locate multivendor cloud platforms at the edge of RAN to overcome the challenges faced by centralized cloud computing paradigm for giving higher speed and low latency. Taking this into consideration, in 2017 for expanding the advancement of MEC application to HetNet, that is, for 4G, 5G, Wi-Fi, and fixed access technology, the word "Mobile" is removed from Mobile Edge Computing and ETSI ISG renamed MEC as Multi-access Edge Computing (Dong et al., 2020). The diversification of MEC helps its server to deploy at different locations within RAN set aside with different component of network edge such as BSs, that is, eNB for 4G and gNB for 5G, optical network unit, radio network controller sites and Wi-Fi access points. At this point, the intelligence is transferred from center to network edge which facilitates communication along with computation; caching and service control and satisfy its name as MEC. Since last few years, researchers concentrate on integration of MEC in 5G communication technology and its applications.

5.3 Convergent network architecture for 5G with MEC

In today's era 5G and beyond network deployment we may encounter heterogeneity of network technologies and even protocol stack implementation. The heterogeneity in communication technologies are attributed by various access networks like 5G-NR, Wi-Fi/Wireless LAN (IEEE 802.11), BLE (IEEE 802.15.1), Zigbee (IEEE 802.15.4), LoRaWAN even satellite access networks for various application-specific access network deployment scenarios. Convergent network architecture

Table 5.1 Comparison of different computing technologies.

Technology	Advantages	Limitations/challenges	Introduced by	Context awareness	Latency
MEC	• It gives benefit to mobile network operator (MNO), application service provider (ASP) and the end user by providing computational offloading, low latency, storage, energy efficiency and high bandwidth.	• Security and privacy, resource optimization, robustness of MEC servers, compatibility issue of web interface, transparency in migration of user applications.	ETSI (2014)	High	About 1 ms
MCC	• Resource management on demand such as network, server, application, storage and computation in a mobile environment.	• The farthest distance of centralized cloud servers from the end user increases the demand for a compute-intensive environment. • Increase latency and network disconnection.		No	30–100 ms
Cloudlet	• Overcome latency and cellular energy consumption problem. • Provide cloud technology nearer to the end user.	• Privacy and security issue of data.	Carnegie Mellon University (2013)	Low	>100 ms
Fog computing	• It allows the single processing devices to collect data from different sensors and take appropriate action. • It provides low latency as comparison to cloud computing.	• Dependency on wireless connection. • Limitations in performing complex action.	Cisco (2011)	Medium	<30 ms

provides a seamless operation in HetNet deployment scenario. A convergent network is one which provides framework for seamless interoperability and required translation at the gateway devices. In a nutshell, Convergent network architecture provides required communication framework to integrate multiple HetNet clusters (segments) into a single functional network to operate seamlessly. Different network technologies are integrated together to work as a single logical as well as operational network realizing a converged network.

Fig. 5.4 depicts a technical visualization of convergent network architecture with 5G and MEC. The main idea behind convergence is to provide interoperability of cooperative communication framework for multiple dissimilar network clusters with heterogeneity in network technologies and protocol stack (TCP/IP) implementation.

In a real-world network deployment scenario, it is inevitable to have multiple HetNet clusters. As we can see in Fig. 5.4, different applications require different kind of access technology (Physical as well as Data-Link layer). Specifically, if we analyze the TCP/IP stack implementation of the Physical Media dependent protocol (PMDP) and physical layer convergence protocol of Layer-I along with the Media Access Control sublayer of Layer-II, we observe that different access technologies like IEEE 802.11 (Wi-Fi), 802.15.1 (Bluetooth), 802.15.4 (Zigbee), LoRaWAN, LTE/5G-NR are needed. Many IoT-based application use IEEE 802.11 as access network. WSNs applications uses Zigbee (IEEE 802.15.4) access

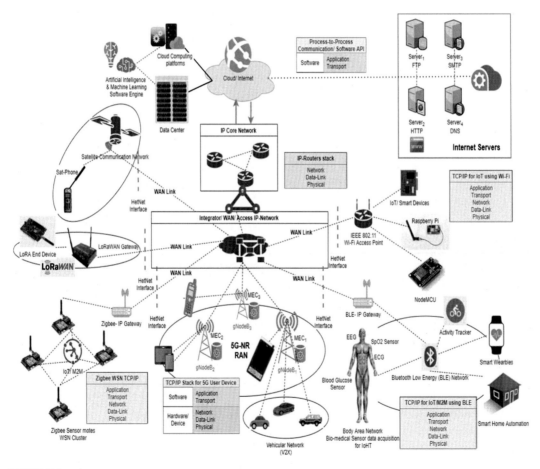

FIGURE 5.4

Convergent network architecture of 5G and MEC.

network. Many smart healthcare wearable devices use BLE IEEE 802.15.1 access network as Wireless Personal Area Network (WPAN). Vehicular Network is often implemented using IEEE 802.11p Wireless Access in Vehicular Environments (WAVE) and also cellular VANET implementation. Some of the IoT and M2M applications are popularly using Long-Range Wide-Area Network (LoRaWAN) by harnessing low data-rate and long range communication designed for IoT and M2M communication. Even satellite communication based access network can be used in remote areas where cellular or other network infrastructure is unavailable. All these application-specific network clusters can be integrated together and work like a converged network with TCP/IP stack and gateway devices. The gateway and clustering techniques are proven to be a successful strategy in dealing with heterogeneity and interoperability. All the different segment/cluster of access network are integrated with existing 5G cellular network using IP-based gateways and WAN access routers making a convergent architecture for 5G with MEC enabled services.

5G Communication is based on eMBB, URLCC and mMTC not delivery (4G) functions (Li, Huang, Li, Yu, & Shu, 2019). It provides advanced communication platform for many advance applications to be developed such as VR/AR, autonomous mobile, Tactile IoT. To meet the communication as well as computation demands, several additional technologies have been developed for 5G including radio access, network resource management, network architecture and applications scenarios, energy saving and system performance. The above technologies includes dense HetNets, cloud radio access network (C-RAN), nonorthogonal multiple access (NOMA), unmanned aerial vehicle (UAV), wireless power transfer (WPT), energy harvesting (EH), IoT and ML.

With the increased use of mobile devices and IoT, data traffic on communication network exponentially increases. Hence, to optimize the network use and minimize device numbers, convergence is an emerging solution. However, MEC can be used in three categories such as for consumer oriented services, third party operator's services and QoE improvement of the network. MEC platform should create a supporting environment for all these services at the edge of the network. The users can be benefited through the computation offloading at the edge. Therefore, user-centric applications such as AR\VR, gaming, big data, location tracking, security and privacy, latency sensitive services (Remote surgery through tactile internet) and QoE (network performance) in 5G need convergence with MEC to support new architecture and advancement in emerging technologies. The 5G and MEC convergence scenarios includes: NOMA, EH and WPT, UAV Communications, IoT, H-CRAN, ML, VM, SDN, NFV and network slicing. Apart from these technologies, block-chain can be integrated with MEC to give various benefits related to context of cloud of thing.

In the Fig. 5.4, we have considered the convergence of 5G with MEC, where the BS of 5G and Mobile Internet world are merged and the edge servers are located in such a way that BS can directly control it and simultaneously BS can access the internet data and the IoT sensors data.

5.4 Current research in 5G with MEC

Advancements in the field of 5G and MEC have been phenomenal since last few years. In this section, we discuss about the research perspectives on the same. Table 5.2 shows a detailed analysis on the current research work on various technical aspects of 5G and MEC. Here, we present an extensive study on few prominent research works pertaining to some emerging field of research which are relevant with 5G, MEC, and Convergent networks. This study focuses on these following

Table 5.2 Current surveys and research works.

Technology	References	Authors contributions	Findings	Research issues
MEC in healthcare	Sodhro et al. (2019)	Window based rate control algorithm (w-RCA) is proposed to develop the medical quality of service (m-QoS) in MEC based healthcare.	From the comparison of performance evaluation with battery smoothing (BSA) algorithm, w-RCA shows better performance.	Steady and inefficient feature of traditional method unable to optimize QoS properly for healthcare application.Rate control algorithm is only used for multimedia and big data application and there is delay in w-RCA.
	Dong et al. (2020)	The two subparts of IoMT, that is, (1) intra WBANs is modeled using bargaining game to handle the wireless channel resource allocation issues and in (2) faraway WBANs the offloading decision issue developed game.	Investigation of edge computing oriented structure in IoMTs with minimizing system wide cost.	Machine learning and block-chain based electronic health data for maintenance of privacy and distributed intelligent spectrum allocation.
	Li et al. (2019)	Proposed a safe and systematic information management system called "EdgeCare" for mobile medical systems. A hierarchical architecture is designed to implement EdgeCare.	EdgeCare provide solution for protection of medical information and support well organized data trading.	Confidentiality, integrity and authentication, traceability, user centric access control.
	Ray, Dash, and De (2019)	Provide taxonomical classification and review of IoT and edge computing, use cases for urban smart living problems and edge-IoT-based architecture proposed for healthcare.	Minimization of dependency on IoT cloud analytics.	Edge gateway, Edge analytics, edge architecture, sensor actuator integration, context awareness and unified framework of integration.
	Hartmann, Hashmi, and Imran (2019)	Survey of latest and appearing edge computing architectures and methodology for healthcare application with requirements and challenges.	A detailed survey on edge computing data functioning.	Low latency communication of voluminous data with maintaining a trade-off between the accuracy and complexity of AI.

(Continued)

Table 5.2 Current surveys and research works. *Continued*

Technology	References	Authors contributions	Findings	Research issues
	Oueida, Kotb, Aloqaily, Jararweh, and Baker (2018)	An integrated architecture of resource preservation net (RPN) framework and edge computing has been proposed for emergency department in healthcare.	Improvement in performance measures such as patient length of stay (LoS), resource usage and patient waiting time.	New privacy policies should be introduced and quality of experience (QoE) of service should be considered.
	Yu and Choi (2020)	Proposed a landmark-free and pose-estimation-free frontal-face synthesis system.	Pose-unconstrained face recognition is improved using MEC for human identification in medical systems.	Improvement in accuracy.
Integration of MEC and 5G	Hu, Patel, Sabella, Sprecher, and Young (2015)	The scientific and technical advantages from mobile edge computing.	Proposed proof-of-concept (PoC) program to exhibit the feasibility of MEC performance.	Internet of things gateway, augmented reality, intelligent video acceleration, deployment framework.
	Kekki et al. (2018)	Presents the benefit of deploying MEC on the N6 reference point of the 5G system.	The user plane function (UPF) remain is same and independent of the deployment of MEC and it integrates the MEC application traffic in the 5G bearer stratum.	Standardization of the application programming interfaces (APIs) to expose the required 3GPP system capabilities.
	Tomaszewski, Kukliński, and Kołakowski (2020)	5G slicing architecture and MEC integration.	Proposed architecture with 5G, APIs and MEC.	MEC using services of APIs, for mobility and roaming application in 5G.
	Lin, Hu, Gao, and Tang (2019)	Optimize the end to end latency with authenticity of user's priorities.	The proposed algorithm reduces end to end delay by allocating communication and computation and resources in healthcare application.	Network delay and users priority in authenticity.
	Gavrilovska, Rakovic, and Denkovski (2018)	Resource scaling in MEC as a favorable perspective for 5G.	Container based virtualization is appropriate for resource scaling more suitable for deployment of 5G-MEC.	Latency, reliability and throughput.
	Tran, Hajisami, Pandey, and Pompili (2017)	Studied three use cases, that is, (1) mobile edge orchestration, (2) collaborative caching and process, (3) multilayer interference cancellation.	Describe the benefits of the proposed approaches for the evolution of 5G.	Resource management, interoperability, service discovery, mobility support, fairness and security.

(Continued)

Table 5.2 Current surveys and research works. *Continued*

Technology	References	Authors contributions	Findings	Research issues
Integration of MEC, 5G and IoT	Sarrigiannis et al. (2019)	Latency based embedding mechanism and an online scheduling algorithm evaluated through a MEC enabled 5G platform.	Maximizes the number of users by utilizing the online allocation of edge and core resources.	Virtual Network Function (VNF) allocation and placement, overhead due to live migration process.
	Pham et al. (2020)	Surveyed an overview of use cases and application of MEC and outline researches in integration of MEC and 5G technology.	Summarized test beds and experimental evaluation.	Machine learning based framework implementation in MEC, federated learning and application for MEC.
	Liu, Peng, Shou, Chen, and Chen (2020)	Analyzed the main features of integration of MEC, 5G and IoT.	Described various technologies including SDN/NFV, virtual machines (VM), network slicing, cloud computing and computation offloading.	Mobility management, edge intelligence, pricing, MEC servers' deployment, green MEC, security and privacy.
	Redondi, Arcia-Moret, and Manzoni (2019)	IoT in 5G using pub/subarchitecture with system design and optimization.	Discussed MQTT + protocol supporting IoT application in MEC based 5G networks.	Orchestration strategies create issues in load balancing and resource allocation process.
	Sanchez-Iborra et al. (2019)	Slicing framework for IoT application of in 5G-based MEC.	Quality of service (QoS) and high scalability.	Security and privacy policy.
	Farris, Orsino, Militano, Iera, and Araniti (2018)	Investigate mobile-IoT-federation-as-a-service (MIFaaS) model in 5G technology to satisfy the low latency requirement.	To handle large data traffic, the combination of NB-IoT and LTE is needed to get high data rate and low latency.	Prediction of mobility pattern, power consumption and trustworthy interaction to build a secure system.
	Hsieh, Chen, and Benslimane (2018)	5G technology is implemented using vMEC, SDN and NFV technologies support various IoT devices to obtain low latency and high bandwidth.	The traffic control strategies decrease in IoT delay, increase QoE and less network congestion.	Latency reduction of IoT application services and QoS.
	Porambage, Okwuibe, Liyanage, Ylianttila, and Taleb (2018)	A survey of MEC technology supporting IoT applications.	The role of IoT towards 5G communication system and technical aspect of empowering MEC in IoT and other various integration technologies.	Scalability, mobility management, security, privacy, computation offloading and resource allocation.

Table 5.2 **Current surveys and research works.** *Continued*

Technology	References	Authors contributions	Findings	Research issues
MEC, 5G and IoT in Healthcare	Braeken and Liyanage (2020)	Architecture of remote monitoring system using 5G technology integration with IoT devices and MEC nodes.	Established an authentication phase between MEC nodes and IoT devices to provide required security features.	Handover mechanism between MEC nodes provide dynamicity and low latency for IoT devices and high computation required security.
	Ning et al. (2020)	To set up a MEC based 5G system for IoMT home health monitoring.	Performance evaluation indicated the efficacy of the proposed algorithm for cost effective and patient benefit using MEC.	Medical data management complexity and power consumption.
	Hewa, Braeken, Ylianttila, and Liyanage (2020)	MEC and block-chain based architecture using ECQV (Elliptic Curve Qu Vanstone) to maintain security and privacy of data between IoT, MEC and cloud.	Performance evaluation demonstrate that the system support voluminous transmission of data through MEC nodes and optimized the block-chain storage by offloading.	Scalability and on chain secure multiparty computation.
	Zhang et al. (2020)	MEC and artificial intelligence technique inspired medical e-healthcare system.	Improved patient treatment and cost effective.	Low latency and high accuracy.

areas: (1) MEC in healthcare, (2) integration of MEC and 5G, (3) Integration of MEC, 5G and IoT and (4) MEC, 5G and IoT in Healthcare.

Fig. 5.5 shows the year-wise research trend over the last 6 years (2015−20) as per Google Scholar in various key-technical aspects like 5G, MEC, and Convergence network such as HetNets, IoT, and M2M communication. These charts give a better understanding of the research focus in key areas related to 5G MEC and convergent network. Number of research papers in 5G, HetNet, IoT are consistent throughout last 6 years. Number of research papers on convergence and M2M is gradually decreasing however, there is a substantial growth in number of publications in MEC. MEC being quite popular among the research community to adapt as a computational technology especially in today's era 5G and beyond for IoT enabled heterogonous networks (Ray et al., 2019).

5.5 Challenges and issues in implementation of MEC

There is a paradigm shift in the computing scenario after introduction of 5G and associated convergence with MEC. Edge technology is the key to user specific computing along with cloud

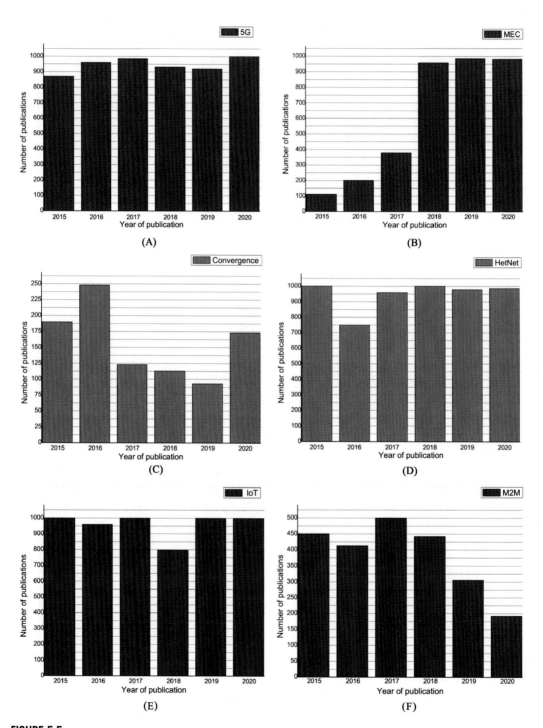

FIGURE 5.5

Research tread in various key-technical aspects: (A) 5G, (B) MEC, (C) convergence, (D) HetNet, (E) IoT and (F) M2M from 2016 to 2020 (past 6 years) * Data acquired from Google Scholar with keyword search.

computing. The convergence with 5G and the recent developments in association with NFV, ICN (Information Centric Networks) and SDN form the basis of MEC implementation (Hartmann et al., 2019; Oueida et al., 2018). Implementation of 5G and MEC for different applications raises many challenges (Yu & Choi, 2020). We discuss the challenges and issues for MEC and 5G so far as their architecture, communication and computing scenarios are concerned.

Use of MEC is becoming increasingly important as there is a steady rise in the host of IoT and mobile applications. These applications need edge specific computations with MEC making cloud computation secondary (Yu & Choi, 2020). MEC servers take on the local computations thereby reducing computational load on cloud servers. In this chapter, we focus on different applications where MEC has taken a key role in enhancing the performance of 5G network. However, we discuss the challenges and issues associated with each of these.

5.5.1 Communication and computation perspective

The use of MEC in 5G is primarily based on communication and computation models. MEC is usually positioned by between the mobile devices and the cloud servers to leverage the device specific computations. The communication model deals with various research issues (Shafi et al., 2017; Zhang et al., 2016).

5.5.1.1 MEC service orchestration and programmability

It should be performed simultaneously by cooperating with each other considering the allocation of VNF and service across the edge cloud platform. The edge cloud platform across various administrative domains creates more challenges for resource aggregation and service mapping. Various service aspects of edge cloud orchestration and programmability includes:

- Resource allocation, service placement, selection of platform and reliability.
- Continuous service and mobility in the edge cloud platform.
- Joint optimization of VNFs and MEC should be performed to execute proper resource utilization and cross layer optimization in edge cloud services and network resources.

The integration of various technologies such as SDN\NFV creates opportunities for deployment of MEC orchestrator by enabling the edge cloud resources control. Currently, different APIs and data models are under development in ETSI MEC for assisting discovery process of millimeter wave (mmWave) and provide RAN and network information for visibility of network towards different applications.

5.5.1.2 MEC service continuity and mobility and service enhancements

It is considered as one of the most important challenging aspect and need further research for service continuity for high speed mobile users. Integration of MEC with MCC encourages the utilization of mobile resources for communication and computation services is considered to be challenging for service orchestration. The multiconnectivity supporting multipath and potentially streaming among various MEC platform encourage further research considering the network scalability and mobile user performance. Quality of Experience (QoE) and Resiliency add direction to the researchers for further research to increase the efficiency and QoE of user. To develop services and application such as resiliency to provide backup features in case of a state of critical parameter and analyze the QoE and other performance related to that state. MEC upgrade the experience of location services, intelligent proximity services by integrating user context information, big data and social applications. From the resource optimization viewpoint it should be studied further to combine network and cloud resources with various services offering QoE. To enhance the efficiency of MEC, the concept of HAEC (Highly Adaptive Energy-efficient Computing) should be considered for further research (Chen & Hao, 2018; Sabella, Vaillant, Kuure, Rauschenbach, & Giust, 2016; Sun & Ansari, 2016). The limitation in the range of cell in a small cell network increases the importance of mobility which need more research exploration for fast migration. Service Discovery is an important research issue where discovery mechanism is introduced to find proper nodes that can be leveraged in a decentralized system. The automatic monitoring and accurate synchronization for multiple devices should consider as important aspect for further research.

5.5.1.3 MEC security and privacy

It is considered very important in MEC as it is deployed at BS and at the areas where it is exposed to physical attack. Therefore, MEC having significant security risk than traditional cloud computing. MEC need more stringent security policies as the third party stake holders can access the information related to user proximity and radio analytics. Hence, authentication needs to be considered for accessing the MEC platform. Isolation between the hosted applications should be taken as challenging issue as security attack on particular application will not affect the other running applications. Therefore, proper encryption technique needs to be studied for secure collaboration and interoperability between heterogeneous resources and different operational parties. Different intrusion detection techniques in cloud computing for large scale geo distributed environment need further exploration.

5.5.1.4 Standardization of protocols

It requires standardized organization to provide a set of universally accepted rules for edge computing in 5G technology. However, the challenges arise to accept the standard due to its flexibility and diversified customization provided by various vendors. At this point, various heterogeneous UEs communication with edge cloud through different interfaces. Addressing heterogeneity and computing technologies in 5G communication faced difficulties in portability across various environments. Programming based models may give a solution to edge node for execution of workload simultaneously in multiple hardware levels (Al-Shuwaili & Simeone, 2017; Dong et al., 2020; Nunna et al., 2015). The parallelism of data and task divide the workload into independent and smaller task to be executed in parallel across various hardware and edge cloud layers.

5.5.1.5 MEC service monetization

It plays an important role for mobile operators to monetize the combined cloud and network resources with third parties. Therefore, resource brokering solution needs further investigation as tariff planning service usage such as video analytics or optimization is becoming a challenging issue on demand. More dynamic pricing models need to be introduced for requirement of advanced accounting and monitoring. Techniques need to be for solving potential economic conflict between various parties for smooth MEC system operation. The collaboration between different network providers for MEC needs common collaboration protocols to access network and context information regardless of their deployment location.

5.5.1.6 Edge cloud infrastructure and resource management

The infrastructure of MEC includes the combination of Telco and IT cloud capabilities within the RAN. For smooth working in highly difficult environment, the edge cloud infrastructure needs to be designed. Web Interface has not been designed for mobile devices so facing compatibility issues and this web interface is not suitable for MEC and arise overhead problem. However, standard protocols require for smooth communication among users, MEC and the cloud (Ren et al., 2019).

It is essential for MEC platform as there is limited computing and storage resources to support constrained number of applications. However, MEC can be considered as a service where operator's resources can be accessed by service provider according to the need. There should a trade-off between fair resource sharing and load balancing to achieve an increased efficiency. It is very important to migrate transparently the user applications to MEC for execution.

5.5.1.7 Mobile data offloading

The consistency in user experience and service continuity should be independent of user location and edge service delivery location creates challenges in mobile data offloading. Another issue is flexibility in diverse forwarding function. Computation offloading for offloading at different level of mobile terminal and edge cloud following changes needed:

- Change in network bandwidth to achieve appropriate latency.
- Requirement of minimal communication overhead.
- Need minimum effort from software developer for rapid development with reduced time and cost.

5.5.2 Application perspective

Convergence of 5G and MEC assures eMBB, URLLC and mMTC along with enhanced performance and scalability. Efficient resource management in the convergence network will enable consumers and network providers to develop applications that require intensive processing closer to the device (data source). This reduces latency for real-time and computation intensive applications. For example, the quality of voice and video over mobile internet will have a enhanced experience (Bai et al., 2020; Giust et al., 2018; Ksentini & Frangoudis, 2020; Ma, Zhao, Gong, & Wang, 2017; Pham et al., 2020). MEC with Intelligent IoT will have computations at edge for variety of devices and applications. IoT-based smart health care will enable improved hospital management and MC applications. Application specific implementations and development on new applications will lead to challenges and issues as discussed in this chapter in details.

5.5.2.1 Industrial IoT application in 5G

5G is capable of providing technological requirement to the modern industries through robotics. The three important factors of 5G such as eMBB, URLLC and MMTC can enhance the performance of industries 4.0. Industry 4.0 deals with the latest automation and data exchanging process in manufacturing sector. IoT is one of the important research domains of industry 4.0. The smart factory concept arises from the industry 4.0 but for 3G and 4G it has some limitations such as end to end delay, energy consumption, reliable wireless communications and to support voluminous devices. However, 5G overcome all these limitations and give high latency and high bandwidth.

At this point, 5G also provides the advancement in time-critical and reliable processes, Nontime-critical communication, and remote control of factories (Althebyan, Yaseen, Jararweh, & Al-Ayyoub, 2016; Ning, Wang, & Huang, 2018).

5.5.2.2 Large scale healthcare and big data management

There is a need of edge computing enabled large scale healthcare system as most of the research related to edge computing healthcare applications are for small scale environment. Althebyan et al. proposed architecture to be fitted for large scale healthcare application (Abdellatif, Mohamed, Chiasserini, Tlili, & Erbad, 2019; Zhang, Weiliang, Fengyi, & Zhouyun, 2018). In this paper, the author consider a large number of user with a decision making model for health worker to observe the trends in disease spread. As in hospitals, large number patients need to be treated and their useful medical data to be recorded and to handle such a large number of data and users, edge enabled large scale healthcare system is required with new analysis technique.

5.5.2.3 Integration of AI and 5G for MEC enabled healthcare application

Scalability encourages converging 5G communication with edge computing and shifts centralized computation to the edge. The deployment of large scale MEC enabled healthcare system need integration of AI to get accurate and in time services by analyzing the factor such as mobility of user and device usage pattern, monitoring medical records. Recently, ML and deep learning are enabling the advancement in medical diagnosis which requires voluminous processing, computation and storage. However, the centralized cloud platform is shifting to the edge keeping URLLC as one of its key factor which requires edge enabled ML trained model for localized data to provide low cost and low latency. The ML categories include (1) supervised learning, (2) unsupervised learning and (3) reinforcement learning and can be used for AI and 5G integrated and MEC enabled healthcare infrastructure deployment. AI gives importance to the context aware health care system to personalize the patient data such as age, gender and recent medical condition which can be taken as input and helps doctor for diagnosis. Considering one AI based model in where the lung cancer has been diagnosed with higher precision than the simple threshold method. Some other research includes voice disorder analysis, prediction of emotion of patient which helps the hospital caretaker to give more attention to the patient (Bellavista, Chessa, Foschini, Gioia, & Girolami, 2018; Bruschi, Bolla, Davoli, Zafeiropoulos, & Gouvas, 2019; Mitra & Agrawal, 2015; Zhang, Di, Wang, Lin, & Song, 2020). However, there are some features such as energy efficiency, computation need further exploration in research through various edge ML training models to get higher system performance (Agiwal, Roy, & Saxena, 2016; Li et al., 2019; Zhang, Wu, Xie, & Yang, 2018).

5.6 Conclusions

MEC and edge computing approach are instrumental in developing application framework in 5G technology and beyond. Further, convergent network architecture is vital in dealing with heterogeneity and interoperability for any real-world deployment scenario using multiple heterogeneous access network technologies with communication stack and computational platforms. In this chapter, we have discussed various key technical aspects of 5G and MEC for convergent networks. The 5G communication standards (3GPP release 15 and beyond) with newer technologies enabled distributed computing (Fog/Edge/MEC) approach are the key technology enabler for a holistic convergent network deployment. For New-gen communication and computation framework, there is the need for modern applications like AR/VR, IoT, M2M, and ultra-reliable and low-latency applications.

Additionally, the hybrid computational approach evolved for proper amalgamation of decentralized edge computing (MEC) and centralized cloud computing would be a smart tailored solution for heterogeneity in computational requirement across different applications scenarios. 5G technology and MEC with integration for HetNet and communication protocols (Convergent Network architecture) would provide a holistic multimodal framework of sense, connect and compute infrastructure required for modern era IoT and Cyber-Physical Systems (CPS). It is needless to mention that these technologies would shape the future beyond imagination with a lot of business potential. IoT-based applications like Smart healthcare would be bolstered using fusion of technologies like 5G and MEC with convergent network architecture. IoHT would harness the technological benefits of 5G network (low-latency, ultra-reliable, high data rate, enhanced mobility support) for bio-medical sensor data acquisition over "internet" (global seamless packet switched data network infrastructure) powered by the hybrid computational model of MEC and cloud computing platforms for better services custom made for edge application and data intensive cloud based analytics engine. Further, emerging technologies like AI, Neural networks, and ML would provide the machine intelligence (perception) for intelligent solutions. We have already discussed 5G and MEC based convergent network technologies, which would provide the required technical feasibility for realizing such intelligent systems. We have also presented a comparative analysis of different computational platforms like MEC, Mobile Cloud Computing (MCC), Cloudlet, and FC and in a nutshell, none of the techniques show technical supremacy over others. Rather, it shows that a perfect blend of different computational approaches harnessing the technical advantages and mitigating the limitations would be the smarter solution to provide a hybrid computational framework for future IoT- and CPS-based systems especially with the aid of communication infrastructure like 5G and beyond. Nevertheless, scenario-specific requirements of data networks always show heterogeneity with access network technology as well as protocol stack implementation in Physical as well as link layer.

We have also presented a convergent network perspective for 5G and MEC with technical overview on heterogeneity and interoperability with the help of clustering and gateways technologies. Further, we have presented a current research perspective on some emerging field of research which are relevant with 5G, MEC and Convergent networks like MEC in healthcare, integration of MEC and 5G, integration of MEC with 5G and IoT also MEC, 5G and IoT in Healthcare. We have also discussed some key-emerging areas of research in this context of HetNet/Convergent network architecture for 5G and MEC with keeping in mind the IoT and M2M application framework. We also mention the challenges and issues in the implementation of 5G with MEC. We have discussed various technical challenges with communication and computational perspective as well as the application perspective. This would be helpful for researchers working on these technical domains of 5G, MEC, and convergent network architecture to focus on challenges to select research objectives and direction and contribute to the scientific community and society at large with their technical innovations and contributions.

References

3GPP the mobile broadband standard. 2021. <https://www.3gpp.org/release-15> Accessed 25.03.20.
3GPP the mobile broadband standard. 2021. <https://www.3gpp.org/release-16> Accessed 25.03.20.

3GPP the mobile broadband standard. 2021. <https://www.3gpp.org/release-17> Accessed 25.03.20.

Abdellatif, A. A., Mohamed, A., Chiasserini, C. F., Tlili, M., & Erbad, A. (2019). Edge computing for smart health: Context-aware approaches, opportunities, and challenges. *IEEE Network*, 33(3), 196−203.

Agiwal, M., Roy, A., & Saxena, N. (2016). Next generation 5G wireless networks: A comprehensive survey. *IEEE Communications Surveys & Tutorials*, 18(3), 1617−1655.

Al-Shuwaili, A., & Simeone, O. (2017). Energy-efficient resource allocation for mobile edge computing-based augmented reality applications. *IEEE Wireless Communications Letters*, 6(3), 398−401.

Althebyan, Q., Yaseen, Q., Jararweh, Y., & Al-Ayyoub, M. (2016). Cloud support for large scale e-healthcare systems. *Annals of telecommunications*, 9, 503−515.

Bai, T., Pan, C., Deng, Y., Elkashlan, M., Nallanathan, A., & Hanzo, L. (2020). Latency minimization for intelligent reflecting surface aided mobile edge computing. *IEEE Journal on Selected Areas in Communications*, 38(11), 2666−2682.

Bellavista, P., Chessa, S., Foschini, L., Gioia, L., & Girolami, M. (2018). Human-enabled edge computing: Exploiting the crowd as a dynamic extension of mobile edge computing. *IEEE Communications Magazine*, 56(1), 145−155.

Braeken, A., & Liyanage, M. (2020). Highly efficient key agreement for remote patient monitoring in MEC-enabled 5G networks. *The Journal of Supercomputing*, 1−24.

Bruschi, R., Bolla, R., Davoli, F., Zafeiropoulos, A., & Gouvas, P. (2019). Mobile edge vertical computing over 5G network sliced infrastructures: An insight into integration approaches. *IEEE Communications Magazine*, 57(7), 78−84.

Chen, M., & Hao, Y. (2018). Task offloading for mobile edge computing in software defined ultra-dense network. *IEEE Journal on Selected Areas in Communications*, 6(3), 587−597.

Chih-Lin, I., Kuklinski, S., Chen, T. C., & Ladid, L. L. (2020). A perspective of O-RAN integration with MEC, SON, and network slicing in the 5G era. *IEEE Network*, 34(6), 3−4.

Dong, P., Ning, Z., Obaidat, M. S., Jiang, X., Guo, Y., Hu, X., . . . Sadoun, B. (2020). Edge computing based healthcare systems: Enabling decentralized health monitoring in Internet of medical Things. *IEEE Network*, 34(5), 254−261.

Farris, I., Orsino, A., Militano, L., Iera, A., & Araniti, G. (2018). Federated IoT services leveraging 5G technologies at the edge. *Ad Hoc Networks*, 68, 58−69.

Gavrilovska, L., Rakovic, V., & Denkovski, D. (2018). Aspects of resource scaling in 5G-MEC: Technologies and opportunities. In: *Proceedings of the IEEE Globecom Workshops (GC Wkshps)*, December 9, 2018, pp. 1−6.

Giust, F., Verin, G., Antevski, K., Chou, J., Fang, Y., Featherstone, W., . . . Purkayastha, D. (2018). MEC deployments in 4G and evolution towards 5G. ETSI White paper 24, pp. 1−24.

Gür, G., Porambage, P., & Liyanage, M. (2020). Convergence of ICN and MEC for 5G: Opportunities and Challenges. *IEEE Communications Standards Magazine*, 4(4), 64−71.

Hartmann, M., Hashmi, U. S., & Imran, A. (2019). Edge computing in smart health care systems: Review, challenges, and research directions. *Transactions on Emerging Telecommunications Technologies*, e3710.

Hewa, T., Braeken, A., Ylianttila, M., & Liyanage, M. (2020). Multi-access edge computing and blockchain-based secure telehealth system connected with 5G and IoT. In: *Proceedings of the 8th IEEE International Conference on Communications and Networking (IEEE ComNet')*.

Hsieh, H. C., Chen, J. L., & Benslimane, A. (2018). 5G virtualized multi-access edge computing platform for IoT applications. *Journal of Network and Computer Applications*, 115, 94−102.

Hu, Y.C., Patel, M., Sabella, D., Sprecher, N., & Young, V. (2015). Mobile edge computing—A key technology towards 5G. ETSI white paper 11 (11), pp. 1−6.

International Telecommunication Union. (2021). <https://www.itu.int/rec/R-REC-M.2083-0-201509-I/en> Accessed 25.03.20.

Kekki, S., Featherstone, W., Fang, Y., Kuure, P., Li, A., Ranjan, A., . . .Wen, K.W. (2018). MEC in 5G networks, ETSI white paper, pp. 1−28.

Ksentini, A., & Frangoudis, P. A. (2020). Toward slicing-enabled multi-access edge computing in 5g. *IEEE Network*, *34*(2), 99−105.

Li, X., Huang, X., Li, C., Yu, R., & Shu, L. (2019). EdgeCare: Leveraging edge computing for collaborative data management in mobile healthcare systems. *IEEE Access*, *7*, 22011−22025.

Lin, D., Hu, S., Gao, Y., & Tang, Y. (2019). Optimizing MEC networks for healthcare applications in 5G communications with the authenticity of users' priorities. *IEEE Access*, *7*, 88592−88600.

Liu, Y., Peng, M., Shou, G., Chen, Y., & Chen, S. (2020). Toward edge intelligence: Multiaccess edge computing for 5G and internet of things. *IEEE Internet of Things Journal*, *7*(8), 6722−6747.

Ma, X., Zhao, J., Gong, Y., & Wang, Y. (2017). Key technologies of MEC towards 5G-enabled vehicular networks. In: *Proceedings of the International Conference on Heterogeneous Networking for Quality, Reliability, Security and Robustness*, pp. 153−159.

Mitra, R. N., & Agrawal, D. P. (2015). 5G mobile technology: A survey. *ICT Express*, *1*(3), 132−137.

Mobile edge computing—A key technology towards 5G. (2021). ETSI white paper. <https://www.etsi.org/images/files/ETSIWhitePapers/etsi_wp11_mec_a_key_technology_towards_5g.pdf> Accessed 25.03.20.

Ning, Z., Dong, P., Wang, X., Hu, X., Guo, L., Hu, B., . . . Kwok, R. Y. (2020). Mobile edge computing enabled 5G health monitoring for Internet of medical things: A decentralized game theoretic approach. *IEEE Journal on Selected Areas in Communications*, 1−6.

Ning, Z., Wang, X., & Huang, J. (2018). Mobile edge computing-enabled 5G vehicular networks: Toward the integration of communication and computing. *IEEE Vehicular Technology Magazine*, *14*(1), 54−61.

Nunna, S., Kousaridas, A., Ibrahim, M., Dillinger, M., Thuemmler, C., Feussner, H., & Schneider, A. (2015). Enabling real-time context-aware collaboration through 5G and mobile edge computing. In: *Proceedings of the 12th International Conference on Information Technology-New Generations*, pp. 601−605.

Oueida, S., Kotb, Y., Aloqaily, M., Jararweh, Y., & Baker, T. (2018). An edge computing based smart healthcare framework for resource management. *Sensors*, *18*(12), 4307.

Pham, Q. V., Fang, F., Ha, V. N., Piran, M. J., Le, M., Le, L. B., . . . Ding, Z. (2020). A survey of multi-access edge computing in 5G and beyond: Fundamentals, technology integration, and state-of-the-art. *IEEE Access*, *8*, 116974−117017.

Porambage, P., Okwuibe, J., Liyanage, M., Ylianttila, M., & Taleb, T. (2018). Survey on multi-access edge computing for internet of things realization. *IEEE Communications Surveys & Tutorials*, *20*(4), 2961−2991.

Ray, P. P., Dash, D., & De, D. (2019). Edge computing for Internet of Things: A survey, e-healthcare case study and future direction. *Journal of Network and Computer Applications*, *140*, 1−22.

Redondi, A.E., Arcia-Moret, A., & Manzoni, P. (2019). Towards a scaled IoT pub/sub architecture for 5G networks: The case of multiaccess edge computing. In: *Proceedings of the IEEE 5th World Forum on Internet of Things (WF-IoT)*, pp. 436−441.

Ren, J., Zhang, D., He, S., Zhang, Y., & Li, T. (2019). A survey on end-edge-cloud orchestrated network computing paradigms: Transparent computing, mobile edge computing, fog computing, and cloudlet. *ACM Computing Surveys (CSUR)*, *52*(6), 1−36.

Sabella, D., Vaillant, A., Kuure, P., Rauschenbach, U., & Giust, F. (2016). Mobile-edge computing architecture: The role of MEC in the Internet of Things. *IEEE Consumer Electronics Magazine*, *5*(4), 84−91.

Sanchez-Iborra, R., Covaci, S., Santa, J., Sanchez-Gomez, J., Gallego-Madrid, J., & Skarmeta, A.F. (2019). MEC-assisted end-to-end 5G-slicing for IoT, *Proceedings of the IEEE Global Communications Conference (GLOBECOM)*, December 9, pp. 1−6.

Sarrigiannis, I., Ramantas, K., Kartsakli, E., Mekikis, P. V., Antonopoulos, A., & Verikoukis, C. (2019). Online VNF lifecycle management in an MEC-enabled 5G IoT architecture. *IEEE Internet of Things Journal*, *7*(5), 4183−4194.

Shafi, M., Molisch, A. F., Smith, P. J., Haustein, T., Zhu, P., De Silva, P., ... Wunder, G. (2017). 5G: A tutorial overview of standards, trials, challenges, deployment, and practice. *IEEE Journal on Selected Areas in Communications, 35*(6), 1201–1221.

Sodhro, H., Luo, Z., Sangaiah, A. K., & Baik, S. W. (2019). Mobile edge computing based QoS optimization in medical healthcare applications. *International Journal of Information Management, 45*, 308–318.

Sun, X., & Ansari, N. (2016). EdgeIoT: Mobile edge computing for the Internet of Things. *IEEE Communications Magazine, 54*(12), 22–29.

Tomaszewski, L., Kukliński, S., & Kołakowski, R. (2020). A new approach to 5G and MEC integration. In: *Proceedings of the International Conference on Artificial Intelligence Applications and Innovations*, Jun 5, pp. 15–24.

Tran, T. X., Hajisami, A., Pandey, P., & Pompili, D. (2017). Collaborative mobile edge computing in 5G networks: New paradigms, scenarios, and challenges. *IEEE Communications Magazine, 55*(4), 54–61.

Yu, W., & Choi, J. (2020). Human identification in health care systems using mobile edge computing. *Transactions on Emerging Telecommunications Technologies, 31*(12), e4031.

Zhang, J., Weiliang, X. I., Fengyi, Y. A., & Zhouyun, W. U. (2018). 5G mobile/multi-access edge computing integrated architecture and deployment strategy. *Telecommunications Science, 34*(4), 109.

Zhang, J., Wu, Z., Xie, W., & Yang, F. (2018). MEC architectures in 4G and 5G mobile networks. In: *Proceedings of the 10th International Conference on Wireless Communications and Signal Processing (WCSP)*, October 18, pp. 1–5.

Zhang, K., Mao, Y., Leng, S., Zhao, Q., Li, L., Peng, X., ... Zhang, Y. (2016). Energy-efficient offloading for mobile edge computing in 5G heterogeneous networks. *IEEE Access, 4*, 5896–5907.

Zhang, Y., Di, B., Wang, P., Lin, J., & Song, L. (2020). HetMEC: Heterogeneous multi-layer mobile edge computing in the 6 G era. *IEEE Transactions on Vehicular Technology, 69*(4), 4388–4400.

Zhang, Y., Chen, G., Du, H., Yuan, X., Kadoch, M., & Cheriet, M. (2020). Real-time remote health monitoring system driven by 5G MEC-IoT. *Electronics, 9*(11), 1753.

Zhang, K., Leng, S., He, Y., Maharjan, S., & Zhang, Y. (2018). Cooperative content caching in 5G networks with mobile edge computing. *IEEE Wireless Communications, 25*(3), 80–87.

An efficient lightweight speck technique for edge-IoT-based smart healthcare systems

Muyideen AbdulRaheem[1], Idowu Dauda Oladipo[1], Alfonso González-Briones[2,3,4], Joseph Bamidele Awotunde[1], Adekola Rasheed Tomori[5] and Rasheed Gbenga Jimoh[1]

[1]*Department of Computer Science, University of Ilorin, Ilorin, Kwara State, Nigeria* [2]*Research Group on Agent-Based, Social and Interdisciplinary Applications, (GRASIA), Complutense University of Madrid, Madrid, Spain* [3]*BISITE Research Group, University of Salamanca, Salamanca, Spain* [4]*Air Institute, IoT Digital Innovation Hub, Salamanca, Spain* [5]*Computer Services and Information Technology, University of Ilorin, Ilorin, Kwara State, Nigeria*

6.1 Introduction

The high expenses of healthcare, the increase in the number of new diseases, and the imbalance in population growth compared with the rate of medical professionals create a wide gap in the use of the conventional healthcare model for the treatment of patients. This necessitates the use of smart sensor Internet of Things (IoT) for diagnosing, monitoring, and treatment of some diseases. The concept of IoT is the global interconnection and integration of every device on the Internet network for the possible exchange of information among the spectrum of heterogeneous devices. This concept has been applied in various fields of human life such as Smart Energy Network, Smart Home, Smart Building, Smart Healthcare, Smart City among others (Beheshti-Atashgaha, Reza Aref, Barari, & Bayat, 2020).

Smart healthcare is emerging as a solution for healthcare challenges in which patients need not travel to the health center to visit a doctor for medical health care services, but rather communicate with a distant physician using IoT devices and advanced features of Information and Communication Technology (ICT). The physicians then remotely analyze the patients' data and recommend drugs and treatments accordingly. Patients need not travel outside their community for medical treatment in expectation of the right application of smart healthcare. Smart healthcare provides benefits capable of turning around the challenges of healthcare services, especially for those people who have uneasy access to healthcare facilities as well as aged peoples. Though smart healthcare shows promising potentials, successful implementation is a source of concern. Several challenges require attention to fully utilize the potential of smart healthcare. Smart healthcare being a networked system with sensitive data that must not be mishandled requires as the foremost priority, the security, confidentiality, and privacy of patients' data as the major concerns. Smart healthcare professionals, heavily rely on the patients' digital health data for diagnosis, therefore the integrity and reliability of this information are essential for error-free medical analysis. Smart healthcare networked systems are vulnerable to a different form of security threats that exist in every stage of a smart healthcare system, such as, data collection, transmission, storage, and access levels.

5G IoT and Edge Computing for Smart Healthcare. DOI: https://doi.org/10.1016/B978-0-323-90548-0.00005-X

One of the main objectives of IoT technology in the healthcare system is to improve the quality of service provided for patients. Various healthcare devices like smart wearables are worn to monitor the vital organs of the human body, such as the kidney, heart, or lung, and once any dangerous symptom of a person's health is present, a report of the danger is generated and communicated to the relevant health authorities. This in turn provides significant usage of health information dissemination for improved, correct, and timely diagnosis. However, such a report is mostly personal information and findings of the patient and needs to be secured against several security loopholes and various malicious attacks on confidentiality, integrity, etc. In the smart healthcare realm, the smart sensor devices used are resources constraint with short battery life, low processing power and small storage capabilities hence conventional security protocols are not efficient to secure such an environment (Noshina, Ayesha, Muhammad, & Farrukh, 2020).

Edge computing is playing a vital role in mining valued and suitable information out of these data for healthcare professionals to take an efficient and effective decision. The edge computing realm, in combination with 5G speeds and contemporary computing procedures, are tending to meet with an energy-efficient and latency requirement for timely collection and analysis of health data (Hartmann, Hashmi, & Imran, 2019), however, security remains an issue. Smart Healthcare or s-Healthcare requires the broad deployment of healthcare services connected across large domains and communicate through networked sensors, controllers' platforms, and servers between patients, physicians, and other healthcare providers. Wireless Sensor Networks (WSNs) structure enables transportation of health data accurately among participating entities thus providing varied series of applications such as patient monitoring, emergency, and immediate response intervention. These WSN technologies such as Radio Frequency Identification (RFID), Bluetooth, LoRa, etc., are required to be equipped with necessary encryptions to secure the health data transported among the entities.

Lightweight encryption is a type of cryptography strictly for resources constrained devices and WSNs. Thus resource-constrained devices are limited in power, energy, and memory as a result of their small size. These devices communicate in an insecure channel in a difficult and critical environment. Conventional cryptography algorithms such as Rivest–Shamir–Adleman (RSA) and Advanced Encryption Standard (AES) use a lot of power and energy not available in smart sensor devices. Therefore a need for new alternatives, referred to as lightweight encryption (Ayo, Folorunso, Abayomi-Alli, Adekunle, & Awotunde, 2020), is to provide security for resource restricts devices. Generally, encryption is divided into two categories: symmetric and asymmetric.

In symmetric encryption, the identical key is shared for both encoding and decoding of data. However, in asymmetric encryption, two different but related keys are used for encryption and decryption of the codes (Jana, Bhaumik, & Maiti, 2013; Mohd, Hayajneh, & Vasilakos, 2015). In encryption operation, a single key is used as a public key available to everyone while the private key is used for decryption operation known secretly to the user that generates it. Lightweight symmetric ciphers are further divided into two categories: block and stream ciphers. In this paper, an asymmetric lightweight block cipher is proposed for encryption of medical data for smart senor devices in the healthcare system. The remainder of this work is organized as follows. Section two reviews related work, section three discusses the model while section four explains the result, and section five presents the conclusion. Fig. 6.1 displayed the architecture of a smart healthcare system.

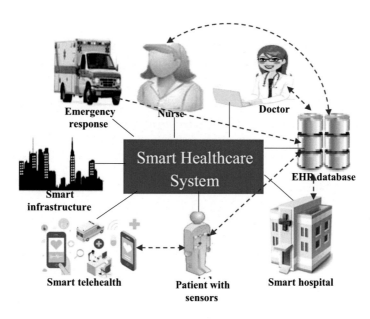

FIGURE 6.1

The architecture of a smart healthcare system.

The architecture of IoT-enabled wearable health care has basic four layers. The foremost layer of the IoT architecture is the layer referred to as the device layer or the sensing layer. It provides, depending on the application, physical objects and sensor devices such as wireless sensors, Radio Frequency Identification (RFID), barcode, and infrared. These devices identify objects and gather desired valuable information in the form of orientation, vibration, location, chemical changes, acceleration, humidity, temperature, and emotional condition with the help of sensors. The information collected is transmitted securely to the communication layer. The communication layer also referred to as the transmission layer processes and transmits the sensor data gathered from the sensor devices using the wireless or wired medium of transmission based on the available technologies such as Wi-Fi, 4G, Zigbee, infrared, and Bluetooth. The data processing layer also called the "middle-ware layer" analyzes and processes the information gathered from the communication layer. The layer establishes and manages the connection with the database using the background technology to process a huge bulk of data in the database. Relying on the knowledge obtained from the three layers, the application layer provides judgment and suggestions on actions to be taken.

6.2 The Internet of Things in smart healthcare system

The modern wave of information technology such as artificial intelligence, big data, the IoT, and cloud computing is turning the conventional medical system into a smart healthcare system that is more convenient, efficient, and more interactive (Hassanalieragh et al., 2015). Smart healthcare is an infrastructure of health service that uses technologies such as IoT, mobile Internet, and wearable devices to access information dynamically, link individuals, healthcare-related materials, and organizations, which then intelligently handle and respond effectively to the needs of the medical system (Bhatt & Bhatt, 2017).

Smart healthcare fosters participation in the healthcare system among the stakeholders, ensures that participants access the needed treatment required, assists medical personals to make an informed decision, and promote the responsible distribution of resources. In this way, smart healthcare is a higher level of information building in the health sector (Tian et al., 2019). Smart healthcare consists of various users, like medical professionals, health facilities, and medical research organizations that integrate and provides different services, including prevention and control of diseases, diagnosis and recovery, administration of health facilities, decision-making on wellbeing, and medical science (Ullah, Shah, & Zhang, 2016).

The components of information technology like 5G, microelectronics, IoT, mobile Internet, cloud computing in addition to modern biotechnology form the foundation of intelligence broadly applied in all areas of smart healthcare (Hassani et al., 2020). Information technology provides patients with technologies such as wearables, to from time to time, tracking their health status, seeking electronic emergency care, and integrating smart healthcare remote facilities from their respective home (Majumder et al., 2017).

To assist and optimize diagnosis, health providers use a range of intelligent clinical decision support tools. Utilizing an interconnected knowledge network such as the Laboratory Information Management System (LIMS), Picture Archiving and Communication Systems (PACS), Electronic Medical Record (EMR), and other knowledge-based networks, health officers can therefore handle medical information in smart healthcare (Fernandes et al., 2020). More effective surgery can be performed with smart healthcare using surgical robotics and mixed-reality technologies. Radiofrequency identification (RFID) technologies can be used from the viewpoint of smart healthcare hospitals to control staffing, materials, and the supply chain, to gather information, and assist in decision-making with automated management platforms (Jokanović, 2020).

From the perception of scientific research agencies, the use of health platforms will improve the experience of patients, using methods like machine learning rather than manual drug monitoring, and using big data for identifying ideal topics (Roski, Bo-Linn, & Andrews, 2014). Smart healthcare assists to effectively minimize the expense and risk of medical operations with the usage of these tools, increase the quality of the use of medical equipment, facilitate exchanges and collaboration in various areas, foster the growth of telemedicine and personal medical care, and eventually make customized medical services everywhere (Zeadally & Bello, 2019).

6.2.1 Support for diagnosis treatment

In line with the application of technology such as artificial intelligence, surgical robotics, and telepresence in smart healthcare to develop the clinical decision support architecture, the treatment and diagnosis of diseases have turned smarter. Important outcomes have been achieved with the successful diagnosis and treatment of such diseases (Islam, Kwak, Kabir, Hossain, & Kwak, 2015). Machine-based computing algorithms have been found to function more effectively than trained physicians and have correctly outperformed the findings of seasoned doctors as a consequence of artificial intelligence diagnosis (Froomkin, Kerr, & Pineau, 2019).

Physicians may offer expert guidance based on the application of the medical decision support system to increase the accuracy of diagnosis, minimize the rate of failed biopsy and medication errors, and allow patients to obtain prompt and effective medical care (Saposnik, Redelmeier, Ruff, & Tobler, 2016). Experts confirmed that the state of the patient and disease status was defined more reliably using technologies to better create a customized care plan and the method of therapy is being more specific (Monje, Foffani, Obeso, & Sánchez-Ferro, 2019). Where the surgeon is allowed greater versatility and consistency with technology and easy deployment, the surgeon has reported high performance relative to the conventional method (Peters, Armijo, Krause, Choudhury, & Oleynikov, 2018).

6.2.2 Management of diseases

chronic diseases contribute to the continuum of human illness and eventually become a new phenomenon with a long path of disease that is incurable and expensive to support; thus it is especially necessary to control and give adequate attention to it (Franceschi et al., 2018). The conventional paradigm of practitioner- and hospital-centered systems of managing health, however, continues to be unable to handle the growing number of patients and diseases adequately. One way out of this situation is the advent of implantable and portable smart health information systems, smart homes, and mobile devices related through Technology in the IoT.

More focus has been paid to patient self-control with the modern health management paradigm of smart healthcare. It stresses self-monitoring of patients in real-time, direct clinical data input, and prompt medical action (Brew-Sam, 2020). Wearable and implantable systems with sophisticated sensors, microprocessors, and wireless technologies intelligently identify and monitor numerous physiological metrics in patients on an ongoing basis, minimizing power usage, improving convenience, and making data accessible for tracking health information (Majumder et al., 2017). This method requires a switch to simultaneous understanding and coordinated treatment from contingency tracking. The related complications posed by the disease are further minimized thus making it possible for healthcare centers to track the disease's survival rate (National Academies of Sciences, Engineering, & Medicine, 2020). For tracking purposes, the advent of mobile phones, smartwatches, etc., offers a new tool, that attempts to incorporate nanomaterials into smartphones to enhance portability further. With this, medical personnel can more effectively track patients globally and their bodies using a super high smartphone (Tian et al., 2019).

Smart healthcare supports the aged and the disabled with health aid in smart homes. Smart homes are individual dwellings or buildings with smart objects in the apartment buildings facilities that track the physical signs and atmosphere of the occupants (Mshali, Lemlouma, Moloney, & Magoni, 2018). Smart homes further carry out events that enhance the performance of living in smart healthcare. The function of artificial intelligence incorporated in smart healthcare is segmented into two facets: home integration and health surveillance (Pramanik, Lau, Demirkan, & Azad, 2017). When gathering health data, these tools offer certain basic resources, allowing individuals who need treatment to lessen their dependency on healthcare professionals and alleviate anguish at home.

By way of software and a health information portal, patients can self-manage their disease using a wearable medical sensor provided by the Stress Detection and Alleviation system to constantly track levels of patient body pressure and give assistance to relieve the body stress automatically (Greiwe & Nyenhuis, 2020). To generate a hierarch for a patient decision support structure that would allow full utilization of the data collected for successful illness analysis, health data from different handheld systems may also be integrated into a hospital information system (Abdel-Basset, Manogaran, Gamal, & Chang, 2019). It will forecast possible issues concern patients and providing advice across the simulator cloud and data analytics ahead of time while helping clinical decision-making (Galetsi, Katsaliaki, & Kumar, 2020).

One other proposal is to create a conducive medical health system that encourages medical stakeholders through lowering entry barriers, to meet and exchange ideas with colleagues. This helps patients to receive advisory services quickly from healthcare services, while physicians can track patient's condition continuously with professional consultants and analysts' support (Vutha, 2020). Mobile technologies like m-Health help mitigate patient complications, eliminate issues with emergency care, enhance the timeliness of surgical procedures, and offer an economic alternative to health services (Albahri et al., 2020).

6.2.3 Risk monitoring and prevention of disease

Health authorities' effort yields the standard disease risk prediction to compile patient data, equate the recommended data from the authorizing organization, and eventually issue the findings of the prediction public. This technique has a certain time delay and this does not offer adequate information to the professionals (Ayo, Awotunde, Ogundokun, Folorunso, & Adekunle, 2020). Disease severity forecasting is complex and specific under smart healthcare. However, it helps patients and clinicians to engage, control the risk of infection proactively, and perform prevention based on its effects of testing (Albahri et al., 2020).

The new cardiovascular disease predictive algorithm collects data through the use of smart apps wearable devices, automatically syncs it through a network to the cloud, and evaluates the outcomes using analytics data algorithms to provide users the forecast results in no time through instant messages (Dias, Marques, & Bhoi, 2021). It has been shown that these interventions are successful. They support physicians and patients at all times to reconsider their habits of treatment and ways of life and even give assistance to health managers in implementing policies relating to community health to achieve the goal of lowering the risk of disease (Poitras et al., 2020). For example, scientists employed algorithms to combined blood glucose parameters, dietary patterns, anthropometry, physical exercise, intestinal microbiota, and other variables to successfully predict changes in glycemic resistance in a diabetes prevention study estimating the glycemic response after monitoring 800 volunteer glucose levels resistance for 46,898 meals per week (Marques, Bhoi, Albuquerque, & Hareesha, 2021).

6.2.4 Virtual support

Robotic assistance is considered an algorithm instead of being an object. Using methods like recognition of voice, mobile apps can connect users, depend solely on big data technology to obtain data, and respond after measurements based on user needs or desires (Abiodun et al., 2021). Digital supports generally function as a gateway to link smart healthcare and medical institutions, patients, and physicians (Rahmani et al., 2018). Healthcare facilities are easily accessible by digital assistants. Medical terms can be translated to familiar, ordinary language for patients using a digital assistant through the mobile device to seek the required medical treatment more precisely (Farahani et al., 2018).

For health personnel, the digital assistant responds instantly to identified details of patient history, thus ensuring adequate and effective treatment of patients and allowing physicians to plan for surgical treatments to save time for other assignments (Bhoi, Sherpa, & Khandelwal, 2018). The deployment of digital support system significantly improve manpower and material costs of medical facilities and adapt more appropriately to the challenges of both parties (Marques et al., 2020). The Nuance technologies are also used to create a conversation between various virtual assistants, in particular between common assistants and highly skilled assistants, thus significantly enhancing the knowledge of stakeholders in the healthcare treatment (Mishra, Mallick, Tripathy, Bhoi, & González-Briones, 2020). Digital assistants may be employed also to help with the management of illnesses, such as using digital assistants to enhance individual mental health, and therefore improve the limited availability of human psychotherapists to provide more patients with emotional wellbeing (Ebert et al., 2018).

6.2.5 Smart healthcare hospitals support

Smart healthcare components comprise of three main items: clinic, geographical, and community. To optimize practices of current patient care and implementation of new features, smart healthcare focuses on environments driven by information and communication technology, concentrating primarily on effective and efficient IoT technology and process automation (Papa, Mital, Pisano, & Del Giudice, 2020). In smart hospitals, there are three fundamental categories of facilities: medical personnel, patient, and admin services. In-hospital management decisions support systems, it is paramount that these healthcare categories be adequately addressed (Jamil, Ahmad, Iqbal, & Kim, 2020).

Mobile computers, smart cities, and personnel are interconnected in-hospital administration by the digital network that integrates many systems of digital IoT devices. This system makes possible the identification and documentation of patients in hospitals, regularly oversee medical staff, and track equipment and biotic samples (Oniani, Marques, Barnovi, Pires, & Bhoi, 2021). In the pharmaceutical sector, intelligent healthcare is also being utilized for manufacturing and distribution of the drug, control of warehouse, anticounterfeiting, and additional required procedures. To realize a reliable, secure, safe, and efficient distribution of hospital materials utilizing RFID technology, a specific RFID tag allocation for each individual can be issued and the data can be transmitted to a device that easily tracked and retrieved the data via the portable apps (Mishra, Tripathy, Mallick, Bhoi, & Barsocchi, 2020).

The creation of an operational control framework assists in carrying out tasks such as resource selection, quality monitoring, and success analysis in terms of decision-making, and can minimize patient costs, optimize resource usage, and help hospitals make recommendations to management (Pramanik, Pradhan, Nandy, Bhoi, & Barsocchi, 2021). Patients can control various features in terms of healthcare services, such as physical exam services, digital consultations, and doctor-patient experiences. Such computerized systems allow the medical care procedures of patients acceptable and more straightforward. In a timely way, patients wait and receive increasingly lifelike care. In summary, the outcome of intelligent healthcare is transformation, development, and modernization (Marks, 2020).

6.3 Application of edge computing in smart healthcare system

The Artificial-IoT (AIoT) systems, along with edge computing, forward to the cloud servers data for deciding on smart healthcare. Quality of Life (QoL) is thus enhanced, protected, and effective with the provision of healthcare services and recorded growth in popularity after combining Intelligence edge computing and IoT-enabled heterogeneous networks for reliable and timely delivery of medical data (Atitallah, Driss, Boulila, & Ghézala, 2020). The progress of the implementation of diverse networks IoT-enabled mobile technology, AI, edge computing, and IoT has aroused the attention of scientists and industrialists in such a way that edge computing was identified as one of the most common technologies currently used in recent times and research efforts are more intensive in this field of research (Faheem et al., 2018). For the science community, edge computing in healthcare networks is very important to addressing big challenges at a global level (Greco, Percannella, Ritrovato, Tortorella, & Vento, 2020).

AIoT increases healthcare's operating productivity by reducing the medical professional nontechnical activities, thus healthcare personnel provided with more time to attend to patients in a patient-centric environment as needed (Luk, 2018). Edge computing technology is smart enough to understand the sharing of real-time services, supplying medical practitioners with continuous computing services, and improving the quality of experience (QoE) of smart healthcare professionals (Chen, Li, Hao, Qian, & Humar, 2018). In smart healthcare, edge computing offers cognitive health monitoring with simple patient data and knowledge on the extent of disease risk in real-time for urgent attention. It also generates complex network resource knowledge for resource allocation in a cost-constraint setting and applies the best network resources to the user with the highest priority which then makes the full edge computing resources accessible to patients with the highest risk of disease and thereby increases the patient's health probability (Omoniwa, Hussain, Javed, Bouk, & Malik, 2018).

Advanced data processing functions required in the data sources and the cloud are implemented by edge computing. It combines clinical and nonmedical data from various sources, processes the data collected on the network, classifies and notifies emergencies, gathers the required information, and transmits the processed data or the data extracted to the cloud (Abdellatif, Mohamed, Chiasserini, Tlili, & Erbad, 2019). Remarkably, with edge computing management of long-term chronic diseases, different healthcare-related apps can be implemented which actively assist patients to participate in administering their treatment and to communicate with their physicians everywhere and at any time (Aceto, Persico, & Pescapé, 2018).

Also, with edge computing, a specialized context-aware processing application can be run with multiple connected data sources and used to easily manage the patient, while optimizing context-based data delivery, such as data type, supported application, and overall conditions of the patient (Porambage, Okwuibe, Liyanage, Ylianttila, & Taleb, 2018). The well-known category of remote monitoring apps is heart tracking apps that detect vital signs associated with the heart to reveal multiple types of heart attacks, such as cardiac arrhythmia, chronic heart disease, ischemia, and myocardial infarction (Ayo, Ogundokun, Awotunde, Adebiyi, & Adeniyi, 2020). Bansal et al. (2015) introduce a real-time heart

Table 6.1 Basic features of IoMT-enabled edge computing in smart healthcare system.

Requirements	IoMT-based edge computing
Mark user	Mobile users
Location of servers	Edge nodes
Service type	Localized information services
Geographical distribution	Distributed
Distance between client and server	Single hop
Delay jitter	Low
Latency	Low
Data processing	Fast
Reliability	Low
Type of connectivity	Wireless
Location awareness	Yes
Server nodes	Large
Computing cost	Low
N/W bandwidth	Low
Reply time	Milliseconds, subseconds
Security	Very secure

wearable sensor device capable of transmitting extracted patients' medical data through Bluetooth to an Android-based listening port with onward transmission to a web server the data for analysis. Table 6.1 shows the basic features of IoMT-enabled edge computing in smart healthcare system.

Edge and cloud computing work together to provide a scalable solution focused on each organization's data collection and analysis requirements. For certain workloads, the edge is suitable for real-time selection and analysis. Simultaneously, the cloud can serve as a centralized place for large-scale analytics. They have real-time and long-term perspectives into efficiency and power interventions such as ML and asset performance management when they work together. Rising computing power at the edge lays the groundwork for autonomous systems, allowing smart healthcare system to boost operational efficiency while freeing up physicians to concentrate on higher-value activities.

Edge computing is critical because it gives smart healthcare system new and better ways to increase operating efficiency, enhance performance and safety, automate all core medical processes, and ensure "always on" availability. It is a leading method for achieving digital transformation of the healthcare operations. To generate actionable healthcare intelligence, edge computing optimizes data collection and analysis at the edge. Edge computing makes technology, systems, and smart healthcare model processes more versatile, scalable, stable, and automated.

Hence, to enable remote cardiac control utilizes a smartphone Android to capture a patient's details captured by wearable sensors and forward it to a web server. The mobile, in this case, is only used, as a contact pivot in the networks, to transmit captured data to the cloud (Shah, Bhat, & Khan). As a result of the heavy energy workload it requires, constant data transfer seems infeasible. Therefore proposed Multiaccess Edge Computing (MEC) MEC architecture in such systems closely monitoring the operating condition of the equipment and its data transmission at the Mobile/Infrastructure Edge Node (MEN) to substantially improved energy saving (Mehrabi et al., 2019). The combination of deriving background knowledge and localization strategies can be effectively and efficiently used by a computing edge to connect the geographical location of the patient with the closest suitable caregivers for emergency attention (Oladele, Ogundokun, Awotunde, Adebiyi, & Adeniyi, 2020).

One of the rapidly increasing ways to collect physiological signals un-interrupting the behavior of a patient is to calculate the heart rhythm using digital camera sensors from facial videos. However, it is not advisable to move big data generated using traditional cloud-based architecture from these camera sensors and some of these implementations could be deemed impractical considering the restricted availability of bandwidth (Abdellatif et al., 2019). The quantity of digital information obtained from a small standard camera will exceed 40 GB in a single day. Reading, compressing, and removing the most relevant information from the data obtained by edge computing dramatically decreases the volume of data to be transmitted to the server, hence the use of bandwidth, and even allows local processing of the data (Kazeem Moses, Joseph Bamidele, Roseline Oluwaseun, Misra, & Abidemi Emmanuel, 2021).

One of the successes of smart healthcare is predictive surveillance of patients at high risk. To incorporate prevention interventions to minimize morbidity and mortality associated with patients at high risk, Ali et al. (2020) proposed to enhance emergency prediction/detection technique of optimal machine learning model. The fast transfer of data to the server is a must in real-time prediction/detection systems. In certain cases, this means that results be processed and even a diagnosis is made as closely as possible to the patient. IoT devices are developed with AI embedded with improved edge computing to provide IoT devices with information (Calo, Touna, Verma, & Cullen, 2017). The use of AI in IoT-enabled networks at the network edge is made better since IoT device-generated data empowers AI embedded in the IoT devices and edge servers to improve knowledge.

In IoT-enabled healthcare networks, before data transfer, IoT devices need to be validated, next, for fast processing, the sensor data needs to be downloaded to high configuration edge computing. With the system, which can build complete network programmability, the offload to edge should be performed intelligently. In terms of resource utilization and load balancing, information needs to satisfy the criteria of edge computing, while security is supported by a lightweight encryption scheme (Li et al., 2020). Data processing, availability, edge virtuosity, and fast and highly efficient data transfer are the essential characteristics of a healthcare system (Dhayne, Haque, Kilany, & Taher, 2019).

Edge computing which is close to data sources is designed with the ability to transport, easy process, and store data generated by numerous wireless sensors, cameras, and controllers at the network edge (Khan, Parkinson, & Qin, 2017). In addition to offering low latency, reduced energy usage for battery-operated systems, and network bandwidth saving, edge computing assists smart healthcare devices with automated diagnosis tracking, proper drug administration to patients, and intensive real-time surveillance of vital signs and early warning of clinical degradation of patents (Pisani et al., 2019).

6.4 Application of encryptions algorithm in smart healthcare system

In smart healthcare-based applications, the main concern is ensuring improved and secured data security and confidentiality in the transmission. Prevention of illegal access to information and cyber-attacks on data security remains the major challenge in the IoT healthcare systems. Although numerous methods have been proposed to secure the health data in the smart healthcare environment, ensuring the safety of information, and cyber-attacks protection against medical data is a vital problem concern that requires adequate attention. Adeniyi, Ogundokun, and Awotunde (2021) proposed a Random Coefficient Selection and Mean Modification Approach (RC-SMMA) for a strong multilevel security solution that focused on hiding information and chaotic theory to protect data privacy and secrecy in an insecure multimedia exchange environment among dual IoT devices.

The approach indicates a promising solution for entire signal processing and the region-oriented attacks, acting as standalone or in a network. As a result of timely communication and the ability of integrated IoT sensors to decide into day to day activities, smart sensors are becoming more significant. Therefore securing the sensors is one of the major concerns as a result of the universal state of the sensors and the high sensitivity of health data transmitted. Also, the power-constrained status of the sensors stresses the need for lightweight security that needs to be tailored to the constraint resource requirements need of the sensors. Sankaran, Abikoye, Ojo, Awotunde, and Ogundokun (2020) propose a lightweight security framework with the development of a level-wise security architecture for IoT devices and further build protocols for secured communication of IoTs using identity-based cryptography. The proposed mechanism was assessed using simulation tools and proved to reduce overhead.

To provide end-to-end secured devices communication, Abdulraheem, Awotunde, Jimoh, and Oladipo (2021) proposed lightweight asymmetric encryption to ensure the protection of IoT gateway services via Asymmetric Key Encryption (AKE) to session share key among the devices and used lattice encryption with low energy consumption sensors to protect broker gateways devices and cloud services. Distributed Denial of Service Attacks Protection (DDSAP), attack by eavesdroppers, and Quantum algorithm attacks were achieved. Varshney & Gupta (Varshney & Gupta, 2017) proposed a blockchain backbone technique to protect and manage data securely among smart devices on the cloud. The technique achieved a scalable and robust result that addressed the identity and security concerns of IoT devices.

Al-Muhtadi et al. (2020) gave an account of the main cyber-attacks in smart environments and proposed a secured lightweight encryption framework to maintain and authenticate the senders and receivers in the cloud environment. The framework was implemented for authentication, role assignment, and access control at the user layer and evaluated for efficiency, sustainability, and secure communication. Beheshti-Atashgaha et al. (2020) proposed an efficient lightweight security authentication framework to protect patients' privacy, identity, and information about their healthcare records and satisfies essential security requirements.

Using the Two-Fish (TF) method of symmetric key encryption, Deebak & Al-Turjman (2020) proposed routing and monitoring protocol to expose and guard against eavesdroppers in the wireless sensor network. The proposed method demonstrates a high percentage of success compared to other existing techniques. Lightweight cryptography (LWC) is a new cipher whose main purpose is to satisfy the criteria provided for the use of restricted objects. The term "lightweight" is for a family of smaller code size cryptographic ciphers, low processing capacity, and low energy consumption. As a result of these resource constraints' difficulty, security solutions focused on LWC built according to IoT specifications are increasingly required (Tausif, Ferzund, Jabbar, & Shahzadi, 2017).

Efforts have been intensified to propose numerous lightweight security algorithms and protocols for smart healthcare support. In general, cryptography is used to safeguard networks with authentication, secrecy, the integrity of knowledge, and control of access. However, standard cryptographic protocols may not be appropriate in IoT system environments because of the resource restriction design of IoT devices (Meneghello, Calore, Zucchetto, Polese, & Zanella, 2019). IoT incorporates a huge variety of devices that communicate with one another and collect a large quantity of data with the availability of powerful technologies, creating the need for IoT imperative protection (Tariq et al., 2019).

By gathering different data from a patient's body, continuous transmitting is infeasible on power constraint AIoT systems. To address these inadequacies, it is preferred to use edge computing and a lightweight authentication device to process data at the edge of the network, near the wearables (Ren, Zhang, He, Zhang, & Li, 2019). The effectiveness of cryptographic algorithms is concerned with LWC, and it is a relatively recent subfield connecting cryptology, computer science, and computer engineering (Al-Garadi et al., 2020). The constrained resource design of IoT devices and individual security are the basic problems for most IoT-based applications. Hence, AIoT applications are becoming common by using the limited resources of IoT devices safely and effectively, especially in healthcare (Hassan, 2019). IoT sensors continuously track patients in healthcare systems and constantly relay the necessary data that needs to be shielded from harmful activities (Majumder et al., 2017).

Services such as medical identification, healthcare personnel, medication inventory, management of drugs, electronic health control, timely monitoring, and interpretation are provided by LWC. Real-time smart healthcare tracking and analysis involve constant data transfer and protection that are crucial for stakeholders, so lightweight authentication mechanisms are the perfect way to protect a low-energy networked enable IoT (Pramanik, Upadhyaya, Pal, & Pal, 2019). Healthcare is among how IoT technology is used to deliver benefits to people with health conditions in which the health of the patient is tracked at all times and everywhere without the patient's need to see the doctor at all times. In the same way, for data protection adversaries, smart health care is now one of the targeted sectors (Kadhim, Alsahlany, Wadi, & Kadhum, 2020).

The result of targeting medical sensors is extremely risky and life-threatening, leading to dangerous issues with life and death. The problem of medical fraud may lead to incorrect care that causes loss of life and even fraudulent rises in financial costs (Kadhim et al., 2020). Any of the main security problems with Medical IoT include allowing an unauthorized party to access the confidential records, medication details, and other details of the patient, such as prescription details that make it vulnerable for patients and even mishandle the patient data (Hossain, Islam, Ali, Kwak, & Hasan, 2018).

6.4.1 Speck encryption

Speck belongs to a family of lightweight symmetric block ciphers having a pair of varying block and key sizes. A block cipher is an encryption algorithm that enables users to have a common key secretly and securely encrypt/decrypt blocks of data. Lightweight encryption is built for efficient implementation of extremely constrained platforms, such as RFID tags, smart sensors, microcontrollers, and other resource-limited devices.

The term "lightweight" does not mean how secured the algorithm is, but rather refers to its suitability for use on highly constrained devices. Thus a secure lightweight block cipher having a given block and key size pairs offers the equivalent level of security as any other secure block cipher with that same block and key size.

Lightweight encryption addresses the request to secure resource constraint devices which the conventional encryptions could not adequately attend to, using algorithms and protocols designed to perform well on platforms for devices with resources constrained.

Speck has support for a block of various sizes such as 32, 48, 64, 96, and 128 bits, and up to three different key sizes to go along with each size of a block. Speck family has ten different algorithms with various block and key sizes, as shown in the table. All values are in bits (Table 6.2).

Speck has operations that are highly efficient for software platforms of varying requirements. There are a quite few numbers of the operation such as NOT, OR, AND, XOR, rotations, modular addition, and subtraction use to achieve nonlinearity as against S-Boxes used in conventional encryptions. Since Speck is not using S-Boxes, then it is not Substitution-Permutation Networks (SPNs). Rather than being SPNs, Speck is an Add−Rotate−XOR (ARX) cipher with Feistel Structure round functions, which provides an adequate equilibrium between operations of nonlinear confusion and linear diffusion. Using the bit permutation rotation operation may not achieve sufficient diffusion, but with additional functional rounds, an adequate level of security is achieved like in the S-Boxes permutation of SPN. Speck nonlinearity is achieved with modular addition. Thus its functions are secured cryptographically and effectively suitable for IoT constraint devices software implementations.

The proposed model

The notations used in the definition of the proposed model are given in the table below

n	A word size
$2n$	A block size
m	Number of words
R	Number of rounds
PT	$2n$-bit input plaintext
CT	$2n$-bit output ciphertext
ki	n-bit round subkey for round i

Table 6.2 Speck parameters.

Block size	Key sizes	n	m
32	64	16	4
48	72	24	3
48	96	24	4
64	96	32	3
64	128	32	4
96	96	48	2
96	144	48	3
128	128	64	2
128	192	64	3
128	256	64	4

K	mn-bit Master key from which round subkeys are generated
$\oplus$	Bitwise exclusive OR operation
Sj	Left cyclic shift by n bits
$S\text{-}j$	Right cyclic shift by n bits
$+$	Mod 2n addition operation

Speck block cipher denoted as Speck2n has an n-bit word, where n represents either 16, 24, 32, 48, or 64. Speck2n having an m-word (mnbit) key is denoted as Speck2n/mn. For instance, Speck64/96 denotes the Speck version having plaintext blocks of 64-bit and with a key having 96-bit ($n = 32$ and $m = 3$) (Fig. 6.2).

The round function of Speck2n encryption is denoted by the mapping

$$Rk(x, y) = ((S - \alpha x + y) \oplus k, S\beta y \oplus (S - \alpha x + y) \oplus k),$$

where x and y are n-bit halves of 2n-bit plaintext, and k is round key.

x = RCS(x, α)	Right shift x by α and assign the result to x
x = x + y	modulo 2n addition of x and y and assign the result to x
x = x $\oplus$ k	XOR x and round key k and assign the result to x
y = LCS(y, β)	Left shifty by β and assign the result to y
y = y $\oplus$ x	XOR y and x and assign the result to y

For decryption, the inverse of the round function is used with modular subtraction instead of modular addition and is given by

$$R - 1k(x, y) = (S\alpha((x \oplus k) - S - \beta(x \oplus y)), S - \beta(x \oplus y)).$$

The parameters α and β are 7 and 2 respectively, for Speck32/64 and they are 8 and 3 for other variance of Speck.

Key schedule

Speck uses key schedules of 2-, 3-, and 4-word depending on Speck variance. Key schedules for Specks use around function, as given below.

Suppose m is the number of words for a key, and the key K can be written as (lm − 2, l0, k0).

Then, to generate sequences of two words ki and li by

$$li + m - 1 = (ki + S - \alpha li) \oplus i$$

and

$$ki + 1 = S\beta \, ki \oplus li + m - 1.$$

The value ki is the ith round key, for i ≥ 0.

The table below shows the Speck rounds and the required parameters (Table 6.3).

The Speck key schedules receive an input key K and produce from it a sequence of T key words k0, kT − 1, where T is the number of rounds. Encryption is the composition $R_{k_{T-1}} \circ \cdot \cdot \cdot \cdot \circ R_{k_1} \circ R_{k_0}$, read from right to left (Fig. 6.3).

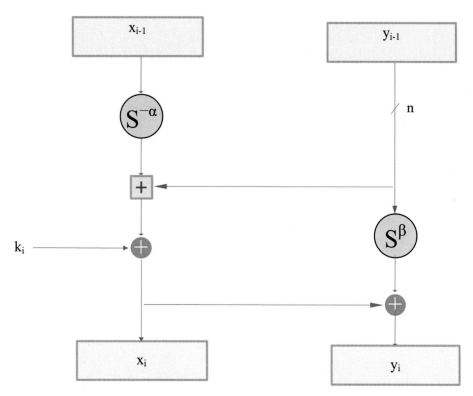

FIGURE 6.2

Speck round function.

Table 6.3 Speck rounds.						
Block size 2n	**n**	**M**	**key size mn**	**Speck rounds**	**α**	**β**
32	16	4	64	22	7	2
48	24	3	72	22	8	3
48	24	4	96	23	8	3
64	32	3	96	26	8	3
64	32	4	128	27	8	3
96	48	2	96	28	8	3
96	48	3	144	29	8	3
128	64	2	128	32	8	3
128	64	3	192	33	8	3
128	64	4	256	34	8	3

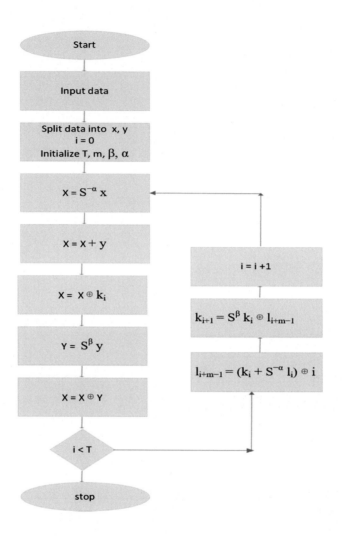

FIGURE 6.3

Speck encryption flow.

Algorithm 6.1 below presents the pseudo-code of the Speck encryption procedure. The key scheduling function forms the encryption key $K = lm - 2, l0, k0$ and generates the T −rounds keys sequence $k0, kT - 1$ using the procedures presented by Algorithm 6.2.

6.5 Results and discussion

The proposed encryption method was developed on the microcontroller (MCU) AVR 8-bit RISC (Reduced Instruction Set Computing) architecture that has programmable on-chip flash memory,

ALGORITHM 6.1 Speck Encryption

input: plaintext
split plaintext into n-bit x, y
input: K encryption key
      initialize T, β, α
      generate round keys k0, kT $-$ 1
      for i = 0 to T-1
         x = (S $-$ α x + y) $\oplus$ ki
         y = Sβ y $\oplus$ x
      end for loop

ALGORITHM 6.2 Round Key generation

initialize m;
generate lm $-$ 2, l0, k0
for I = 0 to T-2
      li + m $-$ 1 = (ki + S $-$ α li) $\oplus$ i
      ki + 1 = Sβ ki $\oplus$ li + m $-$ 1
    end for loop

Table 6.4 Features of the device.

Name	Value
MCU	ATmega328PB
Program memory type	Flash
Program memory size (KB)	32
CPU speed (MIPS/DMIPS)	20
SRAM (B)	2048
Data EEPROM/HEF (bytes)	1024
Digital communication peripherals	2-UART, 2-SPI, 2-I2C
Capture/compare/PWM peripherals	3 Input Capture, 3 CCP, 10PWM
Timers	2 $\times$ 8-bit
Number of comparators	1
Operating voltage range (V)	1.8 to 5.5
Pin count	32
Low power	Yes

SRAM, IO storage space, and EEPROM, as shown in Table 6.1. Different block size data was processed for encryption in KB and our experimental findings are seen in Table II. The execution time tests the amount of encryption and decryption block cycles used. This is the contrast between the beginning and the end of the process (Table 6.4).

Table 6.5 Performance analysis of the proposed model.

File size (Kb)	Encryption time (MM)	Encryption time (MM)
0.82	0.121	0.120
1.65	0.216	0.215
12.32	0.893	0.891
36.50	2.014	2.012
50.2	3.142	3.138
100.7	5.461	5.459

Table 6.6 Comparison results of our proposed model with another model.

Algorithm	Speed (Clockcycle /bytes)	Memory usage (bytes)	Encryption time (ms)	Decryption Time (ms)
AES	24695	1709	650	124
DES	28401	1608	797	281
SIMON	39313	755	700	300
Speck	30957	1364	719	236

Table 6.7 Coefficients of extracted features for dataset A.

Coefficient	Variance	Standard deviation	Energy
D_1	25.2164	5.0216	2.8564e + 04
D_2	587.553	24.2395	3.0435e + 05
D_3	5.3957e + 03	73.4555	1.4426e + 06
D_4	9.9058e + 03	99.5279	1.9874e + 06
A_4	1.5439e + 04	124.2539	4.0502e + 06

Other block cipher techniques such as AES, DES, SIMON were compared to evaluate the analysis of memory use and speed. Table III presents the findings of the comparison. The findings as shown in this table indicate that the memory use of AES, DES, and SIMON is greater than the proposed memory usage of the proposed system. In comparison, the time to encrypt and decrypt is less in the proposed system than in most of the lightweight ciphers (Table 6.5).

The table uses various algorithms to display the different encryption times of the same input data. This illustrates that the suggested encryption solution takes the same amount of time as other regular block ciphers (Table 6.6).

The proposed system was still tested using the EEG datasets taken from Bonn database a widely used dataset (Andrzejak et al., 2001). The dataset contains five various datasets labels as A, B, C, D and E. The dataset A, D, and E was analyzed to measure the latency in both cloud and edge computing using computational time and transmission delay. Dataset A was taken from a

Table 6.8 Coefficients of extracted features for dataset E.

Coefficient	Variance	Standard deviation	Energy
D_1	1.4426e + 03	37.9819	1.8934e + 06
D_2	6.4382e + 04	253.736	4.8707e + 07
D_3	7.0151e + 05	837.560	3.0676e + 08
D_4	6.9684e + 05	834.769	1.887e + 08
A_4	1.7177e + 06	1.310e + 03	4.0854e + 08

Table 6.9 Comparison of cloud-based and edge-based computational time.

Smart healthcare	Latency (s)
Cloud-IoT framework	2.503
Edge-IoT framework	1.432

stable individual, while dataset D was taken during an interracial situation. During seizure activity, dataset E was collected from the epileptegenic region (ictal state). Each dataset contains 100 EEG epochs, each of which contains 4097 samples. The EEG data was collected at a sampling rate of 173.61 Hz. Tables 6.7 and 6.8 demonstrate the statistical parameters that were derived from the subbands (Sayeed, Mohanty, Kougianos, Yanambaka, & Zaveri, 2018). The values of all statistical parameters are clearly higher for dataset E. The derived attribute estimates for datasets A and D are extremely similar. Dataset A, D, and E versus Dataset E were used to assess identification in this study.

The SimulinkR was used to calculate the proposed system's latency. The cloud-based IoT had a latency of 2.503 seconds, while the latency of 1.432 seconds was generated by the edge-based IoT system. Both computational cost and processing delay are included in the latency. In real-world various devices are connected to the cloud server in their millions, the cloud-based IoT applications and excess data in the transmission line causes higher network congestion and system latency. Therefore from the edge-based IoT framework there is 46% significant decrease in latency, thus help in the reduction of latency during computational time and reduced transmission delay during processing. These features are very critical and crucial in smart healthcare applications (Table 6.9).

6.6 Conclusions and future research directions

The IoT-based system is an upcoming information and communication science technology. A large number of IoT technologies have been developed in the healthcare field to remotely detect, track, predict, and manage chronic diseases. The data generated by these devices is huge and the processing and transmissions need typical infrastructure to cope with them. Edge computing can be applying to overcome the challenges of IoT-cloud-based systems. The edge computing technology with lower latency, high computational paradigm, and scalability will help IoT-based system devices in real-time data collection and processing. The deployment of cloud data is far away from the network; this causes the response time delay in real-time. Moreover, the cloud may cause significant overhead on the backbone network to the user application due to the huge amount of big data sent to the cloud. Hence, the application of edge computing will bring the storage resources and computational closer to the end-user devices, thus reduce the burden on the cloud. The edge computing architecture is heterogeneous devices, and geographically distributed ubiquitously connected at the end of a network to provide collaboratively variable and flexible communication, storage services, and computation. The information transmitted by the smart sensor in smart healthcare systems (SHS) is personal, requiring secrecy, trustworthiness, and usability for healthcare practitioners to take timely and accurate decisions, so the data required to be transmitted and stored safely. For the IoT-based Smart Healthcare system, a lightweight ciphering technique was implemented. The proposed approach is based on basic operations of ARX Addition, Rotation XORing, swapping, slicing, etc. The outcome of the implementation reveals that the memory occupation is limited while the speed for producing a successful cipher is important. In different constrained devices, the proposed technique can be applied and tested. It is also possible to perform differential and linear crypto-analysis of this algorithm in the future to ensure the cipher's robustness.

References

Abdel-Basset, M., Manogaran, G., Gamal, A., & Chang, V. (2019). A novel intelligent medical decision support model based on soft computing and IoT. *IEEE Internet of Things Journal, 7*(5), 4160–4170.

Abdellatif, A. A., Mohamed, A., Chiasserini, C. F., Tlili, M., & Erbad, A. (2019). Edge computing for smart health: Context-aware approaches, opportunities, and challenges. *IEEE Network, 33*(3), 196–203.

Abdulraheem, M., Awotunde, J. B., Jimoh, R. G., & Oladipo, I. D. (2021). An efficient lightweight cryptographic algorithm for IoT security. *Communications in Computer and Information Science, 2021*(1350), 444–456.

Abikoye, O. C., Ojo, U. A., Awotunde, J. B., & Ogundokun, R. O. (2020). A safe and secured iris template using steganography and cryptography. *Multimedia Tools and Applications, 79*(31–32), 23483–23506.

Abiodun, M. K., Awotunde, J. B., Ogundokun, R. O., Misra, S., Adeniyi, E. A., Arowolo, M. O., & Jaglan, V. (2021, February). Cloud and big data: A mutual benefit for organization development. *Journal of Physics: Conference Series, 1767*(1), 012020.

Aceto, G., Persico, V., & Pescapé, A. (2018). The role of information and communication technologies in healthcare: Taxonomies, perspectives, and challenges. *Journal of Network and Computer Applications, 107*, 125–154.

Adeniyi, E. A., Ogundokun, R. O., & Awotunde, J. B. (2021). *IoMT-based wearable body sensors network healthcare monitoring system. IoT in healthcare and ambient assisted living* (pp. 103–121). Singapore: Springer.

Albahri, A. S., Alwan, J. K., Taha, Z. K., Ismail, S. F., Hamid, R. A., Zaidan, A. A., ... Alsalem, M. A. (2020). IoT-based telemedicine for disease prevention and health promotion: State-of-the-Art. *Journal of Network and Computer Applications, 173*, 102873.

Al-Garadi, M. A., Mohamed, A., Al-Ali, A. K., Du, X., Ali, I., & Guizani, M. (2020). A survey of machine and deep learning methods for internet of things (IoT) security. *IEEE Communications Surveys & Tutorials, 22*(3), 1646–1685.

Ali, F., El-Sappagh, S., Islam, S. R., Kwak, D., Ali, A., Imran, M., & Kwak, K. S. (2020). A smart healthcare monitoring system for heart disease prediction based on ensemble deep learning and feature fusion. *Information Fusion, 63*, 208–222.

Al-Muhtadi, J., Saleem, K., Al-Rabiaah, S., Imran, M., Gawanmeh, A., & Rodrigues, J. J. P. C. (2020). A lightweight cybersecurity framework with context-awareness for pervasive computing environments. *Sustainable Cities and Society*, 102610. Available from https://doi.org/10.1016/j.scs.2020.102610.

Andrzejak, R. G., Lehnertz, K., Mormann, F., Rieke, C., David, P., & Elger, C. E. (2001). Indications of non-linear deterministic and finite-dimensional structures in time series of brain electrical activity: Dependence on recording region and brain state. *Physical Review E*, *64*(6), 061907.

Atitallah, S. B., Driss, M., Boulila, W., & Ghézala, H. B. (2020). Leveraging Deep Learning and IoT big data analytics to support the smart cities development: Review and future directions. *Computer Science Review*, *38*, 100303.

Ayo, F. E., Awotunde, J. B., Ogundokun, R. O., Folorunso, S. O., & Adekunle, A. O. (2020). A decision support system for multi-target disease diagnosis: A bioinformatics approach. *Heliyon*, *6*(3), e03657.

Ayo, F. E., Folorunso, S. O., Abayomi-Alli, A. A., Adekunle, A. O., & Awotunde, J. B. (2020). Network intrusion detection is based on deep learning model optimized with rule-based hybrid feature selection. *Information Security Journal: A Global Perspective*, 1–17.

Ayo, F. E., Ogundokun, R. O., Awotunde, J. B., Adebiyi, M. O., & Adeniyi, A. E. (2020, July). Severe acne skin disease: A fuzzy-based method for diagnosis. Lecture Notes in Computer Science (including subseries Lecture Notes in Artificial Intelligence and Lecture Notes in Bioinformatics), 12254 LNCS, pp. 320–334.

Bansal, A., Kumar, S., Bajpai, A., Tiwari, V. N., Nayak, M., Venkatesan, S., & Narayanan, R. (2015). Remote health monitoring system for detecting cardiac disorders. *IET Systems Biology*, *9*(6), 309–314.

Beheshti-Atashgaha, M., Reza Aref, M., Barari, M., & Bayat, M. (2020). Security and Privacy-preserving in e-health: a new framework for patient. *Internet of Things*. Available from https://doi.org/10.1016/j.iot.2020.100290.

Bhatt, Y., & Bhatt, C. (2017). *Internet of things in healthcare. Internet of things and big data technologies for next generation HealthCare* (pp. 13–33). Cham: Springer.

Bhoi, A. K., Sherpa, K. S., & Khandelwal, B. (2018). Arrhythmia and ischemia classification and clustering using QRS-ST-T (QT) analysis of electrocardiogram. *Cluster Computing*, *21*(1), 1033–1044.

Brew-Sam, N. (2020). *App use and patient empowerment in diabetes self-management: Advancing theory-guided mHealth research*. Springer Nature.

Calo, S. B., Touna, M., Verma, D. C., & Cullen, A. (2017, December). Edge computing architecture for applying AI to IoT. In *Proceedings of the IEEE international conference on big data (Big Data)* (pp. 3012–3016). IEEE.

Chen, M., Li, W., Hao, Y., Qian, Y., & Humar, I. (2018). Edge cognitive computing based smart healthcare system. *Future Generation Computer Systems*, *86*, 403–411.

Deebak, D., & Al-Turjman, F. (2020). A hybrid secure routing and monitoring mechanism in IoT-based wireless sensor networks. *Ad Hoc Networks*, *97*, 102022. Available from https://doi.org/10.1016/j.adhoc.2019.102022.

Dhayne, H., Haque, R., Kilany, R., & Taher, Y. (2019). In search of big medical data integration solutions-a comprehensive survey. *IEEE Access*, *7*, 91265–91290.

Dias, R. M., Marques, G., & Bhoi, A. K. (2021). Internet of Things for enhanced food safety and quality assurance: A literature review. *Advances in Electronics, Communication and Computing*, 653–663.

Ebert, D. D., Van Daele, T., Nordgreen, T., Karekla, M., Compare, A., Zarbo, C., ... Kaehlke, F. (2018). Internet-and mobile-based psychological interventions: Applications, efficacy, and potential for improving mental health. *European Psychologist*.

Faheem, M., Shah, S. B. H., Butt, R. A., Raza, B., Anwar, M., Ashraf, M. W., ... Gungor, V. C. (2018). Smart grid communication and information technologies in the perspective of Industry 4.0: Opportunities and challenges. *Computer Science Review*, *30*, 1–30.

Farahani, B., Firouzi, F., Chang, V., Badaroglu, M., Constant, N., & Mankodiya, K. (2018). Towards fog-driven IoT eHealth: Promises and challenges of IoT in medicine and healthcare. *Future Generation Computer Systems*, *78*, 659–676.

Fernandes, M., Vieira, S. M., Leite, F., Palos, C., Finkelstein, S., & Sousa, J. M. (2020). Clinical decision support systems for triage in the emergency department using intelligent systems: A review. *Artificial Intelligence in Medicine*, *102*, 101762.

Franceschi, C., Garagnani, P., Morsiani, C., Conte, M., Santoro, A., Grignolio, A., ... Salvioli, S. (2018). The continuum of aging and age-related diseases: Common mechanisms but different rates. *Frontiers in medicine*, *5*, 61.

Froomkin, A. M., Kerr, I., & Pineau, J. (2019). When AIs outperform doctors: confronting the challenges of a tort-induced over-reliance on machine learning. *Arizona Law Review*, *61*, 33.

Galetsi, P., Katsaliaki, K., & Kumar, S. (2020). Big data analytics in health sector: Theoretical framework, techniques and prospects. *International Journal of Information Management*, *50*, 206−216.

Greco, L., Percannella, G., Ritrovato, P., Tortorella, F., & Vento, M. (2020). Trends in IoT based solutions for health care: Moving AI to the edge. *Pattern Recognition Letters*, *135*, 346−353.

Greiwe, J., & Nyenhuis, S. M. (2020). Wearable technology and how this can be implemented into clinical practice. *Current Allergy and Asthma Reports*, *20*, 1−10.

Hartmann, M., Hashmi, U. S., & Imran, A. (2019). Edge computing in smart health care systems: Review, challenges, and research directions. *Transactions on Emerging Telecommunications Technologies*. Available from https://doi.org/10.1002/ett.3710.

Hassan, W. H. (2019). Current research on Internet of Things (IoT) security: A survey. *Computer networks*, *148*, 283−294.

Hassanalieragh, M., Page, A., Soyata, T., Sharma, G., Aktas, M., Mateos, G., ... Andreescu, S. (2015, June). Health monitoring and management using Internet-of-Things (IoT) sensing with cloud-based processing: Opportunities and challenges. In *Proceedings of the IEEE international conference on services computing* (pp. 285−292).

Hassani, F. A., Shi, Q., Wen, F., He, T., Haroun, A., Yang, Y., ... Lee, C. (2020). Smart materials for smart healthcare−moving from sensors and actuators to self-sustained nanoenergy nanosystems. *Smart Materials in Medicine*.

Hossain, M., Islam, S. R., Ali, F., Kwak, K. S., & Hasan, R. (2018). An internet of things-based health prescription assistant and its security system design. *Future Generation Computer Systems*, *82*, 422−439.

Islam, S. R., Kwak, D., Kabir, M. H., Hossain, M., & Kwak, K. S. (2015). The internet of things for health care: A comprehensive survey. *IEEE Access*, *3*, 678−708.

Jamil, F., Ahmad, S., Iqbal, N., & Kim, D. H. (2020). Towards a remote monitoring of patient vital signs based on IoT-based blockchain integrity management platforms in smart hospitals. *Sensors*, *20*(8), 2195.

Jana, S., Bhaumik, J., & Maiti, M. K. (2013). Survey on lightweight block cipher. *International Journal of Soft Computing and Engineering*, *3*(5), 183−187.

Jokanović, V. (2020). Smart healthcare in smart cities. Towards Smart World: Homes to Cities Using Internet of Things, 45.

Kadhim, K. T., Alsahlany, A. M., Wadi, S. M., & Kadhum, H. T. (2020). An overview of patient's health status monitoring system based on Internet of Things (IoT). *Wireless Personal Communications*, *114*, 2235−2262.

Kazeem Moses, A., Joseph Bamidele, A., Roseline Oluwaseun, O., Misra, S., & Abidemi Emmanuel, A. (2021). Applicability of MMRR load balancing algorithm in cloud computing. *International Journal of Computer Mathematics: Computer Systems Theory*, *6*(1), 7−20.

Khan, S., Parkinson, S., & Qin, Y. (2017). Fog computing security: a review of current applications and security solutions. *Journal of Cloud Computing*, *6*(1), 1−22.

Li, J., Cai, J., Khan, F., Rehman, A. U., Balasubramaniam, V., Sun, J., & Venu, P. (2020). A secured framework for SDN-based edge computing in IoT-enabled healthcare system. *IEEE Access*, *8*, 135479−135490.

Luk, C. Y. (2018). *The impact of digital health on traditional healthcare systems and doctor-patient relationships: The case study of Singapore. Innovative perspectives on public administration in the digital age* (pp. 143−167). IGI Global.

Majumder, S., Aghayi, E., Noferesti, M., Memarzadeh-Tehran, H., Mondal, T., Pang, Z., & Deen, M. J. (2017). Smart homes for elderly healthcare—Recent advances and research challenges. *Sensors*, *17*(11), 2496.

Marks, M. (2020). Emergent medical data: Health Information inferred by artificial intelligence. *UC Irvine Law Review.*

Marques, G., Bhoi, A. K., Albuquerque, V. H. C. d., & Hareesha, K. S. (Eds.), (2021). *IoT in healthcare and ambient assisted living.* Springer.

Marques, G., Miranda, N., Kumar Bhoi, A., Garcia-Zapirain, B., Hamrioui, S., & de la Torre Díez, I. (2020). Internet of things and enhanced living environments: Measuring and mapping air quality using cyber-physical systems and mobile computing technologies. *Sensors, 20*(3), 720.

Mehrabi, M., You, D., Latzko, V., Salah, H., Reisslein, M., & Fitzek, F. H. (2019). Device-enhanced MEC: Multi-access edge computing (MEC) aided by end device computation and caching: A survey. *IEEE Access, 7,* 166079–166108.

Meneghello, F., Calore, M., Zucchetto, D., Polese, M., & Zanella, A. (2019). IoT: Internet of threats? A survey of practical security vulnerabilities in real IoT devices. *IEEE Internet of Things Journal, 6*(5), 8182–8201.

Mishra, S., Mallick, P. K., Tripathy, H. K., Bhoi, A. K., & González-Briones, A. (2020). Performance evaluation of a proposed machine learning model for chronic disease datasets using an integrated attribute evaluator and an improved decision tree classifier. *Applied Sciences, 10*(22), 8137.

Mishra, S., Tripathy, H. K., Mallick, P. K., Bhoi, A. K., & Barsocchi, P. (2020). EAGA-MLP—An enhanced and adaptive hybrid classification model for diabetes diagnosis. *Sensors, 20*(14), 4036.

Mohd, B. J., Hayajneh, T., & Vasilakos, A. V. (2015). A survey on lightweight block ciphers for low-resource devices: Comparative study and open issues. *Journal of Network and Computer Applications, 58,* 73–93.

Monje, M. H., Foffani, G., Obeso, J., & Sánchez-Ferro, Á. (2019). New sensor and wearable technologies to aid in the diagnosis and treatment monitoring of Parkinson's disease. *Annual Review of Biomedical Engineering, 21,* 111–143.

Mshali, H., Lemlouma, T., Moloney, M., & Magoni, D. (2018). A survey on health monitoring systems for health smart homes. *International Journal of Industrial Ergonomics, 66,* 26–56.

National Academies of Sciences, Engineering, and Medicine. (2020). *Opportunities to improve opioid use disorder and infectious disease services: Integrating responses to a dual epidemic.*

Noshina, T., Ayesha, Q., Muhammad, A., & Farrukh, A.K. (2020). Blockchain and smart healthcare security: A survey. *Procedia Computer Science.*

Oladele, T. O., Ogundokun, R. O., Awotunde, J. B., Adebiyi, M. O., & Adeniyi, J. K. (2020, July). Diagmal: A malaria coactive neuro-fuzzy expert system. *Lecture notes in computer science (including subseries lecture notes in artificial intelligence and lecture notes in bioinformatics),* 12254 LNCS, pp. 428–441.

Omoniwa, B., Hussain, R., Javed, M. A., Bouk, S. H., & Malik, S. A. (2018). Fog/edge computing-based IoT (FECIoT): Architecture, applications, and research issues. *IEEE Internet of Things Journal, 6*(3), 4118–4149.

Oniani, S., Marques, G., Barnovi, S., Pires, I. M., & Bhoi, A. K. (2021). *Artificial intelligence for Internet of Things and enhanced medical systems. Bio-inspired neurocomputing* (pp. 43–59). Singapore: Springer.

Papa, A., Mital, M., Pisano, P., & Del Giudice, M. (2020). E-health and wellbeing monitoring using smart healthcare devices: An empirical investigation. *Technological Forecasting and Social Change, 153,* 119226.

Peters, B. S., Armijo, P. R., Krause, C., Choudhury, S. A., & Oleynikov, D. (2018). Review of emerging surgical robotic technology. *Surgical Endoscopy, 32*(4), 1636–1655.

Pisani, F., de Oliveira, F. M. C., de Souza Gama, E., Immich, R., Bittencourt, L. F., & Borin, E. (2019). Fog computing on constrained devices: Paving the way for the future IoT. *Advances in Edge Computing: Massive Parallel Processing and Applications, 35,* 22–60.

Poitras, M. E., Hudon, C., Godbout, I., Bujold, M., Pluye, P., Vaillancourt, V. T., ... Légaré, F. (2020). Decisional needs assessment of patients with complex care needs in primary care. *Journal of Evaluation in Clinical Practice, 26*(2), 489–502.

Porambage, P., Okwuibe, J., Liyanage, M., Ylianttila, M., & Taleb, T. (2018). Survey on multi-access edge computing for internet of things realization. *IEEE Communications Surveys & Tutorials*, *20*(4), 2961−2991.

Pramanik, M., Pradhan, R., Nandy, P., Bhoi, A. K., & Barsocchi, P. (2021). Machine learning methods with decision forests for Parkinson's detection. *Applied Sciences*, *11*(2), 581.

Pramanik, M. I., Lau, R. Y., Demirkan, H., & Azad, M. A. K. (2017). Smart health: Big data enabled health paradigm within smart cities. *Expert Systems with Applications*, *87*, 370−383.

Pramanik, P. K. D., Upadhyaya, B. K., Pal, S., & Pal, T. (2019). *Internet of things, smart sensors, and pervasive systems: Enabling connected and pervasive healthcare. Healthcare data analytics and management* (pp. 1−58). Academic Press.

Rahmani, A. M., Gia, T. N., Negash, B., Anzanpour, A., Azimi, I., Jiang, M., & Liljeberg, P. (2018). Exploiting smart e-Health gateways at the edge of healthcare Internet-of-Things: A fog computing approach. *Future Generation Computer Systems*, *78*, 641−658.

Ren, J., Zhang, D., He, S., Zhang, Y., & Li, T. (2019). A survey on end-edge-cloud orchestrated network computing paradigms: transparent computing, mobile edge computing, fog computing, and cloudlet. *ACM Computing Surveys (CSUR)*, *52*(6), 1−36.

Roski, J., Bo-Linn, G. W., & Andrews, T. A. (2014). Creating value in health care through big data: opportunities and policy implications. *Health Affairs*, *33*(7), 1115−1122.

Saposnik, G., Redelmeier, D., Ruff, C. C., & Tobler, P. N. (2016). Cognitive biases associated with medical decisions: A systematic review. *BMC Medical Informatics and Decision Making*, *16*(1), 1−14.

Sayeed, A., Mohanty, S. P., Kougianos, E., Yanambaka, V. P., & Zaveri, H. (2018, December). A robust and fast seizure detector for IoT edge. In *Proceedings of the IEEE international symposium on smart electronic systems (iSES)(Formerly iNiS)* (pp. 156−160). IEEE.

Shah, J. L., Bhat, H. F., & Khan, A. I. (2020). Integration of cloud and IoT for smart e-healthcare. In *Healthcare paradigms in the Internet of Things ecosystem* (pp. 101−136). Academic Press.

Tariq, N., Asim, M., Al-Obeidat, F., Zubair Farooqi, M., Baker, T., Hammoudeh, M., & Ghafir, I. (2019). The security of big data in fog-enabled IoT applications including blockchain: A survey. *Sensors*, *19*(8), 1788.

Tausif, M., Ferzund, J., Jabbar, S., & Shahzadi, R. (2017). Towards designing efficient lightweight ciphers for internet of things. *KSII Transactions on Internet and Information Systems (TIIS)*, *11*(8), 4006−4024.

Tian, S., Yang, W., Le Grange, J. M., Wang, P., Huang, W., & Ye, Z. (2019). Smart healthcare: making medical care more intelligent. *Global Health Journal*, *3*(3), 62−65.

Ullah, K., Shah, M. A., & Zhang, S. (2016, January). Effective ways to use Internet of Things in the field of medical and smart health care. In *Proceedings of the international conference on intelligent systems engineering (ICISE)*, (pp. 372−379). IEEE.

Varshney, G., & Gupta, H. (2017). A security framework for IoT devices against wireless threats. In *Proceedings of the second international conference on telecommunication and networks (TEL-NET)*.

Vutha, A. Z. (2020). *Dealing with performance-based health technology assessment outcomes for medical devices: A South African perspective*. (Doctoral dissertation), University of Pretoria.

Zeadally, S., & Bello, O. (2019). Harnessing the power of Internet of Things based connectivity to improve healthcare. *Internet of Things*, 100074.

Deep learning approaches for the cardiovascular disease diagnosis using smartphone

Abdulhamit Subasi[1], Elina Kontio[2] and Mojtaba Jafaritadi[2,3]

[1]*Faculty of Medicine, Institute of Biomedicine, University of Turku, Turku, Finland* [2]*Faculty of Engineering and Business, School of Information and Communications Technology, Turku University of Applied Sciences, Turku, Finland* [3]*Faculty of Technology, Department of Computing, University of Turku, Turku, Finland*

7.1 Introduction

Machine learning (ML) is an artificial intelligence (AI) subdivision, which allows computer algorithms to learn from input / training data. In the development of ML algorithms in AI, neural networks (NN) play a crucial role. NNs have been inspired by the function of biological neurons in the human brain. Each neuron is a part of a processing element, and an interconnection of those neurons contributes to tremendous computational capacity that can solve complex problems. The computational model for NN opened the way for studies on NN to follow two paths, one for biological processes in the brain and the other for the use of NN to AI. NN were designed specifically to solve complex problems, which could not fit with rule-based methods and algorithms. Some information may be absent in real world problems, or all the rules may not be known. In AI and, thus, in NN, heuristics play a crucial role (Vasuki & Govindaraju, 2017).

NN are mostly utilized for the classification of signals and images and are trained with the learning capacity to diagnose disease. There is a collection of training data available and relationships between training inputs and the desired outputs (or pattern classes) must be fed into the network. On these known data, the network is trained and learns to recognize/interpret new data. ML methods have been developed to make learning occur, with the concept being learned by the algorithm through training. The data or patterns are supplied as inputs, the outputs are defined, and the algorithm learns to explore the connection between inputs and outputs. More hidden layers are needed when the problem is complicated, such as in signal and image recognition for disease diagnosis, and this makes the neural network "deep." The performance of the classification is enhanced by hundreds of hidden layers and learning becomes "deep learning" (DL) (Vasuki & Govindaraju, 2017).

DL that is a sub-branch of ML, is one of AI's most commonly employed techniques. Usually, DL is a technique for extracting informative features automatically by integrating several linear and nonlinear processing elements in a deep architecture (Deng & Yu, 2014). Artificial neural networks (ANNs) made quick strides in the last decade and the name of DL began to be employed. Numerous DL models are employed based on the type of application. The leading and most popular architectures in medical image processing are Convolutional Neural Networks (CNNs), which is a form of DL, and its derivatives (Kharazmi, Zheng, Lui, Jane Wang, & Lee, 2018; Premaladha & Ravichandran, 2016; Wang et al., 2018). Hybrid deep networks refer to systems, which are built by merging various DL models to achieve better performance. CNN and its variations, autoencoders, deep belief neural networks, Recurrent neural networks (RNN), and Long Short Term Memory (LSTM) are the most utilized and popular DL models in the diagnosis and treatment of diseases. Traditional ML methods are primarily focused on predefined hand-crafted features and are normally developed for particular issues only. However, DL algorithms do not need an explicit description of features, but instead employ data and incorporate features that are high-dimensional and difficult to understand for achieving results. Several approaches, such as k nearest neighbor (k-NN), support vector machines (SVM), random forest, and Gaussian mixture models that are among the traditional ML algorithms, were replaced by DL algorithms because of the superior performance and achievements of DL techniques. These traditional approaches are not adjustable, robust, and computationally intense, since they are manually developed for the diagnosis and treatment of diseases (Pacal, Karaboga, Basturk, Akay, & Nalbantoglu, 2020).

The healthcare industry is facing different concerns associated with disease diagnosis and cost-effective delivery of services (Awan, Ali, Aadil, & Qureshi, 2018). One of the crucial needs of a healthcare system is to provide the patient with adequate services by the utilization of medical records, lifestyle habits, and any variability of molecular traits. These systems have faced numerous challenges relevant to data processing, the identification of information, data retrieval and decision making with the rapid growth of emerging technology. Several intelligent instruments have been developed based on AI and data-driven methods to overcome these challenges. In order to create mutual information for further exploration and quantitative analysis, these methods include an effort to connect several data sources. In addition, for particular diseases such as brain tumors and coronary heart diseases, multiple prediction-based approaches have been designed (Alanazi, Abdullah, Qureshi, & Ismail, 2018). However, because of their temporal dependencies, inconsistencies, interpretability, high-dimensionality and diversity, there are many difficulties to address in the sense of biomedical data analysis for disease diagnosis and treatment. For disease diagnosis and treatment, modern and automated technology-enabled Smart Healthcare Platforms have been employed. Data are relevant to demographic details of patients, laboratory test results, image-based radiological documents, prescriptions, medication dose records, and dietician appointment records in these Smart Healthcare Platforms. Healthcare services are delivered by Smart Healthcare Platforms utilizing smart phone/watch and connected technologies. Patient information is gathered in a file for the purpose of disease diagnosis and treatment and offering timely care, in particular for the areas where urgent intervention is a challenge. Despite the existence of these modern communication technologies, because of the scalability and various technology issues, digital development in the healthcare sector also does not fulfill all requirements. A more efficient Smart Healthcare infrastructure has to be built to make the service simpler for providers, patients and healthcare workers. The vast volume of associated medical data is contained in the cloud, so it is a major challenge to decide if we can utilize those data to achieve better forecasts (Qureshi, Din, Jeon, & Piccialli, 2020). It is desirable for forecasting models to examine the data and forecast patient statistics. This chapter introduces and explores the use of AI approaches to detect diseases such as heart attacks, brain cancers, diabetes, cardiac disease, and prediction of epileptic seizures in this perspective.

In modern societies, healthcare is one of the most critical issues, since the life quality of people depends mainly on it (Bagga & Hans, 2015). The healthcare industry, though, is extremely diverse, broadly spread and decentralized. The provision of sufficient medical care from a professional point of view needs access to reliable patient records that is rarely accessible when and where it is required (Grimson et al., 2001). In addition, the large variance in the test-ordering for diagnostic aims implies the need for an adequate and suitable range of test data (Daniels & Schroeder, 1977; Wennberg, 1984). Smellie, Galloway, Chinn, and Gedling (2002) expanded this claim by implying that the broad discrepancies found in the request for general disease diagnosis and treatment are primarily due to individual changes in clinical practice and are thus theoretically prone to changes by making doctors' decisions more reliably and better informed (Stuart, Crooks, & Porton, 2002). Medical data thus also includes a vast variety of heterogeneous factors, obtained from multiple databases, such as populations, history of illness, medications, biomarkers, medical records, allergies, or genetic markers, each of which gives a distinct partial view of the status of the patient. In addition, among the above references, statistical properties are fundamentally distinct. This causes contribute to complications and factual errors in the identification, diagnosis and treatment of the disease, thereby preventing patients from providing proper care (Dick, Steen, & Detmer, 1997). There is also a strong need for an efficient and rigorous approach that allows for the timely identification, diagnosis and treatment of diseases and can be utilized by physicians as a decision support system (Zhuang, Churilov, Burstein, & Sikaris, 2009). The statistical, computational and medical fields are now facing the challenge of exploring new methods for modeling disease diagnosis and treatment, as existing paradigms struggle to answer any of this situation (Huang, Chen, & Lee, 2007). This necessity is closely connected with advances in other areas, such as ML, AI or Big Data. As the quantity of medical data recorded and stored digitally is immense and rapidly expanding, data management and interpretation is similarly evolving to transform this massive data into information and knowledge, which assists them to accomplish their goals (Caballé, Castillo-Sequera, Gómez-Pulido, Gómez-Pulido, & Polo-Luque, 2020).

Advanced methods for explaining these structural variations in data need to be used. Within an area known as AI, several techniques were invented. AI is a component of computer science which intends to make computers smarter. Learning is one of the fundamental criteria for any intelligent behavior. Presently, most scholars believe that without learning, there is no intelligence. A learning challenge can be identified as the complexity of enhancing certain measures by some level of training experience while performing certain tasks. In turn, ML has appeared within AI as a tool for creating dataset analysis algorithms (Jordan & Mitchell, 2015). Today, ML offers many important methods for intelligent information processing. Furthermore, its technology is presently well adapted for the study of medical data and, particularly, a wide variety of medical diagnostic work has been carried out on small-specialized diagnostic problems (Kononenko, 2001), where initial implementations of ML have been found. ML algorithms, which are helpful instruments in clinical diagnosis, have been successfully used for the disease diagnosis and treatment (Sriram, Rao, Narayana, & Kaladhar, 2016). Evidently, several ML methods perform very well on a wide range of critical issues. Nevertheless, as they get incredibly complicated and the curse of dimensionality in the data is high, they failed in tackling the key issues in AI. DL then appeared as a special form of ML. The development of DL was therefore inspired and planned to circumvent the inability of conventional algorithms to deal with high-dimensional data and to learn complex functions in high-dimensional spaces (Caballé et al., 2020; Goodfellow, Bengio, Courville, & Bengio, 2016).

Every year, cardiovascular diseases (CVDs) claim more than 17 million death worldwide (Cardiovascular diseases, 2021), so early detection of CVDs for faster recovery becomes an essential task. One-lead portable electrocardiograph (ECG) instruments such as chest patches and wristbands were invented, implemented and commonly used to achieve this role. ECG is also a non-invasive technique that cardiologists usually use in their clinical practice. It helps individuals to track themselves in real time, supplying the signal that holds the data about whether or not CVDs are occurring. A natural challenge then emerges about how to recognize heart rhythm abnormalities immediately, because arrhythmias can need early diagnosis. It has been receiving more and more scientific interest in the area of medical diagnosis due to the high mortality rate of CVDs (Li et al., 2020).

The most frequent cardiac arrhythmia is atrial fibrillation (AFib) (Colilla et al., 2013), but its paroxysmal nature and low relationship with symptoms make it extremely challenging to diagnose and treat. Actually, for one-quarter of patients with asymptomatic AFib, stroke is the first clinical appearance of AFib (Jaakkola et al., 2016). In response, a range of remote tracking systems have been more commonly used for diagnosis and treatment of AFib. These include wearable heart sensors, medical wearables, direct-to-consumer applications, and smartphone apps (Hickey, Riga, Mitha, & Reading, 2018). Turchioe et al. (2020) focused on testing free mobile apps so access to health services are not obstacles to their acceptance and use. There is growing awareness that photoplethysmography (PPG) can be used for heart disease tracking, an affordable and readily available technology used in all smartphones. PPG sensors sense variations in the concentration of tissue blood originating from peripheral pulses (Allen, 2007). As a light source flash from a mobile camera illuminates body tissues in the finger, PPG waveforms are generated inside smartphones, and a photodetector, which is camera, measures changes in light intensity across the tissue (Allen, 2007). PPG might be utilized to scan big portions of the population for irregular heart rate detection without any cost (Turchioe et al., 2020).

To avoid cardio-embolic strokes, a prompt diagnosis of AFib is important. Appropriate and cost-effective screening methods for asymptomatic AFib are yet to be adopted. Mechanical heart activity with gyroscopes and accelerometers of smartphones can be recorded in mechanocardiography (MCG). Seismocardiography is a method previously established in which mechanical cardiac activity is measured with accelerometers (Inan et al., 2015). Gyrocardiography is a novel modality introduced by Jafari Tadi et al. (2017) in which gyroscopes are utilized to measure mechanical cardiac signals. Jointly, these two methods constitute the concept of MCG. The potential of an advanced mobile MCG application to distinguish AFib from sinus rhythm (SR) was tested by Jaakkola et al. (2018a, 2018b). From a total of 300 hospitalized subjects (150 in AFib and 150 in SR), a 3-minute MCG recording was acquired with a smartphone put on their chest during the tracking of rhythm with telemetry electrocardiography (ECG). The rhythm of the MCG recordings as either AFib or SR was classified by proposed algorithm. Compared to ECG interpretation, the precision of the MCG algorithm to distinguish AFib from SR was tested by two independent cardiologists. Without any external hardware for accurate and usable AF scanning, the smartphone MCG accurately detects AFib (Jaakkola et al., 2018a, 2018b).

This chapter is organized as follows: disease diagnosis and treatment are discussed in Section 7.2. Section 7.3 includes DL approaches for the disease diagnosis and treatment, which includes a description of the DL algorithms. Section 7.4 delivers details of the case study for detection of atrial fibrillation using smartphone. Section 7.5 is the discussion. Finally, in Section 7.6, conclusions are given.

7.2 **Disease diagnosis and treatment**

In precision medicine, interpretation of healthcare data can play a crucial role. For instance, by considering many forms of patient data, including environment, genomics variants, imaging genomics, existing medications, and lifestyle, customized cancer care aims to offer the right treatment for the right patient. New technology such as medical imaging, genomics, and lifetime tracking technologies have created vast amount of healthcare data over the past decade, allowing researchers to provide patients with improved therapies. Our knowledge of diseases, and how we can handle patients, is still inadequate, considering this massive volume of data. In order to interpret such a big volume of data, due to the data complexity, the use of ML and AI techniques (Kalantari et al., 2018), like the DNN, has become more desirable. Particularly, it is a key issue to implement accurate diagnostic tools focused on data-driven techniques and ML approaches to define the connections between all various forms of patient data. A wide variety of AI and ML methods have been utilized over the past decade to accurately analyze huge healthcare data. For instance, in cardiac disease detection, a logistic regression-based predictive approach was developed for automatic detection of heart disease (Kumar & Gandhi, 2018). In medical imaging, ML has been introduced to deliver an automated discovery of object characteristics (Ravì et al., 2016). DNN-based methods are appealing a lot of interest among various ML models, particularly in the analysis of large datasets. DL approaches are feature learning methods where data is filtered through a cascade of multiple layers. As they process large-scale data, DNN models become more and more precise, helping them to surpass many traditional models of ML. These methods are becoming stimulating methods to evaluate healthcare data, provided the success of DL approaches in various fields and their steady continuous methodological development. Thorough the utilization of DL approaches on healthcare and biomedical data, a broad range of projects have been carried out (Shamshirband, Fathi, Dehzangi, Chronopoulos, & Alinejad-Rokny, 2020).

A CNN for region classification of semantically coherent tissues was designed by Dubrovina, Kisilev, Ginsburg, Hashoul, and Kimmel (2018) and has been utilized for mass detection of mammograms. The findings suggest the computation and classification's high accuracy. Jiao, Gao, Wang, and Li (2018) suggested the classification of DL-masses where CNN layers organize simple discriminatory representation to improve the classification accuracy. Misclassified cases are placed in an "error book" during each training period. If the validation error is not minimized in a loop, a stored example is sent to the training set. A CNN was also used for other breast cancer tomosynthesis studies (Samala et al., 2016). For brain-MRI using the multi-scale CNN, a tissue segmentation approach was suggested by Moeskops et al. (2018). With varying degrees of abnormality, the procedure was tested. The findings indicate that it can correctly segment brain tissues. In the clinical sector, model transparency is an important topic impacting real world medical decision making and patient care in terms of predictions (Nie et al., 2015). A new DL technique for brain tumor segmentation by combining Fully Convolutional Neural Networks (FCNN) and Conditional Random Fields (CRF) was suggested by Zhao et al. (2018). The purpose of such integration was to enhance the robustness of the scheme. The approach includes four phases, which include preprocessing, segmenting image slices with embedded FCNNs and CRF-RNN, feature extraction and classification using DL models. The system of CRF was also formulated as RNN. In three steps, the model was educated. First, FCNN was educated in the use of image patches. Second, through the use of image slices, CRF-RNN was trained. The image slices were eventually used to fine-tune the whole network (Shamshirband et al., 2020).

Diabetic retinopathy (Dr) is one of the major reasons of blindness in the working-age population. It is one of the diabetes problems that is most anticipated. The fundamental issue of Dr is that, at advanced stages, it becomes incurable, so early diagnosis is critical. However, owing to the number of patients and the limited number of skilled specialists, this involves remarkable complexity in the health care sector. This motivated the requirement to build an automatic diagnostic tool to help with early diagnosis of Dr. Several efforts in this direction have been made, and several methods have been proposed based on hand-crafted features that have revealed encouraging competence in identifying Dr regions in retinal fundus images. For traditional ML approaches used for Dr diagnosis, hand-crafted features are widely utilized. Expert experience, however, is a requirement for hand-crafted features, and it involves extensive investigation into different options and exhaustive parameter adjustments to select the right features. In comparison, approaches built on hand-crafted features cannot generalize well. The existence of massive datasets and the immense computational power provided by graphics processing units (GPUs) have inspired research on DL algorithms that demonstrated excellent success in different computer vision tasks and achieved a significant success over conventional hand-crafted-based approaches in recent years. Several DL-based algorithms were also developed to analyze retinal fundus images for different tasks in order to implement automated computer-aided diagnostic systems for Dr (Asiri, Hussain, Al, & Alzaidi, 2019).

Neuroscientists and physicians also use noninvasive imaging methods in current neuroscience and clinical practice to test hypotheses and DL models, analyze brain functions, and identify brain disorders. Functional magnetic resonance imaging (fMRI) is one of the most widely utilized imaging technique, which can be employed to explain the functions of the human brain and to identify and diagnose brain diseases. In order to better understand fMRI data, developments in AI and the improvements of DL methods achieved remarkable performance. In order to analyze fMRI images, DL approaches quickly became the state of the art and have achieved better performance in different fMRI applications. DL is typically viewed as an end-to-end learning mechanism, which can mitigate the needs of feature engineering and thereby diminish the requirements for domain awareness to some degree. fMRI data may be considered as images, time series, or image series within the paradigm of DL. Therefore, it is possible to create different DL models, such as CNNs, RNN, or a mixture of both, to utilize fMRI data for various tasks (Yin, Li, & Wu, 2020).

An ensemble of DL was used by the authors in Moeskops et al. (2018) to classify three types of cancers, specifically breast invasive carcinoma, stomach adenocarcinoma, and lung adenocarcinoma. They chose the significant genes utilizing distinguished gene expression study to create this model. Five CNN classifiers were then used to train these chosen genes and assembled to achieve the final result (Xiao, Wu, Lin, & Zhao, 2018). Sharma, Zerbe, Klempert, Hellwich, and Hufnagl (2017) recommended a system of automated classification of gastric carcinoma using CNN. Three multichannel ROI-based deep structured algorithms, specifically DBN, CNN, and stacked denoising AE for lung cancer diagnosis, were introduced in another study (Sun, Zheng, & Qian, 2017). In (Timmis et al., 2020), in order to analyze encephalogram signals for epilepsy diagnosis, the authors presented a new computer-aided diagnostic method employing CNN. This CNN approach includes 13 layers to construct a complex and powerful seizure prediction models and is compared to other ML models (Shamshirband et al., 2020). Choi, Schuetz, Stewart, and Sun (2017) employed the most appropriate application of DL in the diagnosis of cardiac disease. They introduced an RNN model of gated time-stamped events.

In the present research scenario, the incidence of chronic kidney disease (CKD) rises per year. The CKD prediction is one of the areas needs for more therapy where, due to their high precision classification capacity, ML methods become crucial in disease diagnosis. In the last decade, the performance of prediction models is contingent on the correct usage of feature extraction and selection algorithms to minimize the size of the results. Ma, Sun, Liu, and Jing (2020) have proposed the Heterogeneous Modified Artificial Neural Network (HMANN) on the Internet of Medical Things (IoMT) platform for the quick identification, segmentation, and identification of chronic renal failure. The developed algorithm operates on the basis of an ultrasound image that is signified as a pre-processing phase to segment the region of interest of kidney. The suggested HMANN approach achieves high precision in kidney segmentation and decreases the time to define the contour considerably.

To boost Parkinson's disease, a DL-based FPCIT SPECT analysis method was suggested by Choi, Ha, Im, Paek, and Lee (2017). In the presented research, it was shown that the inter-observer variability problem can be solved by this approach. Recently, Gunduz (2019) employed DL for the characterization of Parkinson's disease on the range of vocal (speech) characteristics. Nine-layered CNN is utilized to solve this problem and achieved reasonable performance.

In order to create a ML platform, which optimize clinical decision frameworks, an unsupervised deep feature learning algorithm was suggested by Miotto, Li, Kidd, and Dudley (2016). To distinguish red lesions in fundus images, an ensemble of domain information and DL was offered by Orlando, Prokofyeva, Del Fresno, and Blaschko (2018). Hand-crafted features have been fused in the learning phase of CNN. It was recorded that the efficiency of this feature fusion based classifier was higher than that of other individual classifiers (Shamshirband et al., 2020).

One of the worst form of cancer is pancreatic cancer and the prognosis in the current scenario is very low. Computer-aided scanning (CAD), detection and functional analyses of radiology images such as CT and MRI also include automated pancreatic tumor image segmentation. Via these approaches, tumor classification will also help to monitor, anticipate and support personalized therapy as part of successful treatment without cancer incursions. ANN have provided positive results for correct segmentation of pancreatic images today. For pancreatic tumor identification, Xuan and You (2020) suggested a DL-based Hierarchical Convolutional Neural Network (HCNN). A RNN is given to solve the problem of segmentation of spatial difference across slices of neighboring images. By optimizing the smoothness and form, the recurrent neural network produces CNN outcomes and fine-tunes its segmentation. In addition, the targets of HCNN configurations and training for the success of pancreatic tumor image segmentation were demonstrated. The experimental findings revealed that the suggested solution will increase the classifier's efficiency and decrease the cost of the network for IoMT.

Automatic diagnosis of cardiac disease is one of the most important and challenging healthcare issues in the world. Cardiac disease affects the functioning of blood vessels and triggers infections of the coronary artery that damage the patient's body, especially in adults and the elderly. In order to successfully cure cardiac patients before a heart attack or stroke will happen, early detection of cardiac disease is extremely significant. By performing diagnostic examinations and using wearable monitors, cardiovascular disorders may be detected. Extracting useful risk factors for cardiac disease from electronic medical tests, however, is challenging when doctors aim to diagnose patients rapidly and reliably. Wearable sensors are also currently used to monitor regularly the body of patient to identify cardiac failure. Therefore, an intelligent framework is required, which can automatically combine the information collected from both sensor data and electronic medical records and which can interpret the data recorded to detect the hidden signs of cardiac failure and diagnose cardiac failure before a heart attack happens. Now, using AI approaches and hybrid models, numerous schemes have been developed to forecast and diagnose CVD. Ali et al. (2020) suggested a novel, smart heart disease prediction tracking framework for healthcare utilizing feature fusion and ensemble DL as a whole.

A simple and accurate procedure for the diagnosis of cardiac arrhythmias is electrocardiograms (ECGs). In clinical practice, patients with multiple and overlapping arrhythmias diagnosed by ECG are frequently used. ECGs that are inappropriately interpreted can lead to improper clinical decisions. Additionally, it is impossible even for an expert cardiologist to correctly interpret the ECG for a patient with instantaneous arrhythmias. About 300 million ECGs are performed yearly worldwide and delivering the right diagnosis can be extremely problematic in middle-income and low-income countries because skilled cardiologists are limited. There is therefore an immediate requirement for an integrated multilabel ECG based cardiac arrhythmia detection system supported by AI methods. Several experiments have been aimed at detecting arrhythmias using DL approaches over the past two years (Attia et al., 2019a, 2019b; Faust, Hagiwara, Hong, Lih, & Acharya, 2018; Kamaleswaran, Mahajan, & Akbilgic, 2018; Sannino & De, 2018; Zhu et al., 2020) and several benchmark datasets have been revealed to be effective for analyzing DL arrhythmia models on the basis of quantitative performance measurements. Researches on heart arrhythmia detection have demonstrated that complex anomalies, such as atrial fibrillation, may be detected by utilizing a robustly developed DL model (Attia et al., 2019a, 2019b; Yildirim, 2018; Yıldırım, Pławiak, Tan, & Acharya, 2018). Zhu et al. (2020) attempted to solve the issue related to patients with more than one heart disease, which is often identified concurrently during the clinical diagnosis period, by analyzing the ECG diagnosis problem at a multilabel stage. The goal was to establish a general form of automated diagnosis of cardiac arrhythmia disorders using ECG tests and to develop an AI-based multilabel ECG diagnosis model covering nearly all forms of arrhythmias.

7.3 Deep learning approaches for the disease diagnosis and treatment

DL is a subfield of ML, which requires networks able to learn from unstructured/unlabeled data. Deep neural learning or deep neural networks is a subclass of ML techniques, which progressively extract sophisticated features from the raw data using multiple layers. It is founded on ANNs with learning from representation. Learning may be unsupervised, semisupervised or supervised. DL draws from vast amounts of unstructured data. ANN, with neuron nodes joined together, are analogous to the human brain. Traditional programs usually interpret data in a linear fashion, but data can be analyzed in a nonlinear way by DL models. Each level learns to change the input data into a much more complex presentation through DL. Advances in the field of ML have shown a great potential to use DL models in disease diagnosis (Gao et al., 2019).

DL has shown its outstanding success in providing the solution to problems associated with pattern recognition, image processing, disease diagnosis and classification. DL is becoming a critical factor in identifying specific patterns in large volumes of gene expression datasets due to the benefit of large amounts of RNA-Seq data and freely accessible microarray gene expression.

Classifying cancer cells and normal cells on the basis of patterns of gene expression is a huge obstacle. In the field of bioinformatics, supervised ML methods are utilized to solve this problem (Ahn et al., 2018).

7.3.1 Artificial neural networks

A theoretical model, which is inspired by the structure of the human brain is a neural network. A large amount of nerve cells, or neurons, make up the human brain. Neurons have a basic three-part configuration that consists of a cell body, a series of dendrites called fibers, and a single long fiber named an axon. The axons and the dendrites derive from the body of the cell, and the dendrites of one neuron are linked to the axons of other neurons. The dendrites act as input channels to the neuron and receive signals sent from other neurons along their axons. The axon serves as a neuron's output channel, and thus the signals transmitted along the axon are processed as inputs by other neurons whose dendrites are linked to the axon. The neuron transmits an electrical impulse, termed as action potential, through its axon to the other neurons, which are connected to it, if the incoming stimuli are big enough. Therefore, a neuron functions as an all-or-no switch, which takes in a set of inputs and either outputs an action potential or no output. A substantial simplification of biological fact is this description of the human brain, but it captures the key points required to explain the analogy between the human brain structure and the theoretical models called NN. These points of comparison are: (1) the brain consists of a huge number of interconnected basic elements known as neurons; (2) the operation of the brain can be explained as data transmission, encoded as high or low electrical signals, or activation potentials, scattered across the neuron network; and (3) each neuron receives a series of stimuli from its neighbors and maps these inputs to either a low or high value output. All neural network models have these aspects (Kelleher, 2019).

7.3.2 Deep learning

It has been observed that NN are ideally adapted for the application of ML. There is one input layer, one output layer and two to three hidden layers in conventional neural networks. There is one input layer, one output layer and several hidden layers for deep neural networks. The more hidden layers there are, the stronger the network is. The layers are interconnected, with the input of the current layer being the output of the previous layer. The inputs/outputs are weighted, and the network efficiency is determined by the weights. The network training includes obtaining the suitable weights for the different layers. Deep networks need higher computational power, computation speed, a big repository and parallel processing of the suitable software (Vasuki & Govindaraju, 2017).

Neurons are structured in layers with links between the layers. Signals are passed, after processing, through the layers. Signals often propagate in the forward direction from the input layer to the output layer, and such networks are termed feed-forward networks. Signals propagate in the opposite direction in such networks, leading to recurrent networks and the development of the back propagation algorithm. Both layers of neurons other than input and output are named 'hidden' and, number of hidden layers varies based on the problem at hand. It becomes a deep neural network when the number of hidden layer is greater than two or three. These deep neural networks can learn how to solve problems, and the learning is known as DL. These deep neural networks and learning algorithms are built to solve complex tasks such as machine vision, pattern recognition, identification of objects, surveillance, classification of images for disease diagnosis, character recognition, speech/speaker recognition, etc. (Vasuki & Govindaraju, 2017).

DL has evolved from AI and ML science. The last two subjects are focused on the new field of ML: machines that learn from examples, and research on the neural network. ML includes designing and testing algorithms that allow a computer to extract (or learn) features from a dataset (sets of examples). We need to know three words to comprehend what ML means: dataset, algorithm, and function (Kelleher, 2019). DL is an AI subfield, which focuses on developing massive models of neural networks, which are capable of making correct data-driven decisions. DL is specifically developed for the problems in which the data is challenging, and massive databases are available. DL is employed in the healthcare industry to process diagnostic images such as X-rays, CT, and MRI scans and to diagnose diseases. DL is now at the heart of self-driving vehicles, in which it is utilized for localization and navigation, motion planning and steering, and awareness of the environment, as well as driver status monitoring (Kelleher, 2019).

A mathematical model that is (loosely) influenced by the structure of the brain is a DL network. To understand DL, it is also helpful to provide an abstract understanding of what a mathematical model is, how model parameters can be set, how we can merge (or compose) models, and how we can use geometry to explain how data is interpreted by a model. The term DL defines a family of neural network models, which have multiple layers of basic data processing units, known as neurons, in the network. NN, which contain several hidden layers of neurons are DL networks. Two are the minimum number of hidden layers expected to be considered deep. Nevertheless, there are more than two hidden layers in most DL networks. The key argument is that the depth of a network is measured by the number of hidden layers, plus the output layer (Kelleher, 2019).

7.3.3 Convolutional Neural Networks

CNNs are built to operate with grid-structured inputs that have robust spatial needs in local areas of the grid. A 2-dimensional image is the clearest example of grid-structured data. Spatial dependencies are often demonstrated by this type of data, since neighboring spatial positions in an image also have identical pixel color values. The multiple colors are captured by an additional dimension, which generates a 3-dimensional amount of input. Thus, the properties of a CNN are contingent on spatial distances between each other. Specific cases of grid-structured data with different kinds of relations between adjacent items may also be treated as other types of sequential data such as text, time series, and sequences. The vast majority of CNN applications concentrate on imaging data, while these networks can still be used for all forms of spatial, temporal, and spatiotemporal data. A significant characteristic of image data shows a certain degree of invariance in localization that is not the case in different grid-structured data types. CNNs, from geographic regions with identical patterns, tend to generate similar feature values. One strength of image data is that it is always possible to explain the effects of particular inputs on the feature illustrations in an intuitive way. A process that is referred to as convolution is a crucial defining feature of CNNs. A convolution process is a dot-product operation between a weight set structured by a grid and related grid-structured inputs received in the input volume from different spatial locations. This sort of procedure is useful for data, such as image data, with a high degree of spatial or other location. CNNs are thus described as networks, which utilize a convolutional operation in at least one layer, even though this operation is used by most CNNs in multiple layers (Aggarwal, 2018).

CNN is composed of an input layer and hundreds of layers for feature recognition. Each of the following three operations is carried out by feature recognition layers: Convolution, Pooling, Rectified Linear Unit (ReLU). Convolution positions the image in the image by convolution filters, which activate certain characteristics. In order to minimize the volume of data to be dealt, pooling conducts nonlinear down sampling. The ReLU preserves positive values and maps negative values to zero. The classification layer is the one before the output layer. It is an N-dimensional output fully connected layer, N being the number of classes to be classified. This layer generates an N-dimensional vector with the probability that each component of the vector belongs to one of the N classes of the input image. To give the categorized output, the Softmax function is employed at the final output layer. Thousands or millions of images have to be fed into the network to achieve precise outcomes. With multiple GPUs running in parallel, higher computing power is achieved (Vasuki & Govindaraju, 2017).

7.4 Case study of a smartphone-based Atrial Fibrillation Detection

CVD, also known as heart disease, is the leading reason for health problems, claiming approximately 30% of total death worldwide (Timmis et al., 2020). The prevalence of heart diseases is rising, mainly due to the growing population of elderly people and the continued pervasiveness of cardiovascular risk factors (Heidenreich et al., 2011). Heart diseases cause immense health and economic burdens, resulting in high risks of morbidity and mortality worldwide (Levenson, Skerrett, & Gaziano, 2002). Early detection of critical heart disorders via advanced smart monitoring systems could be of potential importance in preventing fatal incidents and reduce expensive interventional treatments. It will also lead to a better quality of patient care and savings for the healthcare sector and the entire global society (Bhavnani, Narula, & Sengupta, 2016).

Increased prevalence of CVD incidents has led governments to resume national efforts to reinforce new accessible and affordable preventive strategies (Schwalm, McKee, Huffman, & Yusuf, 2016). With the new detection and prevention policies, many lives are saved, and the burden of immense health crises will decline, which will lead to a better economy, life quality, and prosperity. In addition to preventive screening, novel technologies may provide improved care of existing diseased patients. Therefore, it is a critically important task to seek more effective and innovative disease or patient management solutions by which the quality of patient care will improve, and the costs of healthcare services minimize (Chaudhry et al., 2006).

Atrial fibrillation (AFib) is a common heart arrhythmia, known for its irregular and randomly frequent appearances (Waktare Johan, 2002). During the normal rhythm, or SR, the heart beats with a regular rate ranging from 60 to 100 beats per minute. However, in the AFib condition, the heart muscle beats randomly and in an uncoordinated fashion (Waktare Johan, 2002). AFib is a chronic heart arrhythmia often diagnosed by interpreting electrocardiography (ECG) signals (Kleiger & Senior, 1974). Prevalence of AFib rises with aging and occurs in almost 3% of people older than 40 years and 70% of individuals between 65 and 85 years of age (Chen, Yi, & Cheng, 2018; Feinberg, Blackshear, Laupacis, Kronmal, & Hart, 1995; Go et al., 2001). About 40% of all strokes are due to AFib, and more than 200,000 deaths per year happen in Europe (Martignani, Massaro, Biffi, Ziacchi, & Diemberger, 2020). AFib detection is a challenging task because it can develop infrequent and asymptomatic. Undiagnosed AFib can cause the formation of a blood clot(s), potentially leading to acute strokes, heart failure (HF), and other heart-related complications (Anter Elad & Callans David, 2009). Persistent AFib can lead to substantial morbidity, mortality, and hospitalization costs (Savelieva & Camm, 2000). Active prevention of brain strokes requires continuous monitoring of atrial fibrillation indications.

Today, remarkable advances have been made in the development of smart monitoring devices as they can leverage the recent improvements in biosensing developments, embedded or pervasive health computing, and physiological data analytics (Onodera & Sengoku, 2018). Modern smartphone devices can offer an easy-to-access non-invasive sensing solution for unobtrusive recording of underlying physiological data to diagnose at-risk patients for early intervention. With the gain in popularity of smartphones, it is possible to deploy the device's built-in motion sensors for AFib detection (Hendriks, Gallagher, Middeldorp, & Sanders, 2018).

Recent developments of telecommunication systems, including Internet of Things (IoT) technology, allows connections of various smart devices, enabling seamless transfer of data, along with advanced data analysis and data storage platforms such as cloud computing and fog computing. Smartphone-based AFib detection is one of the innovative application domains in IoT that induces immense interest from the healthcare sector, the research community, and the public division (Kotecha & Kirchhof, 2017). Smartphone AFib apps sought to improve patients care, satisfaction, and cost-efficiency in the healthcare industry.

Current mobile phone screening solutions are based on ECG (Lau et al., 2013), pulse oximeter/PPG (Lee, Reyes, McManus, Maitas, & Chon, 2013), and microelectromechanical (MEMS) technology (Jaakkola et al., 2018a, 2018b), allowing early detection of heart arrhythmia. These modern sensing technologies intend to simplify and revolutionize personal monitoring by enabling individuals to track their cardiovascular status themselves. Recently, different smartphone and smartwatch modalities such as AliveCore Kardia (Albert & Myles, 2015), Apple Watch (Perez et al., 2019), Samsung Simband smartwatch (Nemati et al., 2016), Withings smartwatch (Tajrishi et al., 2019), as well as many others with similar technologies have been established for AFib detection. These wearable/mobile modalities often require separate dedicated software/applications to facilitate long term or intermittent monitoring. However, they pose two notable limitations. Such modalities are high-priced and always require additional hardware hampering large-scale screening purposes.

CardioSignal (Precordior Oy, Turku, Finland) is a new cardiography platform which deploys multidimensional built-in inertial measurement unit (IMU) sensors of smartphones for recording cardiovascular mechanical activity. Fig. 7.1 shows a CardioSignal app for smartphone detection of cardiac disorders including heart arrhythmias. User-friendly and cost-effective IMU sensors based on MEMS technology together with the advanced signal processing and data analytics are used in CardioSignal app to improve prevention of the cardiac disorders through early-stage detection and rapid determination of the characteristics of an acute cardiac event, e.g., heart arrhythmia, with greater certainty. Measuring precordial vibrations has been explored as a technique for assessing the mechanical condition of the heart (Inan et al., 2015). The generic motion of the heart consists of translation of the center of mass in three orthogonal directions and rotation about the center of mass around three orthogonal axes (Young & Axel, 1992). Small-scale accelerometer and gyroscope sensors allow six degree of freedom (6DoF) cardiac motion sensing in the three-dimensional space (Jafari Tadi et al., 2017).

In this chapter we present biosignal processing and DL techniques to comprehensively assess mechanical status of the heart from smartphone-derived data. The focus of this chapter is to

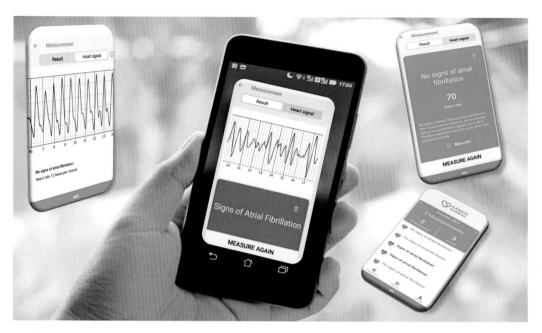

FIGURE 7.1

CardioSignal smartphone detection of atrial fibrillation.

introduce data-driven algorithms to decode cardiac motion pattern solely based on multidimensional IMU data. We present data fusion and AI techniques to characterize and analyze physiological patterns derived from the MEMS sensors. The focus is to show the capability of modern data analytic techniques in the context of smartphone and MEMS technology to overcome existing challenges and improve obtained parameters estimated from mechanical cardiovascular monitoring while introducing a new modality based on motion sensors, called mechanocardiography (MCG). The techniques addressed in this chapter may be extensively used in the detection of other heart disorders such as HF, myocardial infarction, and coronary artery diseases (CAD). Current use case considers AFib detection via retrospective MCG data from MODE-AF study (Jaakkola et al., 2018a, 2018b). We introduced CardioSignal platform as the first medical-grade smartphone MCG product for self-monitoring with the envision that the emergence of 5G telecommunication may foster its wider usability worldwide. The present use case does not reflect the current in use data analytics/processing frameworks for CardioSignal app in any form. The scientific insights proposed in this use case are purely based on academic investigations in the field of smartphone MCG and data analysis.

7.4.1 Smartphone data acquisition

For smartphone MCG technology, built-in IMUs include micro-sized accelerometers and gyroscopes—both offering three channels of motion sensing—are used to measure translational and rotational cardiac movements, typically with a sampling frequency of 200 Hz (Lahdenoja et al., 2018). A dedicated

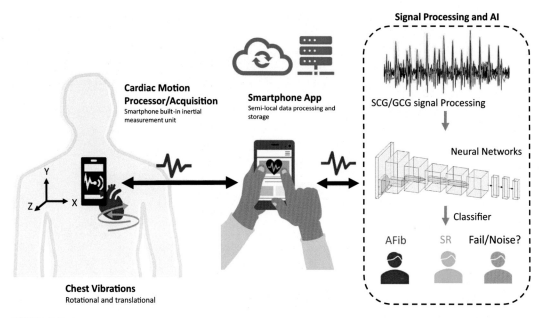

FIGURE 7.2

Overall diagram of data acquisition and processing pipeline.

smartphone application (app) is required, as a user-interface, to collect the raw data, visualize the measurements, and facilitate decision making. Data processing can be performed either through online processors—on a cloud server—or offline—by smartphones' processor. CardioSignal is one of the primary medically approved applications which is commercially available for online data processing where a cloud server is used for data storage, visualization, and arrhythmia detection only. Fig. 7.2 shows the overall pipeline for smartphone MCG starting from data collection, online/offline data preparation, and data-driven analytics based on CNNs. This section describes data acquisition procedure including collection and placement of the smartphone.

Smartphone's IMU sensors are sensitive to the linear and rotational displacements and the sensing mechanism in most of them is capacitance change related to the movement of a spring mass in multiple orientations or axis (Nihtianov & Luque, 2014). The smartphone is placed (longitudinally) on the subject's chest (near to the body of sternum), while the screen is facing upwards and the bottom edge of the phone at the level of the lower chest. Smartphone-derived acceleration and angular velocity data are obtained in three different orientations, x-, y-, and z-axis. The x-axis corresponds to the right-to-left lateral, y-axis to the head-to-foot, and z-axis to the dorsoventrally aligned movements (see Fig. 7.2). During the recording, the subjects are asked to remain calm, silent, and motionless.

7.4.2 Biomedical signal processing

Biomedical signal processing starts with filtering each of the sensors channels, namely 3-axis accelerometer and 3-axis gyroscope. MCG signals include various forms of noise from intentional or unintentional movements. Such noise can be discarded by using signal denoising methods to filter out artifacts and noise. Denoising is one of the key steps of the analysis of MCG signals which allows successive processes without losing relevant information. Signal denoising allows the removal of white noise and signals offset, resulting in increased signal-to-noise ratio. Researchers typically utilize digital filters to remove or decrease the undesirable signal components by transforming the signals into another domain (Akay, 2012).

Another noise removal technique is singular spectrum analysis which decomposes the signal into its favorable and unwanted components (Golyandina & Zhigljavsky, 2013). This approach is consisted of two complementary stages, decomposition and reconstruction, and tends to be useful when the frequency spectrum of noise may not be known a priori.

In addition to denoising, the pulse amplitude signal is obtained by computing the envelope of the accelerometer and gyroscope signals. Several methods based on Hilbert transform (Cizek, 1970), continuous wavelet transform (Rioul & Duhamel, 1992), median filter (Pratt, 2007), moving average filter (Akay, 2012), and Savitzky−Golay (Orfanidis, 1996) filter are used to derived the envelope of the MCG signals. These smoothing filters allow to discover important rhythm patterns in the signals while leaving out unimportant (i.e., noise) information. The goal of envelope detection is to obtain low frequency pulsatile changes in sensors data so that it's easier to see rhythm changes in the MCG signals.

Following filtering and envelope detection, the resulting MCG axes/channels of each sensor are divided into a shorter sequences of discrete segments. These segments or frames are obtained with overlapping windows. A Hamming window is commonly used to facilitate the spectral analysis via fast Fourier transform (FFT), e.g., in terms of spectral leakage, side lobe amplitude, and width of the central peak (Prabhu, 2014). Segmentation is a particularly important part of the data curation and choosing the right window length is very critical as shorter or longer segments can result in different outcomes. A suitable window length preserves the statistical properties of the signals that are almost constant over time. The overlapping windowing phase, although bears more processing load, allows a smoother change of the parameters between the signal frames. This will make the processing in frequency domain more robust to noise (Prabhu, 2014). Fig. 7.3 shows an example of the bandbass and envelope filtered MCG signals during AFib mode.

Motion artifact removal in MCG signals is an incredibly challenging task and different methods have been previously suggested (Tadi et al., 2016) These methods include recognizing distorting patterns from the MCG signals, either by calculting a power envelope of the signal using the root mean quare (RMS) or power spectral density features of the signal with FFT and discard the noisy components or spectra which exceed the determined threshold automatically.

7.4.3 Prediction and classification

Detection of cardiac arrhythmia from mechanical signals is an innovative technology and is often challenging due to the motion artifacts including intentional and unintentional movements. Visual characterization and interpretation of MCG waveforms are not as straightforward as in ECGs, mainly because mechanical waves are inherently complex, varying and less monomorphic as compared to the ECG waves. Advanced analytical techniques are, therefore, essential to thoroughly explore pathological patterns from the mechanical signals.

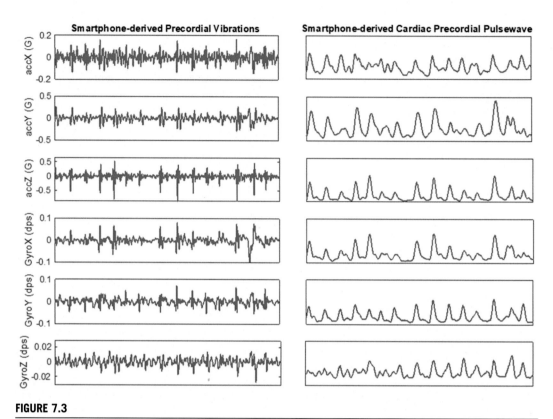

FIGURE 7.3

smartphone-derived MCG signals. The left-side column shows filtered acceleration and angular velocity signals while the right-side columns show the corresponding envelope signals.

Research efforts on arrhythmia detection using MCG technology are mainly focused on showing rhythm irregularities either by rule-based algorithms, e.g., learning recurrent waveforms using auto-correlation (Hurnanen et al., 2017), or ML techniques (Lahdenoja et al., 2018; Mehrang et al., 2019, 2020; Tadi et al., 2019). The latter considers various hand-engineered features such as heart rate variability estimations, median absolute deviation of successive intervals, spectral entropy (Rezek & Roberts, 1998), approximate and sample entropy (Richman & Moorman, 2000), Shannon entropy (Shannon, 1948), turning point ratio, and many other features (Tadi et al., 2019). The feature engineering is tailored to classic ML classifiers such as SVM with radial basis function kernels, random forests (RF) based on Bootstrap-aggregated (bagged) decision trees, XGBoost, robust boosting (RB), multilayer perceptron, and other types of classifiers to determine arrhythmic episodes. These binary classifiers have been used to distinguish AFib from normal SR only; while noisy measurements as one of the major causes of false-positive (FP) detection are investigated by DL methods to avoid premature decision making.

To reduce the processing workload and inconclusive assessments caused by FP readings made by sMCG, data-driven methods based on deep neural networks are proposed to classify SR, AFib, and noisy recordings from a short smartphone recording. DNNs include CNN and RNN to identify dominant spatial and spatiotemporal characteristics of physiological signals mimicking the approach a trained human would take in identifying irregular cardio-mechanical signals. This chapter introduces a DL framework which allows data-driven processing of MCG data, namely acceleration and angular velocity signals. This approach extends previous efforts towards efficient and fast processing of sMCG signals among the full population of the elderly patients.

Deep convolutional-recurrent neural networks are recently developed to characterize arrhythmia pattern from the MCG biosignals. The artificial neural network takes as input the filtered and envelope signals obtained from accelerometer and gyroscope sensors and outputs one prediction. The neural model takes as input only the sensors data, and no other patient metadata or signal-related features. This dual-sensor sequential model concatenates both filtered and envelope signals into one input tensor before feeding to the network.

The network architecture shown in Fig. 7.4 resembles densely connected residual network architecture inspired from three previously designed networks (Hannun et al., 2019; He, Zhang, Ren, & Sun, et al., 2015a, 2015b; Huang, Liu, van der Maaten, & Weinberger, 2018). A residual neural network (ResNet) is an ANN which mimics pyramidal cells in the cerebral cortex. Residual neural networks are based on certain connections, or shortcuts to skip some layers. Deep ResNet models are implemented with multiple shortcut connections that contain nonlinearities (ReLU) and batch normalization in between (He et al., 2015a, 2015b). Networks with several densely connected

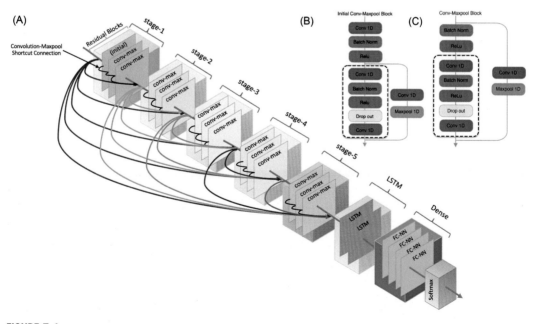

FIGURE 7.4

Dense residual learning architecture (A) for smartphone MCG data processing. Convolutional-max pool (Conv-Max) blocks (B) are sequentially deployed following the initial Conv-Max block (C).

shortcut blocks are referred to as DenseNets (Huang et al., 2018). In the context of residual neural networks, a nonresidual network may be described as a plain network.

The smartphone MCG model of densely connected residual nets has major differences including a new configuration of shortcut connections between the residual blocks, namely a convolutional-max pool block where 1D convolution and max pool operations are sequentially used (see Fig. 7.4B and C). Additionally, densely residual learning is adopted to every three stacked residual blocks in each stage. The baseline architectures are the same as residual nets, except that convolutional layers consider 1-dimensional data plus max pooling, and a shortcut connection is added to every stack of three residual blocks forming a dense residual function.

This network consists of residual blocks with two convolutional layers in each block and one convolution in the shortcut connection. The first block comprises an extra primitive convolution layer as well as a convolution layer in the shortcut connection. Batch normalization and a leaky rectified linear activation (ReLU) are also used at the beginning of each convolutional layer adopting the preactivation block design. Dropout within each block and after the nonlinearity with a probability of .3 is also deployed. Dropout is a common regularization technique for deep neural networks and has proven to be phenomenally successful by giving 1%−2% accuracy boost. Batch normalization, Dropout, and leaky ReLU allow stable learning process and prevent vanishing/exploding gradient.

Following the convolutional layers, a recurrent network consisting of two LSTM layers are used. Batch normalization and leaky ReLU are used on the output of each layer too. LSTM networks are a type of recurrent neural network capable of learning time dependencies. The CNN layers can be highly effective at automatically extracting and learning hidden features from biomedical signal sequence. Convolutional layers can be used in a hybrid mode with recurrent layers such as LSTMs to interpret subsequences of input signals. This hybrid model is called a CNN-LSTM. The output layer of the LSTM block is then fed into fully connected neural network block with four layers. A ReLU activation function is used on the output of each layer. The last fully connected Softmax layer calculates a probability distribution over the three output classes, AFib, SR, or noise. Softmax is used as the last activation function of a neural network and returns decimal probabilities for each class.

The convolutional layers of residual networks are typically initialized with randomly initialization weights. Glorot and He initialization method are commonly used in DL to alleviate the unstable gradients problem (Glorot & Bengio, 2010; He et al., 2015a, 2015b). Model optimization for training deep neural network is critical as it can be sometimes slow due to poor customization of the network. Several ways are then considered to speed up the training phase such as random initialization, proper choice of activation functions, using Batch Normalization, and ad-hoc fine tuning of the hyperparameters. In addition, fast optimizing strategies are important to speed up the training phase, resulting in rapid and stable learning processes. The most popular optimization algorithms used by the DL community are momentum optimization, Nesterov Accelerated Gradient, AdaGrad, RMSProp, and Adam and Nadam optimization (Géron, 2019). The smartphone MCG model deploys Adam (Kingma & Ba, 2017) as the optimizer and considers categorical cross-entropy as the loss function with a mini-batch size of 16. The categorical cross-entropy is preferred due to the nature of MCG multiclass classification problem and the way the labels are provided to the network—e.g., one-hot encoded labeling.

The hyperparameters, such as number of convolutional layers, the size and number of the convolutional filters, and optimization algorithm are chosen in a manual tuning manner in the MCG neural model. However, there are other automated techniques to explore a search space much more efficiently than manually. These methods work like zooming into a region of the space which has higher probability of returning better outcomes in a shorter time. Some Python libraries to optimize hyperparameters DL models are Hyperopt, Keras tuner, Scikit-Optimize (skopt), Hyperband, Spearmint, and many other custom-designed tools available in the DL community (Géron, 2019).

7.4.4 Experimental data

Experimental data for this use case study is consisted of retrospective (de-identified) measurements from 300 patients including 150 patients with AFib as the prevalent heart rhythm. An Android smartphone dedicated for clincal research was used for the trial and each measurement took approximately three minutes. Two sets of measurement senarios including physician-applied and patient-applied were planned to collect data from the patients. Among them, 182 patients (86 AFib) proceeded with two recordings, one physician-applied and one patient-applied. The remaining patients (n = 118) were either nervous, physically in poor condition, or not interested to perform the self-applied measurements. Those measurements in which either the patient or physician failed to obtain a valid recording, for example, due to intentional or unintentional movements, lack of concentration, late/poor placement, and phone drop were labeled as noise. In total, 827 MCG measurements, of which 345 recordings labeled as noise, were considered in this use case study. Data collection was strictly followed according to the Helsinki declaration in all phases and the study was reviewed by the Ethical Committee of the Hospital District of South-Western Finland before data collection. A reference ECG (Philips IntelliVue MX40) was also acquired simultaneously with the sMCG recordings. ECG meaurements allowed accurate assessment of the cardiac rhythm either as SR, AFib, or other as defined by two independent cardiologists. Table 7.1 shows detailed clinical characteristics of the study patients.

DL projects typically require that the dataset is divided into three subsets, namely train, validation, and test sets. The train set is used to fit the parameters (e.g., weights) of the model. The model is trained on the training dataset using a supervised learning method. The validation data is used to tune the models' parameters, while the test data is deployed to assess the error rate of the tuned model at the final stage (Géron, 2019). Random sampling of the biomedical measurements in only one of the subsets (train/valid/test) is required to avoid any data leakage, and overfitting. Therefore, each MCG patient's measurment(s) is put into only one category, regardless of being captured by the physician or the patient him/her-self. This way of data sampling is a standard strategy to prevent overfitting. However, making data subsets will reduce the amount of the labeled training data which can consequently influence the ending classification results. Data augmentation is a convenient step to improve classification performance when large-scale labeled training dataset is not available (Shorten & Khoshgoftaar, 2019). Since the availability of labeled MCG signals for training the DL models is limited to the above retrospective measurements, a data augmentation step is used to leverage data insufficiency by transforming the existing MCG data to create a new one, yet maintaining the correct labels. Researchers typically consider simple augmentation techniques such as adding noise, flipping, rotating, and permutating biomedical signals (Um et al., 2017). Permutation method is a simple form of signal generation in which the temporal location of within-window events is randomly perturbed. Adding Gaussian noise is another technquie which can give

Table 7.1 Clinical characteristics of the subjects with sinus rhythm (SR) or atrial fibrillation (AFib) during the mechanocardiography (MCG) recordings.

Clinical variable	SR (n = 150)	AFib (n = 150)	P-value
Age, y	74.5 (73.0−76.1)	75.0 (73.5−76.6)	.660
Female sex	66 (44.0)	66 (44.0)	1.000
Chest circumference, cm	103 (101−104)	105 (103−107)	.043
BMI, kg/m^2	27.5 (26.7−28.2)	29.0 (28.1−29.9)	.013
History of heart failure	24 (16.0)	73 (48.7)	<.001
Hypertension	101 (67.3)	103 (68.7)	.902
History of ischemic stroke	12 (8.0)	10 (12.7)	0.255
Coronary artery disease	90 (60.0)	54 (36.0)	<.001
History of AMI	68 (45.3)	26 (17.3)	<.001
History of atrial fibrillation	30 (20.0)	150 (100)	<.001
Heart rate, beats/min	70.8 (68.7−73.0)	88.0 (84.6−91.4)	<.001
Respiratory rate	16.9 (16.2−17.5)	19.2 (18.2−20.2)	<.001
Systolic blood pressure, mmHg	145 (141−149)	137 (134−141)	0.003
Diastolic blood pressure, mmHg	70.0 (67.7−72.2)	80.9 (77.7−84.1)	<.001
LBBB configuration in ECG	15 (10.0)	12 (8.0)	.687
RBBB configuration in ECG	10 (6.7)	10 (6.7)	.597
Edema in chest X-ray	38 (37.6)	67 (54.5)	.015
ProBNP (pg/mL)[a]	1095 [3596]	2965 [5603]	<.001
Patients with SVES during MCG	47 (31.3)	0 (0.0)	<.001
Patients with VES during MCG	52 (34.7)	73 (49.0)	.014

AMI, *indicates acute myocardial infarction;* BMI, *body mass index;* ECG, *electrocardiogram;* LBBB, *left bundle branch block;* LVEF, *left ventricular ejection fraction;* proBNP, *pro-brain natriuretic peptide;* RBBB, *right bundle branch block;* SVES, *supraventricular extrasystolia;* VES, *ventricular extrasystolia.*

similar looking biosignals with extra noise components. Rotation simply inverts the sign of the sensor readings without changing the labels. It is also possible to create a label-preserving transformation by which simultaneous combination of rotation, Gaussian noise, and permutation methods are applied on the MCG signals (Um et al., 2017). In total, 28408 augmented MCG segments of 10 s are generated from the two IMU sensors.

7.4.5 Performance evaluation measures

A common way to evaluate the performance of a deep neural network is to look at the error metrics such as F_1 score, accuracy, sensivity, specificity, precision, and other relevant measures. The evaluation metric introduced in this study is based on PhysioNet Challenge 2017 (Clifford et al., 2017) and the goal is to measure how well the model performs in distinguishing biosignals on the test subset. Classification metrics for the assessment of multiclass learning in MCG AFib detection study are calculated according to Table 7.2.

Table 7.2 Definition of parameters for scoring.

	SR	AFib	Noise	Total
SR	S_s	S_a	S_n	$\sum S$
AFib	A_s	A_a	A_n	$\sum A$
Noise	N_s	N_a	N_n	$\sum N$
Total	$\sum s$	$\sum a$	$\sum n$	

F_1 measure for each class is obtained as an average of:

$$F1_{AF} = \frac{2 \times A_a}{\sum A + \sum a}$$

$$F1_{SR} = \frac{2 \times S_s}{\sum S + \sum s}$$

$$F1_{noise} = \frac{2 \times N_n}{\sum N + \sum n}$$

The final score is calculated as an arithmetic mean of the per-class F_1 scores as described in following:

$$Micro_{F1} = \frac{F1_{AF} + F1_{SR} + F1_{noise}}{3}$$

Researchers often consider two-sided 95% confidence intervals (CI) for an aggregate measure of the model performance and network stability, and to be more conservative for accuracy. The CI for the performance metrics is obtained with n times replications of the entire train, validation, and test process. Other metrics such as accuracy, precision, recall, specificity, and Cohen's kappa coefficient are also calculated to evaluate the performance of the DL algorithm in detection of AFib. These classification metrics in a simple form are calculated by counting the total number of true and false predictions with respect to the true labels. True positives is counted when the classifier correctly predicts a disease group, while true negatives happens when the model predicts healthy group correctly. If the model wrongly predicted a negative class as positive, a FP is counted and if predicted as negative when it is actually positive, a false negative is counted. Interested reader can refere to (Géron, 2019) for the detailed information on the model evaluation methods.

7.4.6 Experimental results

The primary goal of this work is to establish a segment-based model for MCG signal classification, which can be implemented using DNNs in automated detection of heart diseases using smartphones. The contribution of this chapter is to find a model which can learn the rhythm and noise characteristics of the MCG signals. The performance of the DL methods was evaluaated and the experimental results are given in the following. The model revealed an overall accuracy of 92% (95% CI: 91−93). We primarily considered 10- and 20-s window lengths to evaluate the performance variability of the model between the two fragments. However, the longer fragment did not show any significant improvement in the performance of the deep model. Therefore, we report the model performance for only 10-s fragments here. The model successfully classified AFib, SR, and noise cases resulting in an averaged sensitivity/recall of 87% (95% CI: 86−88) and successfully in an averaged sensitivity/recall of 87% (95% CI: 86−88) and specificity of 93% (95% CI: 92−94) for the corresponding segment. The respective micro and macro F_1 scores of the algorithm were 87% and 80%, while the precision or positive predictive value was 78% (95% CI: 76−79) and the negative predictive value 92.5% (95% CI: 92−93). The kappa coefficient of the method was 78% showing a substantial agreement in classification between the MCG algorithm and visual interpretations. The performance of current smartphone MCG model-based on deep neuarl networks compares favorably to the various ECG (single-lead) algorithms in detection of AFib. In 2017 Physionet Challenge (Clifford et al., 2017), where a total of 12,186 ECGs were used, the overall macro F_1 score for the top 11 algorithms was 83% ± 2%. Our study scored an average macro F_1 measure of 80%.

7.5 Discussion

With the recent advancements of IoT and smartphone systems, remote processing of health data has become more popular than it was in the past. Smart devices are fast becoming available globally, even among developing countries, and the number of elderly users is also growing at a rapid rate in those countries, where low-cost healthcare solutions are crucial. A personal smart monitoring program such as CardioSignal allows seamless assessment of the health risks by early detection of cardiovascular disorders. The readily available motion sensors in smart devices provide an unprecedented opportunity for the cost-effective screening of heart diseases if they can be diagnosed via early symptoms.

In addition to smartphones, IMU sensors can be embedded into a wearable patch or implantable system for long-term monitoring. These sensors allow seamless measurement of the myocardium's mechanical movement and are not subjected to electrical or implantable stimulating interventions. This allows continuous cardiac function monitoring via the integration of motion signal processing tools in devices like pacemakers and cardioverter-defibrillators.

Various smartphone and smartwatch-based ECG and/or optical sensor recorders are commercially available to check the heart activity and have shown decent performances to detect irregularities. However, they are often expensive and sometimes inconvenient and with the growing prevalence of AFib in an aging population this is an important usability consideration for healthcare systems. Rhythm classification using mechanical signals has been the focus of ML investigations using a variety of techniques such as linear and quadratic discriminant analysis, support vector machines, RB, XGboost, and RF. However, most of those studies relied on the derivation of vibrational pattern features to classify different rhythm types as well as on feature selection techniques for further generalization and performance improvements.

A clear advantage of methods based on NN is that they do not necessarily require direct feature extraction from the physiological signals and perform complex nonlinear operations. One drawback of DL methods is that they are computationally expensive and require large datasets for training. As with other clinical DL applications, the key challenge for MCG analysis is not only computational resources but also the availability of large-scale annotated (labeled) datasets with the required clinical information.

Smartphone as a new modality demands minimum professional knowledge and coordination so that people can quickly collect biosignals in an everyday manner. Self-screening allows patients to collect their vital signals themselves outside the hospital environment. With the development of smart health technologies, the self-screening has become more popular in the elderly patients. Smartphone MCG vision is to better serve the society by facilitating early detection and interventions of CVDs. Smartphones can promptly collect and analyze patient's data, enabling people to get the same quality of diagnostic medical services without shuttling back and forth between hospitals and their homes. Smart healthcare not only reduces the immense burden of social challenges, but it also lowers the economic burden on end users. However, the collection and AI-based processing of massive data still have concerning data privacy risks, which may lead to various ethical problems and endanger the vital interests of users. According to the Ethics Guidelines for Trustworthy AI guidelines, trustworthy AI should be: (1) lawful—respecting all applicable laws and regulations, (2) ethical—respecting ethical principles and values and (3) robust—both from a technical perspective while considering its social environment (AI HLE, 2020). In this chapter, we explore some of the key questions prompted by ethical AI in relation to information privacy.

Even though the use of analytical tools in medicine, especially in cardiovascular medicine, is growing, there still serious challenges about using clinical data and especially protected health information about patients (Mathur, Srivastava, Xu, & Mehta, 2020). In Europe, data privacy laws become more stringent with the establishment of General Data Protection Regulations (GDPR). GDPR defines a body of legal safeguards on digital data protection and privacy for all individuals within the European Union and the European Economic Area. The GDPR outlines the principles associating with the processing of personal data. It defines, for example, the usage and storage principles, the lawfulness of data processing, and several special conditions on personal data usage (REGULATION (EU) 2016/679). With the increasing use of remote data processing in health care, especially with AI tools dealing with electronic medical records of the patients, the GDPR is truly relevant as it recognizes data concerning health as a special category of data and supplies a definition for health data for data protection purposes.

The foundations introduced by the GDPR are consistent and applicable to health data as well. However, specific safeguards for personal health data and for a definitive interpretation of the rules that allow comprehensive protection of such data must yet be addressed promptly. Processes that accelerate innovation and better quality in healthcare, such as AI or mobile health tools, require robust data protections to support the trust and confidence of the users in the rules designed to protect their data.

Under the new laws, personal data extends to the type of data that is collected about people, including online identifiers, economic, cultural, and health information. As recommended by the GDPR, personal data must continue to be processed according to the six data protection principles: (1) Processed legally, fairly, and transparently, (2) Collected for specific legal goals, (3) Adequate, relevant, and limited to necessity, (4) Accurate and up to date, (5) Stored if needed and (6) Ensure security, integrity, availability, and confidentiality (REGULATION (EU) 2016/679).

However, there are several limitations to use AI/ML in healthcare. For example, in cardiovascular medicine, dichotomic and improper calibration are known problems of AI-based ML methods (Johnson et al., 2018). Furthermore, AI-based systems need to address data privacy concerns (Johnson et al., 2018; Powles & Hodson, 2017) and data integrity to prevent faulty data curation, selection bias, and historical stereotypes in data analysis.

7.6 Conclusion

The healthcare industry is facing different concerns associated with disease diagnosis and cost-effective delivery of services (Awan et al., 2018). One of the crucial needs of a healthcare system is to provide the patient with adequate services by the utilization of medical records, lifestyle habits, and any variability of molecular traits. Several intelligent IoT have been developed based on AI and data-driven methods to overcome these challenges. For disease diagnosis and treatment, modern and automated technology-enabled Smart Healthcare Platforms have been employed. Healthcare services are delivered by Smart Healthcare Platforms utilizing smart phone/watch and connected technologies. Patient information is gathered in a file for the purpose of disease diagnosis and treatment and offering timely care, in particular for the areas where urgent intervention is a challenge. This chapter introduces and explores the use of DL approaches to detect diseases such as heart attacks, brain cancers, diabetes, cardiac disease, and prediction of epileptic seizures in this perspective.

Moreover, this chapter addresses the globally prominent issues of DL models in detecting CVDs, for example arrhythmia, with multidimensional motion sensors using a smartphone application. Atrial fibrillation is one of the main reasons for heart-related disabilities, and other comorbidities such as stroke, HF, and other complications. The results of the presented use case study together with a body of earlier investigations in the field acknowledge the clinical potentials of the cardiac motion sensing to reliably find signs of heart disorders, without any external complementary equipment.

Due to the broad accessibility and availability of smart devices, it is possible to develop a universal diagnosis system as a part of efficient global prevention and detection strategies to deal with heart diseases. Smartphone MCG devices may offer a reliable and cost-efficient screening and monitoring possibility for AFib compared with other monitoring modalities.

References

Aggarwal, C. C. (2018). *Neural networks and deep learning*. Springer.

AI HLE (2020, Jul. 17). Assessment List for Trustworthy Artificial Intelligence (ALTAI) for self-assessment. *Shaping Europe's digital future - European Commission*. https://ec.europa.eu/digital-single-market/en/news/assessment-list-trustworthy-artificial-intelligence-altai-self-assessment. Accessed 15.01.21.

Ahn, T., Goo, T., Lee, C., Kim, S., Han, K., Park, S., & Park, T. (2018). *Deep learning-based identification of cancer or normal tissue using gene expression. In 2018 IEEE international conference on bioinformatics and biomedicine (BIBM)*, 1748–1752.

Akay, M. (2012). *Biomedical signal processing*. Academic Press.

Alanazi, H. O., Abdullah, A. H., Qureshi, K. N., & Ismail, A. S. (2018). Accurate and dynamic predictive model for better prediction in medicine and healthcare. *Irish Journal of Medical Science, 187*(2), 501–513. Available from https://doi.org/10.1007/s11845-017-1655-3.

Albert, D.E. & Myles, C.A. (2015, Mar. 26). *Smartphone and ecg device microbial shield*. US20150087952A1.

Ali, F., et al. (2020). A smart healthcare monitoring system for heart disease prediction based on ensemble deep learning and feature fusion. *Information Fusion, 63*, 208–222.

Allen, J. (2007). Photoplethysmography and its application in clinical physiological measurement. *Physiological Measurement, 28*(3), Art. 3.

Anter, E., Jessup, M., & Callans, D. J. (2009). Atrial fibrillation and heart failure. *Circulation, 119*(18), 2516−2525. Available from https://doi.org/10.1161/CIRCULATIONAHA.108.821306.

Asiri, N., Hussain, M., Al Adel, F., & Alzaidi, N. (2019). Deep learning based computer-aided diagnosis systems for diabetic retinopathy: A survey. *Artificial Intelligence in Medicine, 99*, 101701.

Attia, Z. I., et al. (2019a). Prospective validation of a deep learning electrocardiogram algorithm for the detection of left ventricular systolic dysfunction. *Journal of Cardiovascular Electrophysiology, 30*(5), 668−674.

Attia, Z. I., et al. (2019b). An artificial intelligence-enabled ECG algorithm for the identification of patients with atrial fibrillation during sinus rhythm: A retrospective analysis of outcome prediction. *The Lancet, 394*(10201), 861−867.

Awan, K.M., Ali, A., Aadil, F., & Qureshi, K.N. (2018, Feb.). Energy efficient cluster based routing algorithm for wireless sensors networks. In *2018 International conference on advancements in computational sciences (ICACS)* (pp. 1−7). doi: 10.1109/ICACS.2018.8333486.

Bagga, P., & Hans, R. (2015). Applications of mobile agents in healthcare domain: A literature survey. *International Journal of Grid and Distributed Computing, 8*(5), 55−72.

Bhavnani, S. P., Narula, J., & Sengupta, P. P. (2016). Mobile technology and the digitization of healthcare. *European Heart Journal, 37*(18), 1428−1438. Available from https://doi.org/10.1093/eurheartj/ehv770.

Caballé, N. C., Castillo-Sequera, J. L., Gómez-Pulido, J. A., Gómez-Pulido, J. M., & Polo-Luque, M. L. (2020). Machine learning applied to diagnosis of human diseases: A systematic review. *Applied Sciences, 10*(15), 5135.

Cardiovascular diseases. https://www.who.int/westernpacific/health-topics/cardiovascular-diseases. Accessed 18.01.21.

Chaudhry, B., et al. (2006). Systematic review: Impact of health information technology on quality, efficiency, and costs of medical care. *Annals of Internal Medicine, 144*(10), 742−752. Available from https://doi.org/10.7326/0003-4819-144-10-200605160-00125.

Chen, Q., Yi, Z., & Cheng, J. (2018). Atrial fibrillation in aging population. *Aging Medicine, 1*(1), 67−74. Available from https://doi.org/10.1002/agm2.12015.

Choi, E., Schuetz, A., Stewart, W. F., & Sun, J. (2017). Using recurrent neural network models for early detection of heart failure onset. *Journal of the American Medical Informatics Association: JAMIA, 24*(2), 361−370.

Choi, H., Ha, S., Im, H. J., Paek, S. H., & Lee, D. S. (2017). Refining diagnosis of Parkinson's disease with deep learning-based interpretation of dopamine transporter imaging. *NeuroImage: Clinical, 16*, 586−594. Available from https://doi.org/10.1016/j.nicl.2017.09.010.

Cizek, V. (1970). Discrete Hilbert transform. *IEEE Transactions on Audio and Electroacoustics, 18*(4), 340−343. Available from https://doi.org/10.1109/TAU.1970.1162139.

Clifford, G. D., et al. (2017). AF classification from a short single lead ECG recording: The PhysioNet/computing in cardiology challenge 2017. *Computers in Cardiology, 44*, Accessed: Jan. 15, 2021. [Online]. Available. Available from https://www.ncbi.nlm.nih.gov/pmc/articles/PMC5978770/.

Colilla, S., Crow, A., Petkun, W., Singer, D. E., Simon, T., & Liu, X. (2013). Estimates of current and future incidence and prevalence of atrial fibrillation in the US adult population. *The American Journal of Cardiology, 112*(8), 1142−1147.

Daniels, M., & Schroeder, S. A. (1977). Variation among physicians in use of laboratory tests II. Relation to clinical productivity and outcomes of care. *Medical Care*, 482−487.

Deng, L., & Yu, D. (2014). Deep learning: Methods and applications. *Foundations and Trends in Signal Processing, 7*(3−4), 197−387. Available from https://doi.org/10.1561/2000000039.

Dick, R. S., Steen, E. B., & Detmer, D. E. (1997). *The computer-based patient record: An essential technology for health care*. National Academies Press.

Dubrovina, A., Kisilev, P., Ginsburg, B., Hashoul, S., & Kimmel, R. (2018). Computational mammography using deep neural networks. *Computer Methods in Biomechanics and Biomedical Engineering: Imaging & Visualization*, *6*(3), 243−247.

Faust, O., Hagiwara, Y., Hong, T. J., Lih, O. S., & Acharya, U. R. (2018). Deep learning for healthcare applications based on physiological signals: A review. *Computer Methods and Programs in Biomedicine*, *161*, 1−13.

Feinberg, W. M., Blackshear, J. L., Laupacis, A., Kronmal, R., & Hart, R. G. (1995). Prevalence, age distribution, and gender of patients with atrial fibrillation: Analysis and implications. *Archives of Internal Medicine*, *155*(5), 469−473. Available from https://doi.org/10.1001/archinte.1995.00430050045005.

Gao, F., et al. (2019). DeepCC: A novel deep learning-based framework for cancer molecular subtype classification. *Oncogenesis*, *8*(9), 1−12.

Géron, A. (2019). *Hands-on machine learning with Scikit-Learn, Keras, and TensorFlow: Concepts, tools, and techniques to build intelligent systems* (Second edition). Sebastopol, CA: O'Reilly Media, Inc.

Glorot, X. & Bengio, Y. (2010). *Understanding the diffficulty of training deep feedforward neural networks* (p. 8).

Go, A. S., et al. (2001). Prevalence of diagnosed atrial fibrillation in adults: National implications for rhythm management and stroke prevention: The AnTicoagulation and risk factors in atrial fibrillation (ATRIA) study. *JAMA: The Journal of the American Medical Association*, *285*(18), 2370−2375. Available from https://doi.org/10.1001/jama.285.18.2370.

Golyandina, N., & Zhigljavsky, A. (2013). *Singular spectrum analysis for time series*. Springer Science & Business Media.

Goodfellow, I., Bengio, Y., Courville, A., & Bengio, Y. (2016). 2 *Deep learning* (vol. 1). Cambridge: MIT press.

Grimson, J., Stephens, G., Jung, B., Grimson, W., Berry, D., & Pardon, S. (2001). Sharing health-care records over the internet. *IEEE Internet Computing*, *5*(3), 49−58.

Gunduz, H. (2019). Deep learning-based Parkinson's disease classification using vocal feature sets. *IEEE Access*, *7*, 115540−115551.

Hannun, A. Y., et al. (2019). Cardiologist-level arrhythmia detection and classification in ambulatory electrocardiograms using a deep neural network. *Nature Medicine*, *25*(1). Available from https://doi.org/10.1038/s41591-018-0268-3, Art. no. 1.

He, K., Zhang, X., Ren, S., & Sun, J. (2015a, Feb.). *Delving deep into rectifiers: Surpassing human-level performance on ImageNet classification*. ArXiv150201852 Cs. [Online]. Available: http://arxiv.org/abs/1502.01852. Accessed 15.01.21.

He, K., Zhang, X., Ren, S., & Sun, J. (2015b, Dec.). *Deep residual learning for image recognition*. ArXiv151203385 Cs. [Online]. Available: http://arxiv.org/abs/1512.03385. Accessed 15.01.21.

Heidenreich, P. A., et al. (2011). Forecasting the future of cardiovascular disease in the United States: A policy statement from the American Heart Association. *Circulation*, *123*(8), 933−944. Available from https://doi.org/10.1161/CIR.0b013e31820a55f5.

Hendriks, J. M., Gallagher, C., Middeldorp, M. E., & Sanders, P. (2018). New approaches to detection of atrial fibrillation. *Heart (British Cardiac Society)*, *104*(23), 1898−1899. Available from https://doi.org/10.1136/heartjnl-2018-313423.

Hickey, K. T., Riga, T. C., Mitha, S. A., & Reading, M. J. (2018). Detection and management of atrial fibrillation using remote monitoring. *The Nurse Practitioner*, *43*(3), 24.

Huang, G., Liu, Z., van der Maaten, L., & Weinberger, K.Q. (2018, Jan.). *Densely connected convolutional networks*. ArXiv160806993 Cs. [Online]. Available: http://arxiv.org/abs/1608.06993. Accessed 15.01.21.

Huang, M.-J., Chen, M.-Y., & Lee, S.-C. (2007). Integrating data mining with case-based reasoning for chronic diseases prognosis and diagnosis. *Expert Systems with Applications*, *32*(3), 856−867.

Hurnanen, T., et al. (2017). Automated detection of atrial fibrillation based on time-frequency analysis of seismocardiograms. *IEEE Journal of Biomedical and Health Informatics*, *21*(5), 1233−1241. Available from https://doi.org/10.1109/JBHI.2016.2621887.

Inan, O. T., et al. (2015). Ballistocardiography and seismocardiography: A review of recent advances. *IEEE Journal of Biomedical and Health Informatics*, *19*(4), 1414−1427. Available from https://doi.org/10.1109/JBHI.2014.2361732.

Jaakkola, J., et al. (2016). Stroke as the first manifestation of atrial fibrillation. *PLoS One*, *11*(12), e0168010.

Jaakkola, J., et al. (2018a). Mobile phone detection of atrial fibrillation with mechanocardiography: The MODE-AF study (Mobile Phone Detection of Atrial Fibrillation). *Circulation*, *137*(14), 1524−1527. Available from https://doi.org/10.1161/CIRCULATIONAHA.117.032804.

Jaakkola, J., et al. (2018b). Mobile phone detection of atrial fibrillation: The mode-AF study. *Journal of the American College of Cardiology*, *71*(11S), pp. A410−A410.

Jafari Tadi, M., et al. (2017). Gyrocardiography: A new nnon-invasive monitoring method for the assessment of cardiac mechanics and the estimation of hemodynamic variables. *Scientific Reports*, *7*(1). Available from https://doi.org/10.1038/s41598-017-07248-y, Art. 1.

Jiao, Z., Gao, X., Wang, Y., & Li, J. (2018). A parasitic metric learning net for breast mass classification based on mammography. *Pattern Recognition*, *75*, 292−301. Available from https://doi.org/10.1016/j.patcog.2017.070.008.

Johnson, K. W., et al. (2018). Artificial intelligence in cardiology. *Journal of the American College of Cardiology*, *71*(23), 2668−2679. Available from https://doi.org/10.1016/j.jacc.2018.03.521.

Jordan, M. I., & Mitchell, T. M. (2015). Machine learning: Trends, perspectives, and prospects. *Science (New York, N.Y.)*, *349*(6245), 255−260.

Kalantari, A., Kamsin, A., Shamshirband, S., Gani, A., Alinejad-Rokny, H., & Chronopoulos, A. T. (2018). Computational intelligence approaches for classification of medical data: State-of-the-art, future challenges and research directions. *Neurocomputing*, *276*, 2−22.

Kamaleswaran, R., Mahajan, R., & Akbilgic, O. (2018). A robust deep convolutional neural network for the classification of abnormal cardiac rhythm using single lead electrocardiograms of variable length. *Physiological Measurement*, *39*(3), 035006.

Kelleher, J. D. (2019). *Deep learning*. Mit Press.

Kharazmi, P., Zheng, J., Lui, H., Jane Wang, Z., & Lee, T. K. (2018). A computer-aided decision support system for detection and localization of cutaneous vasculature in dermoscopy images via deep feature learning. *Journal of Medical Systems*, *42*(2), 33. Available from https://doi.org/10.1007/s10916-017-0885-2.

Kingma, D.P. & Ba, J. (2017, Jan.). *Adam: A method for stochastic optimization*. ArXiv14126980 Cs. [Online]. Available: http://arxiv.org/abs/1412.6980. Accessed 15.01.21.

Kleiger, R. E., & Senior, R. M. (1974). Longterm electrocardiographic monitoring of ambulatory patients with chronic airway obstruction. *Chest*, *65*(5), 483−487. Available from https://doi.org/10.1378/chest.65.5.483.

Kononenko, I. (2001). Machine learning for medical diagnosis: History, state of the art and perspective. *Artificial Intelligence in Medicine*, *23*(1), 89−109.

Kotecha, D., & Kirchhof, P. (2017). ESC apps for atrial fibrillation. *European Heart Journal*, *38*(35), 2643−2645. Available from https://doi.org/10.1093/eurheartj/ehx445.

Kumar, P. M., & Gandhi, U. D. (2018). A novel three-tier Internet of Things architecture with machine learning algorithm for early detection of heart diseases. *Computers & Electrical Engineering*, *65*, 222−235.

Lahdenoja, O., et al. (2018). Atrial fibrillation detection via accelerometer and gyroscope of a smartphone. *IEEE Journal of Biomedical and Health Informatics*, *22*(1), 108−118. Available from https://doi.org/10.1109/JBHI.2017.2688473.

Lau, J. K., et al. (2013). iPhone ECG application for community screening to detect silent atrial fibrillation: A novel technology to prevent stroke. *International Journal of Cardiology*, *165*(1), 193−194. Available from https://doi.org/10.1016/j.ijcard.2013.01.220.

Lee, J., Reyes, B. A., McManus, D. D., Maitas, O., & Chon, K. H. (2013). Atrial fibrillation detection using an iPhone 4S. *IEEE Transactions on Bio-Medical Engineering*, *60*(1), 203−206. Available from https://doi.org/10.1109/TBME.2012.2208112.

Levenson, J. W., Skerrett, P. J., & Gaziano, J. M. (2002). Reducing the global burden of cardiovascular disease: The role of risk factors. *Preventive Cardiology*, *5*(4), 188−199. Available from https://doi.org/10.1111/j.1520-037X.2002.00564.x.

Li, Y., et al. (2020). CraftNet: A deep learning ensemble to diagnose cardiovascular diseases. *Biomedical Signal Processing and Control*, *62*, 102091.

Ma, F., Sun, T., Liu, L., & Jing, H. (2020). Detection and diagnosis of chronic kidney disease using deep learning-based heterogeneous modified artificial neural network. *Future Generation Computer Systems*.

Martignani, C., Massaro, G., Biffi, M., Ziacchi, M., & Diemberger, I. (2020). Atrial fibrillation: An arrhythmia that makes healthcare systems tremble. *Journal of Medical Economics*, *23*(7), 667−669. Available from https://doi.org/10.1080/13696998.2020.1752220, Jul.

Mathur, P., Srivastava, S., Xu, X., & Mehta, J. L. (2020). Artificial intelligence, machine learning, and cardiovascular disease. *Clinical Medicine Insights: Cardiology*, *14*. Available from https://doi.org/10.1177/1179546820927404, p. 1179546820927404, Jan.

Mehrang, S., et al. (2019). Reliability of self-applied smartphone mechanocardiography for atrial fibrillation detection. *IEEE Access*, *7*, 146801−146812. Available from https://doi.org/10.1109/ACCESS.2019.2946117.

Mehrang, S., et al. (2020). Classification of atrial fibrillation and acute decompensated heart failure using smartphone mechanocardiography: A multilabel learning approach. *IEEE Sensors Journal*, *20*(14), 7957−7968. Available from https://doi.org/10.1109/JSEN.2020.2981334.

Miotto, R., Li, L., Kidd, B. A., & Dudley, J. T. (2016). Deep patient: An unsupervised representation to predict the future of patients from the electronic health records. *Scientific Reports*, *6*(1), 1−10.

Moeskops, P., et al. (2018). Evaluation of a deep learning approach for the segmentation of brain tissues and white matter hyperintensities of presumed vascular origin in MRI. *NeuroImage: Clinical*, *17*, 251−262.

Nemati, S., et al. (2016). Monitoring and detecting atrial fibrillation using wearable technology. *Annual International Conference of the IEEE Engineering in Medicine and Biology Society*, *2016*, 3394−3397. Available from https://doi.org/10.1109/EMBC.2016.7591456.

Nie, L., Wang, M., Zhang, L., Yan, S., Zhang, B., & Chua, T.-S. (2015). Disease inference from health-related questions via sparse deep learning. *IEEE Transactions on Knowledge and Data Engineering*, *27*(8), 2107−2119.

Nihtianov, S., & Luque, A. (2014). *Smart sensors and MEMS: Intelligent devices and microsystems for industrial applications*. Woodhead Publishing.

Onodera, R., & Sengoku, S. (2018). Innovation process of mHealth: An overview of FDA-approved mobile medical applications. *International Journal of Medical Informatics*, *118*, 65−71. Available from https://doi.org/10.1016/j.ijmedinf.2018.07.004.

Orfanidis, S. J. (1996). *Introduction to signal processing*. Englewood Cliffs, N.J: Prentice Hall.

Orlando, J. I., Prokofyeva, E., Del Fresno, M., & Blaschko, M. B. (2018). An ensemble deep learning based approach for red lesion detection in fundus images. *Computer Methods and Programs in Biomedicine*, *153*, 115−127.

Pacal, I., Karaboga, D., Basturk, A., Akay, B., & Nalbantoglu, U. (2020). A comprehensive review of deep learning in colon cancer. *Computers in Biology and Medicine*, 104003.

Perez, M. V., et al. (2019). Large-scale assessment of a smartwatch to identify atrial fibrillation. *The New England Journal of Medicine*, *381*(20), 1909−1917. Available from https://doi.org/10.1056/NEJMoa1901183.

Powles, J., & Hodson, H. (2017). Google DeepMind and healthcare in an age of algorithms. *Health Technology, 7*(4), 351−367. Available from https://doi.org/10.1007/s12553-017-0179-1.

Prabhu, K. M. M. (2014). *Window functions and their applications in signal processing*. Boca Raton, [Florida]: CRC Press/Taylor & Francis.

Pratt, W. K. (2007). *Digital image processing: PIKS scientific inside*. Hoboken, NJ: Wiley.

Premaladha, J., & Ravichandran, K. S. (2016). Novel approaches for diagnosing melanoma skin lesions through supervised and deep learning algorithms. *Journal of Medical Systems, 40*(4), 96. Available from https://doi.org/10.1007/s10916-016-0460-2, Feb.

Qureshi, K. N., Din, S., Jeon, G., & Piccialli, F. (2020). An accurate and dynamic predictive model for a smart M-Health system using machine learning. *Information Sciences, 538*, 486−502.

Ravì, D., et al. (2016). Deep learning for health informatics. *IEEE Journal of Biomedical and Health Informatics, 21*(1), 4−21.

Rezek, I. A., & Roberts, S. J. (1998). Stochastic complexity measures for physiological signal analysis. *IEEE Transactions on Bio-Medical Engineering, 45*(9), 1186−1191. Available from https://doi.org/10.1109/10.709563.

Richman, J. S., & Moorman, J. R. (2000). Physiological time-series analysis using approximate entropy and sample entropy. *American Journal of Physiology—Heart and Circulatory Physiology, 278*(6), H2039−H2049. Available from https://doi.org/10.1152/ajpheart.2000.278.6.H2039.

Rioul, O., & Duhamel, P. (1992). Fast algorithms for discrete and continuous wavelet transforms. *IEEE Transactions on Information Theory, 38*(2), 569−586. Available from https://doi.org/10.1109/18.119724.

Samala, R. K., Chan, H., Hadjiiski, L., Helvie, M. A., Wei, J., & Cha, K. (2016). Mass detection in digital breast tomosynthesis: Deep convolutional neural network with transfer learning from mammography. *Medical Physics, 43*(12), 6654−6666.

Sannino, G., & De Pietro, G. (2018). A deep learning approach for ECG-based heartbeat classification for arrhythmia detection. *Future Generation Computer Systems, 86*, 446−455.

Savelieva, I., & Camm, A. J. (2000). Clinical relevance of silent atrial fibrillation: Prevalence, prognosis, quality of life, and management. *Journal of Interventional Cardiac Electrophysiology: An International Journal of Arrhythmias and Pacing, 4*(2), 369−382. Available from https://doi.org/10.1023/A:1009823001707.

Schwalm, J.-D., McKee, M., Huffman, M. D., & Yusuf, S. (2016). Resource effective strategies to prevent and treat cardiovascular disease. *Circulation, 133*(8), 742−755. Available from https://doi.org/10.1161/CIRCULATIONAHA.115.008721.

Shamshirband, S., Fathi, M., Dehzangi, A., Chronopoulos, A. T., & Alinejad-Rokny, H. (2020). A review on deep learning approaches in healthcare systems: Taxonomies, challenges, and open issues. *Journal of Biomedical Informatics*, 103627.

Shannon, C. E. (1948). A mathematical theory of communication. *The Bell System Technical Journal, 27*(3), 379−423. Available from https://doi.org/10.1002/j.1538-7305.1948.tb01338.x.

Sharma, H., Zerbe, N., Klempert, I., Hellwich, O., & Hufnagl, P. (2017). Deep convolutional neural networks for automatic classification of gastric carcinoma using whole slide images in digital histopathology. *Computerized Medical Imaging and Graphics: The Official Journal of the Computerized Medical Imaging Society, 61*, 2−13.

Shorten, C., & Khoshgoftaar, T. M. (2019). A survey on image data augmentation for deep learning. *Journal of Big Data, 6*(1), 60. Available from https://doi.org/10.1186/s40537-019-0197-0.

Smellie, W., Galloway, M., Chinn, D., & Gedling, P. (2002). Is clinical practice variability the major reason for differences in pathology requesting patterns in general practice? *Journal of Clinical Pathology, 55*(4), 312−314.

Sriram, T., Rao, M. V., Narayana, G., & Kaladhar, D. (2016). A comparison and prediction analysis for the diagnosis of Parkinson disease using data mining techniques on voice datasets. *International Journal of Applied Engineering Research*, *11*(9), 6355−6360.

Stuart, P. J., Crooks, S., & Porton, M. (2002). An interventional program for diagnostic testing in the emergency department. *The Medical Journal of Australia*, *177*(3), 131−134.

Sun, W., Zheng, B., & Qian, W. (2017). Automatic feature learning using multichannel ROI based on deep structured algorithms for computerized lung cancer diagnosis. *Computers in Biology and Medicine*, *89*, 530−539.

Tadi, M. J., et al. (2016). A real-time approach for heart rate monitoring using a Hilbert transform in seismocardiograms. *Physiological Measurement*, *37*(11), 1885−1909. Available from https://doi.org/10.1088/0967-3334/37/11/1885.

Tadi, M. J., et al. (2019). Comprehensive analysis of cardiogenic vibrations for automated detection of atrial fibrillation using Smartphone Mechanocardiograms. *IEEE Sensors Journal*, *19*(6), 2230−2242. Available from https://doi.org/10.1109/JSEN.2018.2882874.

Tajrishi, F. Z., Chitsazan, M., Chitsazan, M., Shojaei, F., Gunnam, V., & Chi, G. (2019). Smartwatch for the detection of atrial fibrillation. *Critical Pathways in Cardiology*, *18*(4), 176−184. Available from https://doi.org/10.1097/HPC.0000000000000192.

Timmis, A., et al. (2020). European Society of Cardiology: Cardiovascular Disease Statistics 2019. *European Heart Journal*, *41*(1), 12−85. Available from https://doi.org/10.1093/eurheartj/ehz859.

Turchioe, M. R., Jimenez, V., Isaac, S., Alshalabi, M., Slotwiner, D., & Creber, R. M. (2020). Review of mobile applications for the detection and management of atrial fibrillation. *Heart Rhythm O2*, *1*(1), 35−43.

Um, T.T. et al. (2017, Nov.). Data augmentation of wearable sensor data for parkinson's disease monitoring using convolutional neural networks. In *Proceedings of the 19th ACM international conference on multimodal interaction* (pp. 216−220). Glasgow UK. doi: 10.1145/3136755.3136817.

Vasuki, A., & Govindaraju, S. (2017). *Deep neural networks for image classification, . Deep Learning for Image Processing Applications* (vol. 31, p. 27). IOS Press.

Waktare Johan, E. P. (2002). Atrial fibrillation. *Circulation*, *106*(1), 14−16. Available from https://doi.org/10.1161/01.CIR.0000022730.66617.D9.

Wang, S.-H., Phillips, P., Sui, Y., Liu, B., Yang, M., & Cheng, H. (2018). Classification of Alzheimer's disease based on eight-layer Convolutional Neural Network with leaky rectified linear unit and max pooling. *Journal of Medical Systems*, *42*(5), 85. Available from https://doi.org/10.1007/s10916-018-0932-7.

Wennberg, J. E. (1984). Dealing with medical practice variations: A proposal for action. *Health Affairs (Millwood)*, *3*(2), 6−33.

Xiao, Y., Wu, J., Lin, Z., & Zhao, X. (2018). A deep learning-based multi-model ensemble method for cancer prediction. *Computer Methods and Programs in Biomedicine*, *153*, 1−9.

Xuan, W., & You, G. (2020). Detection and diagnosis of pancreatic tumor using deep learning-based hierarchical convolutional neural network on the internet of medical things platform. *Future Generation Computer Systems*.

Yildirim, Ö. (2018). A novel wavelet sequence based on deep bidirectional LSTM network model for ECG signal classification. *Computers in Biology and Medicine*, *96*, 189−202.

Yıldırım, Ö., Pławiak, P., Tan, R.-S., & Acharya, U. R. (2018). Arrhythmia detection using deep convolutional neural network with long duration ECG signals. *Computers in Biology and Medicine*, *102*, 411−420.

Yin, W., Li, L., & Wu, F.-X. (2020). Deep learning for brain disorder diagnosis based on fMRI images. *Neurocomputing*.

Young, A. A., & Axel, L. (1992). Three-dimensional motion and deformation of the heart wall: Estimation with spatial modulation of magnetization−a model-based approach. *Radiology*, *185*(1), 241−247. Available from https://doi.org/10.1148/radiology.185.1.1523316.

Zhao, X., Wu, Y., Song, G., Li, Z., Zhang, Y., & Fan, Y. (2018). A deep learning model integrating FCNNs and CRFs for brain tumor segmentation. *Medical Image Analysis*, *43*, 98−111.

Zhu, H., et al. (2020). Automatic multilabel electrocardiogram diagnosis of heart rhythm or conduction abnormalities with deep learning: A cohort study. *Lancet Digital Health*.

Zhuang, Z. Y., Churilov, L., Burstein, F., & Sikaris, K. (2009). Combining data mining and case-based reasoning for intelligent decision support for pathology ordering by general practitioners. *European Journal of Operational Research*, *195*(3), 662−675.

Advanced pattern recognition tools for disease diagnosis

Abdulhamit Subasi[1], Siba Smarak Panigrahi[2], Bhalchandra Sunil Patil[3], M. Abdullah Canbaz[4] and Riku Klén[5]

[1]Faculty of Medicine, Institute of Biomedicine, University of Turku, Turku, Finland [2]Department of Computer Science and Engineering, Indian Institute of Technology Kharagpur, Kharagpur, West Bengal, India [3]Department of Mechanical Engineering, Indian Institute of Technology Kharagpur, Kharagpur, West Bengal, India [4]Computer Science Department, Indiana University Kokomo, Kokomo, IN, United States [5]Turku PET Centre, University of Turku, Turku, Finland

8.1 Introduction

Machine learning (ML) is a subbranch of artificial intelligence (AI) that is extensively employed in medical image analysis to diagnose diseases. Computer-aided diagnosis (CAD) techniques were employed to provide more precise and useful medical imaging interpretation results. In the mid-1980s, CAD algorithms started to evolve and were first employed for mammography and chest radiography to detect and diagnose cancer (Giger, Doi, & MacMahon, 1988). Other approaches, such as ultrasound and computed tomography (CT) (Kawata et al., 1999), were then expanded. CAD algorithms has been generally employed data-driven strategy in the early days, as many deep learning (DL) algorithms utilized today. But early CAD methods were primarily focused on feature extraction. The development of CAD systems was not helped by feature extraction, as it had many shortcomings (LeCun, Bengio, & Hinton, 2015). The feature extraction was substituted by transfer learning and DL to overcome these shortcomings by using a transfer learning approach to boost the performance of the CAD systems (Bengio, Courville, & Vincent, 2013). DL, a ML subfield, is a technique widely employed in research areas, including medical images, natural language processing, speech recognition, and computer vision. Because of the increase in computing capacity with the hardware costs reduction and many new datasets, DL has received much interest in the last decade. In disease detection and diagnosis, as well as in biomedical image segmentation, DL algorithms are fruitful since high-level features from raw images can be automatically extracted using DL to assist clinicians. Furthermore, convolutional neural networks (CNNs), which is a form of DL algorithm, have become crucial with the success in medical image analysis (LeCun et al., 2015). Newly, several studies have begun to utilize DL in new research areas due to the availability of high performance computing hardware components (Pacal, Karaboga, Basturk, Akay, & Nalbantoglu, 2020).

Image recognition is one of the most formidable challenges of the algorithms to learn and solve. The human brain recognizes and categorizes existing as well as new images with almost 100

5G IoT and Edge Computing for Smart Healthcare. DOI: https://doi.org/10.1016/B978-0-323-90548-0.00011-5

percent accuracy when several images are available. ML techniques have been proposed to mimic this behavior of the human brain precisely. Whenever the images are taken under various conditions, such as shift in lighting, translation or rotation of objects, incomplete or hidden objects, different postures in the image, the problem becomes more complicated. These circumstances result in numerous image sets involving the same object, contributing to the recognition or classification problem's complexity. In the past, several image recognition techniques have been developed, such as the k-means clustering, Principal Component Analysis (PCA), Minimum Distance Classifier, Maximum Likelihood Classifier, Bayes Classifier, Support Vector Machines (SVM), etc. The image recognition can be object-based or pixel-based. Each pixel's attributes are extracted in pixel-based recognition to denote it as a member of a specific class. In object-based classification, segmentation is performed to extract objects or regions in the image and evaluate their attributes. In order to be classify medical images in a good manner, image attributes or features must be extracted. The algorithm's efficiency is based on the number of features employed in the procedure. This creates the "curse of dimensionality" problem. Reducing dimensionality, which is similar to reducing features, is needed to reduce the computational burden. The more the number of features, the more data processing and storage is required. This increases the algorithm's time complexity. Algorithms, which categorize objects with a minimum number of features, and less time, are more effective (Vasuki & Govindaraju, 2017).

Classical ML methods were mainly utilized to make predictions and inferences on data until the 1990s. However, it had many disadvantages depending on handcrafted features, which were constrained by human-level precision (Nanni, Ghidoni, & Brahnam, 2017). However, in the case of DL, handcrafted feature engineering is not needed; instead, features are extracted automatically from the data during training. Moreover, with the help of creative techniques, the computational power of modern computers, and the existence of the large volume of datasets, DL can make more precise predictions and classifications. Such a DL model's main capabilities that made them efficient and widely used are their feature extraction, massive parallelism, and nonlinearity capabilities (Goodfellow, Bengio, Courville, & Bengio, 2016). There are different DL algorithms such as Long Short Term Memory (LSTM) (Sak, Senior, & Beaufays, 2014), Recurrent Neural Networks (RNN) (Mandic & Chambers, 2001), CNN (Ghosh, Sufian, Sultana, Chakrabarti, & De, 2020), Generative Adversarial Networks (GAN) (Goodfellow et al., 2014), etc. Following the success of a CNN-based model called AlexNet (Krizhevsky, Sutskever, & Hinton, 2012), several DL models, such as ResNet (He, Zhang, Ren, & Sun, 2016a), VGGNet (Simonyan & Zisserman, 2014), DenseNet (Huang, Liu, Van Der Maaten, & Weinberger, 2017), GoogLeNet (Szegedy et al., 2015), ZFNet (Zeiler & Fergus, 2014), MobileNet have been proposed especially for computer vision problems (Sufian, Ghosh, Sadiq, & Smarandache, 2020; Sultana, Sufian, & Dutta, 2018).

Nowadays, cancer has a high death rate and is one of the leading illnesses that has impacted human health. Cancers are the results of tumors that are malignant. On the other hand, benign tumors, which can be removed easily, are not cancers and can rarely be harmful. In contrast, malignant tumors or cancers are dangerous when they spread irregularly and uncontrollably. Cancer is a main concern of clinicians and scientists who are involved in this area. The researchers introduced a variety of early cancer detection studies since early cancer diagnosis saves human life and is critical to the fight against the disease (Schiffman, Fisher, & Gibbs, 2015). Medical imaging is a powerful application, which plays an important role in the early detection of cancer (Abd El-Salam, Reda, Lotfi, Refaat, & El-Abd, 2014). The analysis of the medical imaging data relevant to the disease progress is exhausting and complicated. Furthermore, if the misdiagnosis of physicians is taken into account, the rate of accuracy declines dramatically, and the early detection can be crucial (Waite et al., 2017). For medical image processing, DL has achieved tremendous progress. In healthcare systems, such as cancer screening and diagnosis, treatment techniques, and disease monitoring, DL practices are utilized. DL is at the forefront of the growing medical data and transforming this data into usable knowledge. In the diagnosis and treatment of colon cancer, the success of DL is truly exceptional. The success of topics such as pathology image analysis or colonoscopy image analysis plays a significant role in diagnosing colon cancer (Pacal et al., 2020).

Parkinson's disease (PD) is a progressive and chronic neurodegenerative disease predominantly influencing balance and movement. Tremor, postural instability, rigidity, and slow movement are the key signs of PD (Durga, Jebakumari, & Shanthi, 2016). The dysfunction and death of essential neurons positioned in the brain cause PD. For contemporary physicians, the exact diagnosis of PD remains a major problem at its early stages. Owing to the similarities of symptoms with other diseases, the challenge of differentiating PD from other neurodegenerative disease is excessive. While the predominant source of clinical diagnosis is the occurrence of traditional clinical symptoms such as tremor, bradykinesia, and other main motor characteristics, PD is related to a variety of nonmotor symptoms that contribute to the total impairment. Thus timely and precise diagnosis of PD is crucial for prompt detection and appropriate diagnosis and for initiating neuroprotective therapies. Neuroimaging methods such as single-photon emission computed tomography (SPECT) presented the diagnosis even at the early stages of the disorder (Booth et al., 2015, p. 1; Cummings et al., 2011; Oliveira, Faria, Costa, Castelo-Branco, & Tavares, 2018). SPECT images are important instruments since PD's progression can be easily seen by the presynaptic dopaminergic deficits in the striatum (Prashanth, Roy, Mandal, & Ghosh, 2016). The consistent evaluation of SPECT images is performed only through visual inspection by skilled nuclear medicine doctors. In certain cases, however, the diagnosis cannot be noticed, particularly in early PD patients. These challenges in the identification of PD patients can be quickly eliminated by incorporating ML techniques. Recent studies have shown that diagnostic methods relying on ML and DL approaches can help clinicians in early diagnosis, clinical preparation, and PD progression monitoring (Orru, Pettersson-Yeo, Marquand, Sartori, & Mechelli, 2012; Prashanth, Roy, Mandal, & Ghosh, 2014). The latest research on the early detection of PD has taken advantage of developments in ML techniques. Particularly, by integrating the feature engineering stage into a learning phase, DL has proved to be an efficient instrument in image recognition (Schmidhuber, 2015). In fact, DL involves a set of data with less preprocessing instead of extracting features in a handcrafted fashion, and then extract the discriminative features through self-learning (Bengio, 2009; LeCun et al., 2015). DL relying on CNNs has proved to be effective instruments for a wide variety of computer vision tasks in medical imaging (Greenspan, Van Ginneken, & Summers, 2016). In order to differentiate PD patients from healthy controls, Martinez-Murcia et al. (2017) proposed CNNs. For PD recognition, Choi, Ha, Im, Paek, and Lee (2017) have introduced a DL based network named PDNet. A DL-based SPECT analysis approach was utilized in their research to enhance the PD diagnosis. The model was trained on the SPECT images taken from PPMI. Their model reported a high performance similar to the evaluation results of the experts. Nevertheless, the DL networks introduced for PD identification are challenging to develop. Since time is of great importance in diagnosing PD, the realistic application of such network solutions is not practical. To fix this issue, Mohammed, He, and Lin (2020) implemented a smaller network and introduced CNNs to use SPECT images to differentiate PD patients from healthy controls. In the detection of PD, the experimental findings revealed high performance. The proposed network architecture outperforms by a significant margin with enhancements in all assessment metrics compared to the previous studies (Choi et al., 2017; Martinez-Murcia et al., 2017). Using a comprehensive repository of SPECT images (2723 images), a CNN-based model was introduced to diagnose PD specifically. With enhancements in precision, sensitivity, and specificity, the developed method surpasses the previous studies. The developed model analyzes whole images and learns features from the images, contributing to maximum efficiency. For the development and implementations, the limited complexity of the network offers a comprehensive benefit. The CNN-based model could change the diagnosis of PD, considering its outstanding performance and lower complexity (Mohammed et al., 2020).

The stage of Alzheimer's Disease (AD) patients was defined by Bringas, Salomón, Duque, Lage, and Montaña (2020) utilizing the DL models and mobility data. This method makes it possible for the condition to be tracked and encourages steps to be taken to ensure ideal treatment and prevent complications. They used data from 35 AD patients obtained from smartphones in a daycare center for a week. Every patient's data sequences recorded accelerometer variations as daily activities were carried out and labeled with the phase of the disorder. To recognize the patterns that define each step, their methodology employs these time series and utilizes a CNN model. Over 90% accuracy was achieved by the CNN-based method, significantly enhancing the traditional feature-based classifiers' results. The results indicate that the data collected from smartphone can be a useful source for the treatment of AD patients and the study of disease progression. Compared to the typical supervised learning models, the proposed CNN-based model enhances the performance of the detection of AD stages.

There has been a sudden increase in the patients affected by COVID-19 (Li et al., 2020), which has increased the load over healthcare systems across the world. But the available hospital beds and personal protective equipment (PPE) (Sawada, Kuklane, Wakatsuki, & Morikawa, 2017) and ventilators are limited. Therefore it is of utmost importance to distinguish patients with severe acute respiratory illness (SARI) (Hatem et al., 2019) who might have COVID-19 infection in order to use the limited resources better. When the number of sick patients is very high at the time of the COVID-19 epidemic, and the disease is still circulating, many study groups use DL methods to screen COVID-19 patients for temperature identification of fever, viral and COVID-19, pneumonia, etc. Furthermore, DL may be employed for other practices, such as patient treatment, systemic breach of social distance identification (Sufian, Jat, & Banerjee, 2020). Wang S et al. (2020) employed a CNN-based DL to scan COVID-19 patients utilizing computed tomography (CT) images with an accuracy, specificity and sensitivity of 89.5%, 88.0%, and 87.0%, respectively. Likewise, Wang, Lin, and Wong (2020) utilized chest X-ray images to diagnose COVID-19 cases with 83.5% accuracy (Sufian, Ghosh et al., 2020).

In this chapter, diagnosis of COVID-19 using chest X-ray images is presented as disease diagnoses example. Using numerous models, one can classify a given X-ray in one of the three classes: normal, COVID-19, and pneumonia. X-ray is advantageous (Hoheisel, Lawaczeck, Pietsch, & Arkadiev, 2005) over most conventional diagnostic tests. Some advantages include cost-effectiveness, easy transport from source to analysis points. Also, unlike CT scans, (Kalender, 2006) portable X-ray machines reduce the utilization of extra personal protective equipment (PPE), as the patient can be tested inside an isolation ward. Further, it reduces the risk of hospital-acquired infection for the patients. With the novel nature of the virus, many radiologists may not themselves be familiar with all the nuances of the infection. They may lack the expertise required to make the diagnosis highly reliable. Therefore this digital method will act as an inspiration for those at the forefront of this research. Nevertheless, we note that both models reflect our current view of this continually changing problem, which is still based on very limited data at the moment.

The rest of the chapter is organized as follows. In Section 8.2, a short introduction to the disease diagnosis is presented. In Section 8.3, pattern recognition tools for the disease diagnosis are given. Section 8.4 presents a case study of COVID-19 detection using an AI approach. Discussion is presented in Section 8.5. Finally, we conclude the chapter in Section 8.6.

8.2 Disease diagnosis

The predictive models can achieve automated disease diagnosis. ML is part of AI, which enables predictive modeling. The idea of ML is to use a large dataset to build a model, predicting the desired outcome for new samples. The general framework for disease diagnosis using AI techniques is shown in Fig. 8.1. ML for disease diagnosis has been applied to medical images in various medical fields including neurodegenerative diseases (Myszczynska et al., 2020), cancer research (Makaju, Prasad,

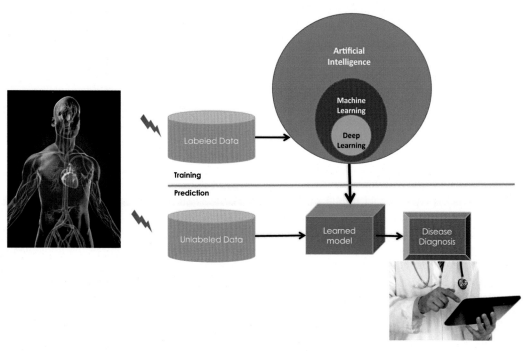

FIGURE 8.1

General framework for disease diagnosis using artificial intelligencetechniques.

Alsadoon, Singh, & Elchouemi, 2018; Yassin, Omran, El Houby, & Allam, 2018), cardiovascular diseases (Wong, Fortino, & Abbott, 2020), COVID-19 (Elaziz et al., 2020), and malaria (Poostchi, Silamut, Maude, Jaeger, & Thoma, 2018). These applications cover various medical imaging techniques such as magnetic resonance imaging (Myszczynska et al., 2020; Wong et al., 2020; Yassin et al., 2018), ultrasound (Wong et al., 2020; Yassin et al., 2018), computed tomography (Makaju et al., 2018; Wong et al., 2020), X-ray (Elaziz et al., 2020), and microscopic images (Poostchi et al., 2018; Yassin et al., 2018).

In ML, the data used for modeling is called training data. Typically, the training data is used to build multiple models. In this case, a separate validation dataset is used for the evaluation of the models. Based on the evaluation, the final model is selected. The final model's performance is estimated using a test dataset, which is again separate from the training and validation datasets. In the ideal case, the training, validation, and test datasets are distinct. In practice, it is not always possible, and the datasets are separated from a larger dataset. Predictive models can be built using supervised or unsupervised learning. Supervised learning means that the model is built using data with known labels. For example, if we want to develop a predictive model for detecting COVID-19, then the data for model building should include a diagnosis for each sample as a label. In unsupervised learning, though, the labels are not used.

Disease diagnosis typically involves the interpretation of clinical tests supplemented with medical images. In the case of COVID-19, the most commonly used test is the reverse-transcriptase polymerase chain reaction or, in short, RT-PCR. However, it is suggested that more sophisticated COVID-19 detection methods are needed (Wu et al., 2020). While reading medical tests is a straightforward task, reading medical images is a more complicated process. In image interpretation software is used to display and analyze the images. In the simplest case, the software is used only to zoom and measure parts of the image. In more complicated cases, the software is used for image analysis, such as segmentation or even diagnosis via AI. In most cases, medical image analysis is manual, or, at best, semiautomatic, and thus automated analysis tools would speed up the analysis and enable reproducible and objective results.

Physicians can use predictive models to decide the condition of a subject. This is called computer-aided decision-making. Predictive models in clinical practice may be used to create, for example, disease risk calculators (Liang et al., 2020; Schalekamp et al., 2020) or automated segmentation tools for images (Shi et al., 2020). The idea of predictive models and computer-aided decision-making is typically not to make the final clinical decision or disease diagnosis but to help the medical experts to do it. While there have been many predictive models for computer-aided decision-making in scientific articles, it is unclear how many of these have been implemented in the software used in everyday clinical work. Moreover, it is unknown how many of the implemented methods have been adopted as a part of clinical practice.

Research on DL's clinical employments has expanded significantly over the past few years, with cancer being the foremost noticeable disease examined and medical images of the predominant data type (Jiang et al., 2017). Two applications of DL for cancer diagnosis are automatic investigation and knowledge discovery. Automatic investigation refers to the need for models for routine clinical diagnostic activities, where success at the expert level was accomplished in many medical fields (Bejnordi et al., 2017; Esteva et al., 2017; Gulshan et al., 2016). In contrast, the knowledge discovery seeks to reveal new patterns in data, which can guide diagnosis, prognosis, response to treatment, or genomic status (Levine et al., 2019).

Segmentation of organs or lesions is often needed in order to promote further analysis and certain therapy methods by making it a key component of computerized systems. Numerous studies were conducted across various organs and forms of pathology (Litjens et al., 2017). Arterys, a startup, recently received FDA approval, the first such approval, for a framework of DL-based oncology image analysis software. Currently, the software concentrates on liver and lung analysis, with permission to eventually expand to all consistent tumors, and can segment lesions, monitoring them, and helping with conventional radiological scoring systems. DL has many other application areas, which can enhance the radiology workflow, in addition to benefits, which directly affect diagnoses, including image quality improvement, content-based retrieval, alignment of multiple images, and research database mining (Liew, 2018; Litjens et al., 2017). A DL-based system treats patients' CT scans for cancer-specific diagnosis by reducing the time to review more urgent images (Titano et al., 2018). In the past, methods focus on the utilization of a handcrafted features described image-based diagnosis in several types of cancer, including breast cancer (Mazurowski, Zhang, Grimm, Yoon, & Silber, 2014), glioblastoma (Gutman et al., 2013), renal cell carcinoma (Karlo et al., 2014), and squamous cell carcinoma of the head and neck (Leger et al., 2017). In research areas such as the prediction of malignant potential in gastrointestinal stromal tumors (Ning et al., 2018) and breast cancer molecular subtype (Zhu et al., 2019) based on imaging incorporating DL techniques have achieved promising results (Levine et al., 2019).

DL offers a substantially different approach for the analysis of histopathology images than handcrafted feature-based methods. DL architectures utilize the original feature extraction step as end-to-end systems. Instead, images are fed straightforwardly into the model after simple preprocessing, including its automatic feature extraction into the network's previous layers under a large parameter space. An essential criterion with these techniques were only tested on metastatic cancer. Further studies in breast cancer also include tumor regions segmentation in breast biopsy slides (Cruz-Roa et al., 2017), discrimination between numerous cancer types histotypes and benign breast changes (Motlagh et al., 2018), and cancer identification based entirely on changes in the local stroma (Bejnordi et al., 2018). The evaluation of prostate biopsies and laparoscopic samples is another task, which is well tailored to computerization (Fricker et al., 2018; Levine et al., 2019).

More recently, a Google team has released its large-scale analysis on prostate cancer scores in prostatectomy samples (Nagpal et al., 2019). The Cancer Genome Atlas (TCGA) digital slide archive already proved to be a rich source of knowledge discovery for integrating histology with clinical and molecular evidence. In order to forecast glioma outcomes, Mobadersany et al. (2018) proposed a CNN model called a survival CNN. Their model could discern findings within molecular subtypes of glioma based on histology alone, thus obtaining increased prognostic precision by integrating histology with typical genomic markers. A CNN-based computational staining technique was employed by Saltz et al. (2018) to trace tumor-infiltrating lymphocyte patterns in more than 5000 slides through 13 types of cancer and compare them with molecular subtypes and survival. DL has been employed in prostate cancer to computerize the detection of the most pathological areas on slides in predicting the status of speckle-type POZ protein (SPOP) (Schaumberg, Rubin, & Fuchs, 2018). Moreover, the authors utilized a groundbreaking approach to overcome the problem of dataset imbalance of unusual mutations by developing an ensemble model trained on subsets of the data with selected number of negative and positive slides (Levine et al., 2019).

Dopaminergic degeneration is a Parkinson's disease (PD) pathological hallmark that can be evaluated by the visualization of dopamine transporters such as FP-CIT SPECT. DL based FP-CIT

SPECT analysis approach was implemented by Choi et al. (2017) to improve the prognosis of PD. This scheme, trained by SPECT images of PD patients and normal controls, achieved a superior classification performance comparable to experts' assessment. Moreover, they found that the proposed automated method could reclassify certain patients clinically identified as PD who had scans without evidence of dopaminergic deficit (SWEDD). The findings showed that the DL-based approach could define FP-CIT SPECT correctly and eliminate human assessment inconsistency. It may assist in imaging prognosis of patients with ambiguous Parkinson's disease and, in more clinical trials, deliver objective patient group classification, especially for SWEDD.

Ear and mastoid disorders are common diseases that can safely be treated with early medical treatment. Nonetheless, if one may not undergo prompt diagnosis and adequate care, sequelae, such as hearing damage, can be left behind. Physical inspection using traditional otoscopy or otoendoscopy and historical examination is the first step in assessing ear and mastoid disease in the clinic. Fortunately, diagnosis utilizing otoscopy or otoendoscopy by nonotolaryngologists is particularly vulnerable to misdiagnosis (Blomgren & Pitkäranta, 2003). The shortage and relatively low diagnostic precision of specialists in the local clinic demands a new type of prognostic approach where ML can play a crucial role. DL were successfully applied to diverse areas of medicine. Most of these studies use a supervised DL process, the CNN. Building CNN from scratch, though, needs a significant volume of dataset and computing capacity that is not feasible in many implementation areas. Instead, a particular application, called transfer learning, may be reused and fine-tuned with existing CNN models pretrained for natural images. Many network layers in a public network model are transferred to a different model during transfer learning, followed by a new fully-connected layer, which classifies certain functions into a new class collection. Medical imaging with transfer learning tests demonstrated a high degree of classification performance equal to, or even higher than constructing a CNN from scratch (Kermany et al., 2018; Shin et al., 2016). Cha, Pae, Seong, Choi, and Park (2019) examined the efficiency of nine models to select the best models. Based on this appraisal, an ensemble classifier was suggested to integrate the classification effects of several models that is supposed to improve the overall classification efficiency relative to a single classifier. Although transfer learning is proven to be effective in a relatively limited dataset, there is still no example of the significance of the classification and model form on the scale of the dataset. Therefore based on the data size, they checked the classifier performance.

COVID-19 is an infection, which affects respiratory symptoms and causes deaths across the globe. In human life, the early prognosis of this infection is crucial. This phase is advancing rapidly with diagnostic experiments relying on DL. A DL model, which can be utilized for early diagnosis of the COVID-19 disease, has been suggested by Canayaz (2020). A data set composed of three classes of healthy, COVID-19, and pneumonia lung X-ray images, with each class comprising 364 images, was employed for this approach. Preprocessing was done on the created data set utilizing the image contrast enhancement algorithm, and a new data set was created. Feature extraction was conducted from this data collection utilizing pretrained models such as VGG19, ResNet, AlexNet, and GoogLeNet. Two metaheuristic algorithms of binary gray wolf optimization and binary particle swarm optimization were employed to choose the best possible features. They were categorized using SVM after combining the elements achieved after the feature selection of the optimized data collection resulting in an average accuracy of 99.38%. The experimental results achieved by two separate metaheuristic algorithms revealed that the proposed methodology would support clinicians during the clinical diagnosis of COVID-19.

To identify the patient diagnosed with COVID-19 from X-ray images, Altan and Karasu (2020) introduced a hybrid model composed of two-dimensional (2D) curvelet transform, chaotic salp swarm algorithm (CSSA), and DL approach. The 2D Curvelet transform is used with the patient's chest X-ray images in the developed framework, and a feature matrix is generated utilizing the coefficients obtained. With the utilization of CSSA, the feature matrix's coefficients are optimized, and COVID-19 disease is identified by the EfficientNet-B0 DL model. Experimental studies indicate that the suggested method can diagnose COVID-19 disease with high precision using chest X-ray images.

8.3 Pattern recognition tools for the disease diagnosis

The classification of medical images refers to the classification of different objects to detect diseases in images such as MRI, X-ray, CT, positron emission tomography, etc. You have to define and classify the numerous objects or regions in the medical image. The classification technique evaluates accuracy based on a single medical image or image set. If the medical image sets are employed, multiple images of the same object with different views and under different circumstances will be included in the set. Compared to classifying single medical images, it can be more efficient in classification, as the algorithm can handle various factors, such as background variations, lighting, or appearances. Image rotation and other transformations can also be invariant. The numerical pixels of the image are the input to the ML algorithm. A value or series of values representing the class would be the output. The ML algorithm is a mapping function that maps the values of the pixels to the appropriate categories, namely, to the classes. It can be a single multivalued function output, or it can be a single-valued function multioutput to show the class. Classification techniques might be supervised or unsupervised for the diagnosis of disease. The number of classes is known in the supervised classification, and a collection of training data with details about their class is given. It is like a teacher's learning. The number of classes is not specified in the unsupervised grouping, and the training data is not available. The interaction (or mapping) must be learned between the data to be categorized and the different groups. Without a teacher, it is like studying. If some knowledge about mapping data to groups is available, it is possible to merge supervised and unsupervised methods to become semisupervised. The most important parameters associated with input data, depending on which the information is categorized, are features. In classification, identifying specific features of the object as characteristics plays a crucial role. For classification, the selection of features from the objects in the medical image is often utilized (Vasuki & Govindaraju, 2017).

Among biologically inspired AI techniques, CNNs are among the most frequently utilized DL architectures, which achieve the best performance. By proposing numerous models to enhance the efficiency of CNNs, several researchers have contributed. Deep CNN's different classical and modern architectures are usually employed as the building block of different architectures for classification, diagnosis, and segmentation. The AlexNet architecture is the architecture that initiated the CNN architectures (Krizhevsky et al., 2012). VGG architecture (Simonyan & Zisserman, 2014) has, however, achieved tremendous popularity. The VGG16 and VGG19 models are the most powerful of the VGG architecture. It is commonly used in CNN architectures for object identification, diagnosis, and classification as a backbone. In addition, with 22 layers, GoogLeNet

(Szegedy et al., 2015) was the competition winning model in 2014. Employing modules called Inception, this architecture minimized the computational complexity and the risk of overfitting. In 2015, the ResNet (He et al., 2016a) architecture, composed of 152 layers, won the ILSVRC competition. Eventually, several new architectures were introduced, such as MobileNets (Howard et al., 2017) and DenseNet (Huang et al., 2017). ResNet, VGG, and Inception are the most widely utilized architectures for image recognition and disease diagnosis. Also, while CNN architectures were initially employed for classification purposes, they were later employed broadly in object recognition, partitioning applications, and disease diagnosis because of their popularity (Pacal et al., 2020).

8.3.1 Artificial neural networks

Artificial neural networks (ANN) is a group of connected input/output units that have a weight relevant to each relation. During the learning process, the network will learn by changing the weights so that the input tuples can predict the appropriate class mark. These need a variety of parameters, which are usually better empirically defined, such as the topology of the network or architecture. The benefits of neural networks involve their high tolerance to noisy data. They can be employed when the relationships between attributes and classes are little known to you. They are well-matched for continuous-valued inputs and outputs when compared to most decision tree algorithms. The most popular ANN architecture is the backpropagation that performs learning on a multilayer feed-forward neural network. Such a network composed of an input layer, one or more hidden layers, and an output layer. Each layer consists of units. The network inputs refer to the attributes for each training tuple being trained. The inputs are fed into the units, which make up the input layer at the same time. All inputs move through the input layer and are then instantaneously weighted and provided to the second layer of "hidden" units. The hidden layer unit outputs can be input into another hidden layer, and so forth. There is no fixed rule to choose the hidden layers to be used, but mostly a single hidden layer is used, and the number of layers is increased if necessary. It is known as the feed-forward network since none of the weight's cycles back to a previous layer. In each layer, various activation functions improve the performance measures with mostly a nonlinear activation function seemingly offers better results. Again, there is no such fixed algorithm to determine which activation function to be used at each layer, but it is determined by tuning each layer with different activation functions. By iteratively processing a data set of training tuples, backpropagation learns how to equate the network's prediction for each tuple with the actual known target value. The target value can be the training tuple's established class mark (for classification issues) or a continuous value (for numerical prediction). The weights are adjusted for each training tuple to minimize the mean squared error between the network's prediction and the actual target value (Han, Pei, & Kamber, 2011).

8.3.2 K-nearest neighbor

K-nearest neighbor (k-NN) finds in the training set a group of k objects closest to the test object and bases the assignment of a label on the predominance of a particular class in the neighborhoods. This method has three main elements: a collection of labeled objects, a distance or similarity metric for measuring distance between objects, and the value of k, the number of nearest neighbors. The distance of this object to the labeled objects is calculated to classify an unlabeled object.

The unknown class label is assigned after the k-nearest neighbors are determined with the help of these identified objects. One crucial aspect is the choice of the distance function. Although various methods can determine the distance between two points, the most appropriate distance measure is a smallest distance between two objects. For example, in identifying documents, using cosine measure instead of the Euclidean distance can be considered better (Wu et al., 2008).

8.3.3 Support vector machines

SVM are supervised learning techniques employed for the classification, regression, and outlier detection. There are particular forms of SVMs that can be employed for specific ML problems, such as regression and classification. SVMs differ from other classification methods since they select a decision boundary, which maximizes the distance from the closest data points of all classes. The decision boundary provided by the SVMs can be referred to as the maximum margin hyperplane or the maximum margin classifier. A basic linear SVM classifier operates by constructing a straight line or a hyperplane between two classes. This means that all data points on one side of a line or a hyperplane can indicate a class. The data points on the other side of the line/hyperplane would be classified in a separate category, with the possibility of an unlimited number of lines/hyperplanes to choose from. What makes a linear SVM learner stronger than any of the other learner, such as the k-nearest neighbors, decides the optimal line/hyperplane to categorize the data points. Choose a line/hyperplane that divides the data and is as far away as possible from the nearest data points (Evgeniou & Pontil, 1999).

8.3.4 Random forests

Random forests (RF) can be constructed using tandem bagging with a random selection of the attributes by growing tree votes and returning the most successful class at ranking. For the determination of the split, the selection of decision trees is based on some aspects. More precisely, for all trees in the forest, each tree depends on the values of an individually and evenly distributed random vector sampled. The trees are grown using the CART methodology. The trees are planted to a full size and are not pruned. For a forest, the generalization error converges as long as there are many trees in the forest. So, overfitting is not a problem. Accuracy entirely depends on the performance of individual classifier trees. Correlation between trees should be avoided to strengthen the individual trees. RF are efficient so that they consider very random and few attributes that are necessarily important for classification. They can be quicker than getting bagged or boosted trees (Han et al., 2011). External calculations measure error, power, and correlation are used to illustrate the answer by increasing the number of features used in the split. Internal estimates are also used to gage variable significance (Breiman, 2001).

8.3.5 Bagging

Different model decisions can be combined into one single prediction. Opinions are taken in classification and the average is calculated for numerical prediction. The models have equal weight in bagging, while weighting is used to impact the more efficient ones depending on the past. In bagging, numerous same-sized training datasets are randomly selected from the question to construct a decision tree for each data set. Such trees are essentially similar to every new test instance, making the same prediction that some trees offer correct results while others fail. Bagging attempts to remove the instability of ML approaches by utilizing random sampling. This sampling technique eventually eliminates some of the instances and duplicates the remainder. The resampling data sets are not independent since they are all created from one dataset. Yet bagging yields a combined model, which achieves considerably better results than the single model generated from the original training data. Bagging may be used for numerical prediction, in which the outcomes of the individual predictions are summed instead of voting (Hall, Witten, & Frank, 2011).

8.3.6 AdaBoost

AdaBoost or adaptive boosting is one of the ensemble boosting classifier that combines multiple weak classifiers to improve the classifiers' accuracy. It is an iterative ensemble method. It constructs a robust learner by combining many weak learners to get a highly accurate and robust learner. The basic idea behind AdaBoost is to adjust the weights of learners and training the data sample in every iteration in a way that it guarantees the precise predictions of unusual observations. Any ML approach can be employed as a base classifier if it takes the weights on the training set.

AdaBoost must provide two conditions:

1. The learner must be trained interactively on numerous weighted training instances.
2. In each iteration, it tries to deliver an outstanding fit for these instances by minimizing training error.

AdaBoost adds learners to the ensemble, iteratively making it better. This algorithm's significant disadvantage is that the model cannot be parallelized since each predictor can only be trained after the previous one has been trained and assessed (Chengsheng, Huacheng, & Bing, 2017).

8.3.7 XGBoost

XGBoost stands for Extreme Gradient Boosting, which applies a Gradient Boosting technique based on decision trees. It constructs short, basic decision trees iteratively. Each tree is termed as a "weak learner" because of its high bias. XGBoost begins by building the first basic tree that has a poor performance. Then it builds another tree, trained to predict what the first tree, which is a weak learner, cannot do. The technique sequentially produces weaker learners, each correcting the previous tree before the stopping condition is met, such as the number of trees (estimators) to be created. XGBoost has additional advantages: training is speedy and can be parallelized/distributed across clusters (Ramraj, Uzir, Sunil, & Banerjee, 2016).

8.3.8 Deep learning

ML includes the design and analysis of algorithms that allow a computer to extract (or learn) functions from a dataset (sets of examples). It is necessary to consider three concepts to understand what ML implies: dataset, algorithm, and function. A dataset is a table in its simplest form, in which each row includes the definition of one example from a domain (Kelleher, 2019). A mathematical model that is (loosely) affected by the human brain is a DL network. Hence in order to understand DL, it is needed to provide an abstract understanding of what a mathematical model is, how model parameters can be tuned, how we can compose models, and how we can use geometry to explain, how a model interprets knowledge. The term DL defines a family of neural network models, which includes multiple layers of basic information processing units, known as neurons, in the network are defined (Kelleher, 2019).

It has been shown that neural networks are ideally adapted for the application of ML algorithms. There is one input layer, one output layer, and two or three hidden layers in conventional neural networks. There is one input layer, one output layer, and several hidden layers for deep neural networks. The more hidden layers there are, the stronger the network is. The layers are interconnected, with the new layer's input being the output of the previous layer. The inputs/outputs are weighted, and the weights determine the network efficiency. The network's training requires having the appropriate weights for the different layers. Deep networks need higher computational power, computation speed, a big dataset, and parallel processing of the appropriate software (Vasuki & Govindaraju, 2017).

DL is AI subfield, which focuses on developing massive neural network models capable of making correct decisions driven by data. DL is specifically tailored to circumstances in which the data is challenging and where big datasets are available. DL is utilized in the medical field to process diagnostic images (X-rays, CT, and MRI scans) and detect health problems. DL is now at the heart of self-driving vehicles. It is used for localization and navigation, motion planning and steering, awareness of the environment, and driver status monitoring (Kelleher, 2019).

8.3.9 Convolutional neural network

CNNs (Albawi, Mohammed, & Al-Zawi, 2017) take, process, and classify an image input into those categories. Computers interpret an input image as an array of pixels, which depends on the image resolution. Based on image resolution it can be height, width, or depth. Regularized models of CNNs are multilayer perceptrons. "Absolute reliability" of such networks makes them vulnerable to overfitting results. Typical methods of regularization involve adding the loss function to some form of weight calculation. CNNs use a particular approach to regularization: they take advantage of the hierarchical structure of data by using smaller and simplified patterns and generate more complicated patterns. Therefore CNNs are at the lower end on the connectivity scale and complexity. Convolution is the first layer for extracting features from an input file. Convolution maintains the relationship between pixels by employing small squares of input data to learn image properties. It is a mathematical operation containing two inputs, such as an image vector and a filter or kernel. Pooling layers will reduce number of parameters once the images become too large. Spatial pooling, also known as down-sampling or subsampling, reduces each map's dimensionality but retains essential information. Fully-connected layers in one layer bind each neuron to each neuron in another layer. The flattened matrix moves through a totally connected row to define the images.

8.3.10 Transfer learning

One of the difficulties encountered in the image analysis is that labeled training data might not be accessible for a specific use. Considering the condition where one has a collection of images, which should be utilized for image retrieval. Labels do not exist in retrieval applications, so the features must be semantically consistent. In certain other instances, one may be willing to perform classification on a data set with a specific set of labels that may be limited in accessibility and different from extensive resources like ImageNet. Since neural networks need a lot of training data to construct from scratch, these settings trigger complications. The categories selected and the wide range of images in the data set are so comprehensive and exhaustive that they can derive image features for general-purpose settings. For instance, an entirely different image data set can be generated by passing the features extracted from the ImageNet data through a pretrained CNN like AlexNet and extracting the multidimensional features from the fully-connected layers. For an entirely different application, such as clustering or retrieval, this new representation may be utilized. This type of methodology is so popular that CNNs are barely trained from scratch. This kind of off-the-shelf feature extraction approach can be used as a kind of transfer learning since a shared resource such as ImageNet can be utilized to extract features in the circumstances in which adequate training data is not accessible to solve various problems (Aggarwal, 2018).

Suppose any extra training data is accessible and can only be utilized to fine-tune the deeper layers. The weights of the early layers are fixed since the previous layers catch only simple features such as edges. In contrast, the deeper layers capture more complex features. In other words, previous layers tend to capture features that are highly generalizable, whereas later layers tend to capture features unique to data (Aggarwal, 2018). In addition, VGG (Simonyan & Zisserman, 2014) highlighted the increasing trend in terms of expanded network depth. Although the best-performing models had 16 or more layers, the tested networks were built with different sizes varying from 11 to 19 layers. VGG's breakthrough is that it decreased filter sizes but boosted depth. It is essential to realize that increased depth requires a reduced filter size. Owing to more rectified linear unit (ReLU) layers, a deeper network would have more nonlinearity and more regularization as the increased depth imposes a structure on the layers through repetitive convolution composition. The VGG still uses 3×3 spatial footprint filters and 2×2 scale pooling. With step 1, the convolution was performed, and a padding 1 was used. In phase 2, the pooling was utilized. Another curious design decision for VGG was that after each max-pooling, the number of filters was always boosted by a factor of 2. Whenever the spatial footprint decreased by a factor of 2, the intention was still to maximize the depth by a factor of 2. Note that after each max-pooling, the number of filters increases by a factor of 2 (Aggarwal, 2018).

ResNet (He et al., 2016a) included 152 layers that was almost an order of magnitude greater than most architectures had historically employed. In the training of such deep networks, the critical problem is that the gradient flow between layers is hindered by the vast number of deep layer operations, which can decrease or increase the gradient size. The basic unit is called a residual module, and by adding together all these basic units, the whole network is formed. An adequately padded filter with a step of 1 is utilized in most layers so that the spatial scale and depth of the input do not change from layer to layer. Some layers, indeed, use stride convolutions to minimize by a factor of 2 per spatial dimension. When employing a more significant number of filters, depth is expanded by a factor of 2. In such a scenario, the identity function over the skipped connection should not be used (Aggarwal, 2018).

Inception models were trained in a partitioned fashion, in which every replica was divided into numerous subnetworks to match the whole model in memory. However, the Inception design is extremely tunable, which ensures several potential adjustments to the number of filters in the different layers that do not impact the fully-qualified network's efficiency. Cheaper Inception blocks than the initial Inception are used for the residual iterations of the Inception networks. Each Inception block is accompanied by a filter-expansion layer used before the inclusion to balance the input's depth to scale up the filter bank's dimensionality. This is required to compensate for the reduction in dimensionality caused by the block of Inception. Another minor technological difference between the residual and nonresidual variants of Inception is that batch-normalization is only used on top of conventional layers in the case of Inception-ResNet, but not on top of summations (Szegedy, Ioffe, Vanhoucke, & Alemi, 2017).

A CNN architecture solely based on depthwise separable convolution layers was suggested by Chollet (2017). In reality, in the role maps of CNNs, the mapping of cross-channel correlations and spatial correlations can be fully decoupled. This theory is a better variant of the Inception architecture hypothesis, and Xception, which stands for "Extreme Inception," is the proposed architecture. To map cross-channel correlations, an "extreme" variant of an Inception module will first utilize a 1×1 convolution and then map each output channel's spatial correlations individually. This extreme type of an Inception module is almost similar to a depthwise separable convolution, an operation employed in the neural network architecture. A depthwise separable convolution is generally referred to in DL systems as "separable convolution," composed of a depthwise convolution. It is also possible to make other intermediate formulations of Inception modules, which lie in between regular Inception modules and depthwise separable convolutions. 36 convolutional layers in the Xception architecture form the network's feature extraction framework. The experimental assessment image description is exclusively explored, and a logistic regression layer would then adopt the suggested convolutional base. Preferably, before the logistic regression layer, one might introduce fully-connected layers. The 36 convolutional layers are organized into 14 modules, all of which have linear residual relations across them except for the first and last modules (Chollet, 2017).

A different connectivity pattern is suggested by Huang et al. (2017) to further optimize the information flow between layers by adding direct links to all subsequent layers from each layer. This network architecture is called the Dense Convolutional Network (DenseNet) because of its dense connectivity. Three sequential operations, batch normalization, followed by a ReLU, and a 3×3 convolution are known as a composite function. When the size of the feature-maps increases, the concatenation operation is not feasible. However, down-sampling layers that alter the size of feature-maps are an integral component of convolutional networks. The network is divided into many densely interconnected dense blocks to allow down-sampling in the proposed architecture. The layers within blocks are referred to as layers of transformation, convolution, and pooling. A batch normalization layer and a 1×1 convolutional layer followed by a 2×2 average pooling layer are the transition layers employed in the experiments. DenseNet has very thin layers, which is the difference from existing network architectures. It has been noted that before each 3×3 convolution, a 1×1 convolution may be incorporated as a bottleneck layer to lessen the quantity of input feature-maps and thereby increase classification performance. The number of feature-maps at transition layers could be minimized to maximize model robustness further. A global average pooling is applied at the end of the last dense block, and then a SoftMax layer is connected. In the three dense blocks, the feature-map dimensions are 32×32, 16×16, and 8×8, respectively.

The MobileNet (Howard et al., 2017) model is focused on depthwise separable convolutions that are a type of factorized convolutions to divide a regular convolution into a depthwise convolution. The depthwise convolution for MobileNets utilizes a single filter to each input channel.

To merge the outputs of the depthwise convolution, the pointwise convolution uses a 1×1 convolution. In one step, a typical convolution filters and merges inputs into a new set of outputs. This is divided into two layers: a depthwise separable convolution, a separate layer for filtering, and a separate layer for merging. This factorization has the potential to reduce computation and model size significantly. MobileNets employ Batchnorm and ReLU nonlinearities for both layers. Compared with regular convolution, depthwise convolution is exceptionally effective. It filters just input channels, though, and it does not merge them to create new features. In order to produce these new features, an extra layer, which calculates a linear combination of the output of depthwise convolution through 1×1 convolution, is therefore required. The fusion of convolution depthwise and pointwise ($1 \times$) convolution is referred to as depthwise separable convolution. MobileNet uses 3×3 depthwise separable convolution that utilizes 8 to 9 times less computing than regular convolutions with just a slight decrease in precision.

8.4 Case study of COVID-19 detection

DL is a model employed to make exceptional advances in retrieving information from images by computers. In certain areas of medicine, mainly radiology and pathology, these techniques have been introduced to tasks and, in some cases, achieved efficiency similar to that of human experts. It could also be employed to provide information on molecular state, prognosis, or sensitivity to treatment (Levine et al., 2019). CNNs are DL architecture suggested in disease diagnosis to solve image processing problems and adapt very well to images. Layers in separate blocks are connected in a CNN architecture instead of being directly connected. The flow of knowledge between these blocks is analogous to the visual cortex and reduces the difficulties of classical ML approaches. It can also automatically extract features from the raw data while eliminating the challenges of manual extraction. The network will allow faster learning in this way and eliminate the overfitting problems. CNNs are called data-hungry DL architectures since it can train millions of parameters to help diagnose diseases by using medical images (Pacal et al., 2020).

Transfer learning is a methodology that efficiently utilizes an already trained model to solve another new problem with minimum re-training or fine-tuning requirements (Pan & Yang, 2010; Zamir et al., 2018). Compared to classical ML approaches, DL requires a significant volume of training data. Therefore the need for a large number of labeled data is a major challenge in solving some critical domain-specific tasks, especially disease diagnosis applications (Altman, 2017). A DL-based transfer learning, DTL, drastically decreases the need for training time and training data for a target domain-specific allocation by selecting a pretrained model for a defined feature extractor (Koitka & Friedrich, 2016) or further fine-tuning (Kumar, Kim, Lyndon, Fulham, & Feng, 2017). A DTL model can be trained for feature extraction utilizing a benchmark dataset. This pretrained model is then further employed to solve a new problem such as the role of COVID-19 by modifying only a few last layers in the architecture head and needed fine-tuning. A VGG-based transfer learning model (Simonyan & Zisserman, 2014) was proposed by Gao and Mosalam (2018). Classification of anomalies in MR images by DTL is proposed by Talo, Baloglu, Yıldırım, and Rajendra Acharya (2019). The developers of that study have used the pretrained fine-tuning ResNet34 model. Their system used adaptation to condition distribution. Q-TRANSFER, which is another DTL model, is suggested by Phan, Sultana, Nguyen, and Bauschert (2020). A DTL-based reinforcement learning technique is employed to alleviate the dataset insufficiency issue in the context of information networking. As the outbreak of COVID-19 disease is alarming worldwide, the detection, quarantine, and adequate care for COVID-19 patients have become a priority. However, with substantial false negative findings, the worldwide standard diagnostic pathogenic laboratory procedure is massively time-consuming and more expensive (Santosh, Parmar, Anand, Srikanth, & Saritha, 2020, p. 19). Around the same time, because of the inadequate facilities and areas compared with many cases, studies are seldom to be carried out in the common health centers or hospitals. Researchers from this area are working hard to build potential DTL models to alleviate these problems to tackle this situation (Apostolopoulos & Mpesiana, 2020; Loey, Smarandache, & Khalifa, 2020). As in the case of Loey et al. (2020), DTL is utilized along with the GAN model to diagnose the COVID-19 disease-based on chest X-ray images. They have three state-of-the-art pretrained models, namely AlexNet (Krizhevsky et al., 2012), GoogLeNet (Szegedy et al., 2015), and ResNet18 (He et al., 2016a). In their experiments, the highest precision is achieved by GoogLeNet among these three pretrained models (Sufian, Ghosh et al., 2020). In this chapter, a transfer learning-based feature extraction technique is utilized to diagnose the COVID-19 using X-ray images. The general framework for COVID-19 detection using AI techniques is shown in Fig. 8.2.

Although there was no broad clinical adoption in these early forays into the computer-aided diagnosis, it must be remembered that they employed technologies that followed the explosion of

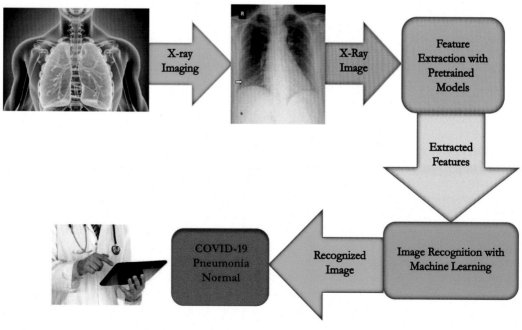

FIGURE 8.2

General framework for COVID-19 detection using artificial intelligence techniques.

DL. Recent head-to-head comparisons shown superior DL performance over other systems (Kooi et al., 2017; Wang et al., 2017). Numerous recent research on big datasets have shown that the efficiency of DL systems is comparable to that of experts conducting traditional diagnostic tasks through several modalities, including spine MRI (Jamaludin et al., 2017), head CT (Merkow et al., 2017), mammography (Kooi et al., 2017), chest X-rays (Rajpurkar et al., 2017), and limb trauma X-rays (Olczak et al., 2017) With the growing evidence that CT chest screening can minimize lung cancer mortality, substantial interest has been created by the automatic identification and assessment of lung nodules (Levine et al., 2019; Setio et al., 2017).

Various DL-based approaches have been developed to identify different thoracic diseases, including pneumonia. Most recently, a lot of research work has been done on the detection of COVID-19 (Makris, Kontopoulos, & Tserpes, 2020; Narin, Kaya, & Pamuk, 2020). In particular, the DL methods that have been employed on the radiography images have shown promising results with high precision. DeTraC (Abbas, Abdelsamea, & Gaber, 2020), a relatively new CNN architecture, is based on transfer learning and class decomposition to enhance the pretrained models' performance on X-ray image classification. It includes three phases. The first phase conducts the deep feature extraction using a pretrained CNN by utilizing ImageNet (Krizhevsky, Sutskever, & Hinton, 2017). The second phase of DeTraC employs the Stochastic gradient descent optimization method (Bottou, 2010) for training the model. The final phase, also known as the Class-composition component, is acquired for the final classification of the images. It uses error-correction measures at the SoftMax layer.

Zhang et al. (2020) developed a new deep anomaly detection model for fast and consistent screening of COVID-19 cases based on chest X-ray images. The model designed in their approach comprises of three components: (1) a backbone network, (2) a module for the detection of anomalies, and (3) a classification module. The backbone network extracts the chest X-ray image characteristics into a d-dimensional feature vector, fed into the classification module. The classification module includes a new deep convolution neural network (DCNN) comprised of a hidden layer, the "sigmoid" activation function, and a single neuron output layer. The anomaly detection module has a similar architecture as the classification module, except it produces the scalar anomaly scores to reveal the anomalous images. This particular model has proven to be a successful approach to reduce the false-positive rate.

In another study focusing on detecting COVID-19 cases from chest X-ray images, namely the COVID-Net (Wang & Wong, 2020), a deep CNN incorporates the human-machine collaboration and machine-driven design exploration stage, employs a lightweight residual projection-expansion-projection-extension (PEPX) design pattern. In addition to COVID-Net, the authors also established an explainability-driven audit approach for decision validation. A similar approach consists of seven DL image classifiers proposed in COVIDX-Net (Hemdan, Shouman, & Karar, 2020) as a framework for integrating DL models altogether. Comparing the state-of-the-art DL modes, the best performance was accomplished by the DenseNet201 and VGG19 classifiers with a classification accuracy of 90%. A recent study, the Monte-Carlo Dropout method, and Bayesian CNN have been investigated to measure this uncertainty (Seoh, 2020). Using COVID-19 X-ray images, the Bayesian DL classifier was trained employing transfer learning on a pretrained ResNet50V2 (He, Zhang, Ren, & Sun, 2016b) model utilizing COVID-19 X-ray images to evaluate the model uncertainty. Their analysis has shown to quantify the uncertainty to enhance the efficiency of human-machine decisions.

8.4.1 Experimental data

We use the public datasets available on Kaggle.com, the Coronahack X-ray Dataset[1], and the chest X-ray pneumonia[2]. The datasets contain 6253 images, of which 1495 are normal lung images, 4371 are X-ray images infected with pneumonia, and 287 images are those infected with COVID-19. We have divided the data into three sets, namely, training, validation, and test sets. Initially, the original data is divided into a 95% set, further divided into 95% and 5% into training and validation sets. The division of the 5% original dataset was labeled as the test set.

8.4.2 Performance evaluation measures

The training set performance measures cannot be assumed similar to that of the test set. The training set may hold the bulk of examples, but the test set is identical to the real-world data. For ConvNets, having a low amount of data is a significant hindrance. The methods used to evaluate the performance measures in such a case are still controversial. The majority of the support is given to the performance measures obtained from the cross-validation sets. The classifier is tuned to provide the best performance measures in the case of the training set. The most important fact to have in mind while training is that none of the test set examples should be used while designing the classifier. Hence the performance measures from the test set can be assumed to be a close comparison to the performance measures on the real-world images. A learner classifies an image into different categories. If the category matches the given class, then it is assumed to be a success. If there is any mismatch, then it is assumed to be an error. The performance measures are mostly based on the error rate of a classifier. The training set holds a bulk of examples. The validation dataset should have a similar distribution to the test set to have the best performance measures. The validation set should be utilized for parameter tuning and the test set for finding the final values of performance measures (Hall et al., 2011).

When vast volumes of data are available, there is no problem: We take a large sample, use it for preparation, and presume that the remainder should be used for processing. In general, the efficiency of a classifier increases with an improvement in the size of the training range. Still, after a certain stage, the efficiency stays constant and does not change further (Hall et al., 2011). In this study accuracy, F1-measure and Kappa statistic is used as a performance evaluation measure.

8.4.3 Feature extraction using transfer learning

DL techniques require many examples or a large set of data. Yet COVID-19 current situation is relatively new, and therefore the number of X-ray images for the classification function is fewer, so we can go for transfer learning. Transfer learning (Hussain, Bird, & Faria, 2018; Pan & Yang, 2010) is about taking features learned on one domain's problem and using them towards another, relatively similar issue. Transfer learning is aimed at improving learning in the target task by incorporating information from the source task. The key facts to test before using transfer learning are that the source task should have been trained on a larger dataset than the target task, and the source and target

---

[1]https://www.kaggle.com/praveengovi/coronahack-chest-xraydataset?select = Coronahack-Chest-XRay-Dataset
[2]https://www.kaggle.com/paultimothymooney/chest-xray-pneumonia

tasks should be identical in nature. Three common measures could improve learning through transfer. First is the initial achievable output in the target task utilizing only the transferred information, compared to an ignorant person's initial output, before any more learning is performed. Second, the amount of time it takes to understand the target mission thoroughly, despite the information transferred, relative to the amount of time it takes to understand it from scratch. Third, the final output level can be reached in the target role close to the last level without transition.

Here we have used different pretrained models like VGG16, VGG19, Inception v3, MobileNet, DenseNet169, DenseNet121, InceptionResNetV2, MobileNetV2, ResNet101. Simonyan and Zisserman (Simonyan & Zisserman, 2014) presented the VGG network architecture (Jaderberg, Simonyan, & Zisserman, 2015). This network is distinguished by its simplicity, using just three universal layers layered on top of each other in rising detail. Max-pooling works to high volume capacity. In comparison to conventional sequential network architectures like AlexNet, OverFeat, and VGG, ResNet is instead a type of "exotic architecture" based on micro-architecture modules. The term microarchitecture refers to the collection of "building blocks" that were used to create the ResNet architecture, which was first presented by He et al. (2016a). They demonstrated incredibly deep networks that can be equipped using normal Stochastic Gradient Decent (and a fair initialization function) by using residual modules. The Inception module's goal is to act as a "multilevel feature extractor" by computing 1 bracket, 3 bracket, and 5 brackets within the same network module—these filter outputs are then stacked along the channel dimension before being fed into the next network layer. GoogleLeNet was the initial implementation of this architecture, but later implementations were simply named Iteration vN, where N corresponds to the version number Google brought out. Xception was developed by François Chollet, founder and chief maintainer of the Keras library. Xception is an extension of the standard Inception architecture.

Transfer learning is one of the most effective approaches, particularly in image recognition and classification for building accurate classifiers when there is a limited supply of target training data. To be more precise, in transfer learning, we use the knowledge gained from rich labeled data in a source domain to improve our predictive models' accuracy. There are various reasons behind the source of training data is being limited, such as data being inaccessible, expensive to harvest and curate, or even rare to find. This approach has been used heavily in ML applications, including image classification (Duan, Xu, & Tsang, 2012; Kulis, Saenko, & Darrell, 2011; Zhu et al., 2011), human motion classification (Harel & Mannor, 2010) sentiment analysis (Wang & Mahadevan, 2011) etc.

8.4.4 Experimental results

For training the algorithm with X-ray images, existing DL models pretrained with the ImageNet database (http://www.image-net.org), capable of classifying 1000 natural objects, were employed. VGG16 and VGG19, ResNet50, ResNet101, MobileNet-V2, MobileNet, Inception-V3, Inception-ResNet-V2, DenseNet169, DenseNet121, and Xception have been used for deep feature extraction among several publicly accessible deep feature extraction models, and many shallow ML techniques are compared because these deep feature extraction and ML models are considered to have higher performance. Various ML algorithms were used on the extracted features obtained from different pretrained models. The ML algorithms used are ANN, KNN, RF, SVM, Bagging, AdaBoost, and XGBoost. The entire dataset of X-ray images was divided into COVID-19, Normal and other types of pneumonia. Accuracy, F1-score, and Kappa were used as performance measures.

If we use ML models on deep features, they give an outstanding result as they have already acquired the necessary features. As we can see from the results below, nearly all the models show significant results. Note that ANN has been run for 100 epochs here.

Random forest and bagging have very high training accuracy as they beat all the models in training accuracy. Still, at the same time, these classifiers present noticeable overfitting to a great extent. ANN and XGBoost can be considered classifiers with less overfitting.

Here too, like in VGG 16, we see that random forest and bagging have a very high training accuracy as they beat all the models in training accuracy, but, at the same time, these classifiers present noticeable overfitting to a great extent, bagging much more than Random Forest. Here most of the classifiers are giving good results with less overfitting.

Random forest and bagging beat all others like usual in the training accuracy, but they present noticeable overfitting again. Moreover, XGBoost shows a very high training accuracy. Surprisingly, SVM shows very poor results in this case.

Random forest, bagging, and XGBoost. have a very high training accuracy as they beat all classifiers and they present obvious overfitting. ANN can be considered as a classifier with less overfitting.

Here again similar to the previous experiments, random forest and bagging present obvious overfitting to a certain extent. Here SVM achieved the best performance and can be considered as a classifier with the least overfitting.

With MobileNet, too, all the classifiers present noticeable overfitting to a certain extent. Also, AdaBoost is seen performing very weakly here compared with the other pretrained models.

We can see that results of InceptionV3 stand out from the other pretrained models as nearly all classifiers have significantly less overfitting and give very high accuracies. Surprisingly SVM performs very poorly with this pretrained model.

Random forest, bagging, and XGBoost have a very high training accuracy as they beat all classifiers, although they present obvious overfitting to a certain extent. Here AdaBoost is seen to show inferior results.

This model, just like the other, faces a problem of overfitting, and random forest and bagging present noticeable overfitting and have the highest training accuracies. Here SVM achieved the best performance and can be considered a model with the least overfitting.

This model shows much more promising results than DenseNet169. Each ML model gives promising results with this pretrained model. Again, random forest and bagging present overfitting. Here XGBoost can be considered as a classifier with the least overfitting.

This model works great with nearly all the classifiers and is also seen to have no or significantly less overfitting. Here too, AdaBoost gives very poor results, whereas XGBoost gives promising results without overfitting.

We would like to underline a few key concepts that make a significant impact in our analysis. The ML models are designed to divide data into three sets, namely, the training, validation and testing set, mainly used to overcome the problem of overfitting. Overfitting is a malfunction in the model when a model perfectly understands (or fits) the training data. In contrast, it cannot make accurate predictions for the testing data that has not been seen before. This is why data is divided into two or more subsets. However, this still might not be enough as data with a lot of noise can lead the ML models to generate false positives, false negatives, and possibly both outcomes. To assess the models' accuracy better, most of the time, the F-measure (or also known as F1 score) and the Cohen's Kappa Statistics is employed. In our analysis, we provided not only the training, validation, and test accuracies; we also included the F1 Scores along with Kappa values. Note that if the model clears out all the weeds and produces perfect results, then the F1 score is considered excellent and gets closer to 1, 1 being the 100% achievement. In contrast, if the model is not successful, the F1 score leans towards the total failure score, which is zero.

Moreover, the accuracy, in all cases (training, validation, and test cases), is the percentage of correctly classified instances out of all instances. Also, Kappa, or as known as Cohen's Kappa Statistics, is the performance measure with a normalization of the results with the baseline of a random selection of the data points picked from the data. Kappa results can be interpreted as if the values are less than or equal to zero (values ≤ 0), indicating no agreement. The values are less than or equal to 1, meaning that there is almost perfect agreement.

Our results seen in Tables 8.1 through 8.11 show some interesting results. Even though all the models seem to have great accuracy ($> 90\%$), the random forest and the bagging are the winners in all deep feature extraction approaches. While random forest had 99.98% accuracy in training accuracy, it falls short on test and validation accuracies. Hence, they present noticeable overfitting. However, considering the F1-measure (over 85%) and Kappa statistics (over 66%), we can assure that, even though the outcome of random forest performs exceptional classification performance in all deep feature extraction techniques, namely, DenseNet121, DenseNet169, InceptionResNetV2, InceptionV3, MobileNet, MobileNetV2, Resnet50, ResNet101, VGG16, VGG19, Xception. Bagging has shown an outstanding performance overall in most deep feature extraction techniques, except, ResNet50 and MobileNet.

Even though the training and testing accuracies were not excellent compared to random forest and XGBoost, Bagging has the highest F1 measure and kappa on both ResNet50 and MobileNet pretrained models. The interesting fact is that, while XGBoost has not performed good in most of the pretrained models, it produced an outstanding performance in MobileNet, InceptionResNetV2, ResNet50, and ResNet101 feature extraction methods.

These results imply that the accuracy combined with F1 measure and kappa gives the most precise picture in terms of the ML models' performance. Depending on the classifier and the pretrained models, the outcome can be significantly different.

Table 8.1 Performance of different classifiers with VGG16 deep feature extraction.

	Training accuracy (%)	Validation accuracy (%)	Test accuracy (%)	F1-Measure	Kappa
Artificial neural networks	92.88	93.60	93.93	0.9386	0.8495
KNN	94.67	91.25	92.65	0.9215	0.8134
Support vector machines	90.27	90.57	92.33	0.9107	0.8020
Random forest	99.98	90.91	92.97	0.9209	0.8148
AdaBoost	86.76	88.89	86.58	0.8607	0.6737
Bagging	99.42	91.92	90.42	0.8984	0.7579
XGBoost	94.88	92.93	94.25	0.9395	0.8539

Table 8.2 Performance of different classifiers with VGG19 deep feature extraction.

	Training accuracy (%)	Validation accuracy (%)	Test accuracy (%)	F1-Measure	Kappa
Artificial neural networks	91.60	89.90	93.29	0.9325	0.8364
KNN	94.15	89.23	90.73	0.9065	0.7716
Support vector machines	89.10	89.56	91.69	0.9141	0.7904
Random forest	99.98	91.58	90.73	0.9051	0.7691
AdaBoost	82.42	81.82	84.46	0.8499	0.6486
Bagging	99.50	87.54	90.10	0.8992	0.7568
XGBoost	94.49	91.25	91.69	0.9161	0.7996

Table 8.3 Performance of different classifiers with ResNet50 deep feature extraction.

	Training accuracy (%)	Validation accuracy (%)	Test accuracy (%)	F1-Measure	Kappa
Artificial neural networks	86.30	87.54	84.66	0.8438	0.6209
KNN	92.66	87.54	85.62	0.8560	0.6469
Support vector machines	71.48	71.72	74.12	0.6451	0.0644
Random forest	99.98	90.24	89.54	0.8529	0.6389
AdaBoost	81.43	83.84	77.64	0.7795	0.4776
Bagging	99.42	91.92	90.42	0.8984	0.7579
XGBoost	99.98	90.24	86.90	0.8651	0.6663

Table 8.4 Performance of different classifiers with ResNet101 deep feature extraction.

	Training accuracy (%)	Validation accuracy (%)	Test accuracy (%)	F1-Measure	Kappa
Artificial neural networks	88.43	88.89	87.54	83.760	0.6936
KNN	92.22	85.86	84.66	84.510	0.6290
Support vector machines	80.84	82.49	82.43	79.390	0.4916
Random forest	99.98	87.88	85.94	0.8569	0.6501
AdaBoost	82.12	84.51	80.19	0.8061	0.5477
Bagging	99.27	89.90	85.30	0.8491	0.6292
XGBoost	99.98	90.57	87.22	0.8689	0.6776

Table 8.5 Performance of different classifiers with MobileNetV2 deep feature extraction.

	Training accuracy (%)	Validation accuracy (%)	Test accuracy (%)	F1-Measure	Kappa
Artificial neural networks	95.52	89.23	92.33	0.9237	0.8169
KNN	93.97	88.89	90.42	0.9012	0.7618
Support vector machines	92.37	91.25	94.25	0.9407	0.8569
Random forest	99.98	87.88	93.29	0.9290	0.8291
AdaBoost	84.99	81.82	84.98	0.8555	0.6647
Bagging	99.31	88.89	90.73	0.9027	0.7666
XGBoost	94.51	89.23	93.61	0.9347	0.8436

Table 8.6 Performance of different classifiers with MobileNet deep feature extraction.

	Training accuracy (%)	Validation accuracy (%)	Test accuracy (%)	F1-Measure	Kappa
Artificial neural networks	97.34	90.91	88.82	0.8871	0.7254
KNN	94.98	91.25	86.58	0.8665	0.6796
Support vector machines	93.71	93.27	88.50	0.8849	0.7224
Random forest	99.98	89.23	87.54	0.8715	0.6826
AdaBoost	81.11	81.82	75.08	0.7631	0.4792
Bagging	99.42	86.87	85.30	0.8511	0.6392
XGBoost	99.98	91.92	90.10	0.8997	0.7548

Table 8.7 Performance of different classifiers with InceptionV3 deep feature extraction.

	Training accuracy (%)	Validation accuracy (%)	Test accuracy (%)	F1-Measure	Kappa
Artificial neural networks	95.45	94.28	95.53	0.9546	0.8909
KNN	96.51	94.28	95.21	0.9516	0.8834
Support vector machines	71.59	75.76	76.04	0.6795	0.1538
Random forest	99.98	94.61	96.49	0.9646	0.9147
AdaBoost	94.72	94.61	95.85	0.9595	0.9002
Bagging	99.63	94.95	95.53	0.9548	0.8916
XGBoost	96.62	95.62	94.89	0.9479	0.8752

Table 8.8 Performance of different classifiers with InceptionResNetV2 deep feature extraction.

	Training accuracy (%)	Validation accuracy (%)	Test accuracy (%)	F1-Measure	Kappa
Artificial neural networks	96.12	95.96	92.01	0.9202	0.8049
KNN	96.77	95.62	90.73	0.9075	0.7770
Support vector machines	92.31	92.93	89.78	0.8734	0.7239
Random forest	99.98	96.30	91.37	0.9138	0.7915
AdaBoost	66.59	70.03	69.01	0.6416	0.1193
Bagging	99.56	95.96	90.10	0.9012	0.7616
XGBoost	99.98	95.62	91.37	0.9140	0.7926

Table 8.9 Performance of different classifiers with DenseNet169 deep feature extraction.

	Training accuracy (%)	Validation accuracy (%)	Test accuracy (%)	F1-Measure	Kappa
Artificial neural networks	96.92	94.28	92.97	0.9290	0.8293
KNN	94.67	93.60	94.25	0.9424	0.8603
Support vector machines	94.28	94.61	96.49	0.9641	0.9139
Random forest	99.98	92.26	92.01	0.9141	0.7892
AdaBoost	84.49	86.53	81.79	0.8213	0.5779
Bagging	99.31	90.24	91.69	0.9086	0.7833
XGBoost	95.52	93.94	93.29	0.9297	0.8279

Table 8.10 Performance of different classifiers with DenseNet121 deep feature extraction.

	Training accuracy (%)	Validation accuracy (%)	Test accuracy (%)	F1-Measure	Kappa
Artificial neural networks	96.35	96.97	96.81	0.9676	0.9216
KNN	96.56	95.96	96.17	0.9610	0.9065
Support vector machines	94.70	96.63	94.57	0.9446	0.8659
Random forest	99.98	96.30	95.53	0.9548	0.8919
AdaBoost	73.93	74.41	75.72	0.7639	0.4449
Bagging	99.68	96.63	95.85	0.9587	0.9018
XGBoost	97.43	97.31	96.49	0.9649	0.9156

Table 8.11 Performance of different classifiers with Xception deep feature extraction.

	Training accuracy (%)	Validation accuracy (%)	Test accuracy (%)	F1-Measure	Kappa
Artificial neural networks	96.35	96.97	96.81	0.9676	0.9216
KNN	96.56	95.96	96.17	0.9610	0.9065
Support vector machines	94.70	96.63	94.57	0.9446	0.8659
Random forest	99.98	96.30	95.53	0.9548	0.8919
AdaBoost	73.93	74.41	75.72	0.7639	0.4449
Bagging	99.68	96.63	95.85	0.9587	0.9018
XGBoost	97.43	97.31	96.49	0.9649	0.9156

8.5 Discussion

In various sectors, AI's rise has undoubtedly become a powerful player and is expected to create much more disruption. Ultimately and naturally, this ability has led to opposing views regarding its possible position and integration into society. AI is supposed to have a major effect on so-called knowledge workers, unlike other previous technical developments that have primarily impacted manual work. There is a wide debate among scientists about the expected future involvement of human specialists and the capacity for human diagnostic capacities to be surpassed by AI (Granter, Beck, & Papke, 2017; Sharma & Carter, 2017). Nevertheless, it is necessary to note the inevitable imprecision in technical prognosis and the role of viewpoints and biases in shaping individual perceptions when considering these issues (Granter, 2016; Levine et al., 2019).

This chapter addressed numerous challenges in typical medical applications and truly approximating a human physician's cognitive processes. Most DL implementations were extremely task-specific, while humans can build relationships, which can enhance performance through many similar tasks. In spite of the limitation of feature-engineered methods, the combination of semantic knowledge with the visual interpretation would likely benefit, particularly in the distinction between uncommon diagnoses with minimal instances available. In addition, pathology and radiology findings are mostly focused not only on a single specimen or scan but also on a connection with the previous ones and other medical histories (Levine et al., 2019; Sharma & Carter, 2017).

In general, DL is greedy for data, considerably more than previous feature-engineered methods, which are less vulnerable to overfitting. The collection of relevant training data in almost all domains is an ongoing problem. Though there are unsupervised and semisupervised learning methods, data sets need manual annotation or curation for most medical tasks (Hosny, Parmar, Quackenbush, Schwartz, & Aerts, 2018). This could be sufficient for qualified research workers or require medical professionals' full input, depending on the assignment's scope. Since the previous layers in DNNs almost invariably learn quite common image characteristics, medical data can be fine-tuned on networks that have been pretrained on broad general image sets, minimizing the amount of data needed total training time (Tajbakhsh et al., 2016; Yosinski, Clune, Bengio, & Lipson, 2014). DL has been criticized for being a 'black box' in contrast to the feature-engineered methods, where it is not completely clear how the algorithm produces outputs from a given input. Although this claim undoubtedly has some significance, strategies for visualizing a network's activation functions and the types of images that activate a given neuron have contributed to characterize the inner workings of these techniques, which continues an active research area (Zeiler & Fergus, 2014; Zintgraf, Cohen, Adel, & Welling, 2017). However, considering the failure of current DL algorithms to clarify their diagnostic mechanism, many challenges, including the degree of physician supervision needed and deciding who is responsible for system error, will need to be solved before their introduction in clinical practice. Medical policymakers and clinicians must proceed carefully and insist on new methods that can be tested extensively in practical circumstances prior to patient treatment, considering the speculation and high standards surrounding DL in medicine (Levine et al., 2019; Topol, 2019).

AI is expected to stay in a diagnostic service position for the near future, where it will accurately define disorders, automate manual operations, and enhance roadmap, but a human being should bear responsibility of final decisions. Nevertheless, considering the current state of the field, precise information remains highly speculative about applying this technology. Regardless of AI's eventual effect directly, as emerging innovations are implemented, the field of diagnostic medicine will continue to evolve. If AI algorithms can generate widespread adoption by clinicians, the expense of the computing technology required to implement AI algorithms is likely to be marginal in the light of total healthcare expenditure (Liew, 2018). If AI can eventually automate a good portion of image recognition, a radiologist or pathologist's role can move to increasingly focusing other activities, such as medical record correlation, report formulation, clinician liaison, and departmental quality control. To better explain the computer techniques employed, this technology would also require a change in diagnostic physicians' education, with the development of an completely new discipline or even a combination of pathology and radiology (Jha & Topol, 2016; Levine et al., 2019; Lundström, Gilmore, & Ros, 2017).

Despite numerous attempts to increase medical accuracy, diseases depend on the procedure and mostly rely on the physician's expertise. X-ray imaging diagnosis needs experience in image diagnosis; radiologists performed slightly more than general practitioners in a sample of video-presented evaluation for diagnosis. In these cases, the new deep network paradigm could support physicians by recommending potential X-ray image-based diagnosis. By integrating clinical knowledge with recommendations, they could achieve greater diagnostic accuracy. COVID-19 was diagnosed with sufficient accuracy by the existing image classification process, with transfer learning feature extraction, which is remarkable in accuracy and diagnostic diversity. It should be remembered that all the clinical X-rays were purposely used, without any selection bias for preparation. Any of the images were used as long as a clinician could get an idea of the given image for diagnosis. We speculate that this realistic database of images makes the model's output dependent on the scale of the database. COVID-19 disease is automatically diagnosed using deep feature extraction based on transfer learning and chest X-ray images.

8.6 Conclusions

In the continued pursuit of computer-aided medical diagnostics, AI is a promising innovation. Development over the past few years reveals its capacity to produce success at human experts' level, but the technology continues to be far away from mainstream clinical application. The practice of diagnostic medicine is likely to be transformed by AI, and we are hopeful that it will eventually contribute to increased patient safety and health care quality (Levine et al., 2019).

In this chapter we presented how AI and ML techniques are utilized in disease diagnosis. Besides an automated COVID-19 diagnosis approach based on deep feature extraction is also presented. Moreover, we have presented some results on detecting COVID-19 positive cases, a pandemic that infected many of the human population in 2020, from chest X-rays using different ML models. After extracting features using DTL, the X-ray images are fed into the shallow ML models to diagnose COVID-19 from X-ray images. Also, we compared different types of transfer learning models for feature extraction using traditional ML models over these pretrained models. Simple transfer learning-based feature extraction gives promising results with models such as VGG16, VGG19, and MobileNet. Finally, the feature extraction method also shows promising results with nearly 90% accuracy in most models and the highest accuracy, reaching 96.49% with Random Forest classifier using InceptionV3 feature extraction.

References

Abbas A., Abdelsamea M.M., & Gaber M.M. (2020). Classification of COVID-19 in chest X-ray images using DeTraC deep convolutional neural network. *arXiv200313815 Cs Eess Stat*, May. <http://arxiv.org/abs/2003.13815>. Accessed: 07.08.20.

Abd El-Salam, M., Reda, S., Lotfi, S., Refaat, T., & El-Abd, E. (2014). *Imaging techniques in cancer diagnosis," In. Cancer biomarkers: Minimal and noninvasive early diagnosis and prognosis* (pp. 19−38). CRC Press.

Aggarwal, C. C. (2018). *Neural networks and deep learning.* Springer.

Albawi S., Mohammed T.A., & Al-Zawi S. (2017). *Understanding of a convolutional neural network*, pp. 1−6.

Altan, A., & Karasu, S. (2020). Recognition of COVID-19 disease from X-ray images by hybrid model consisting of 2D curvelet transform, chaotic salp swarm algorithm and deep learning technique. *Chaos, Solitons, and Fractals, 140*, 110071.

Altman, R. (2017). Artificial intelligence (AI) systems for interpreting complex medical datasets. *Clinical Pharmacology and Therapeutics, 101*(5), 585−586.

Apostolopoulos, I. D., & Mpesiana, T. A. (2020). Covid-19: automatic detection from X-ray images utilizing transfer learning with convolutional neural networks. *Phys. Eng. Sci. Med., 43*(2), 635−640. Available from https://doi.org/10.1007/s13246-020-00865-4.

Bejnordi, B. E., Veta, M., van Diest, P. J., van Ginneken, B., Karssemeijer, N., Litjens, N., . . . Venâncio, R. (2017). Diagnostic assessment of deep learning algorithms for detection of lymph node metastases in women with breast cancer. *JAMA: The Journal of the American Medical Association, 318*(22), 2199−2210.

Bejnordi, B. E., Mullooly, M., Pfeiffer, R. M., Fan, S., Vacek, P. M., Weaver, D. L., . . . Sherman, M. E. (2018). Using deep convolutional neural networks to identify and classify tumor-associated stroma in diagnostic breast biopsies. *Modern Pathology: An Official Journal of the United States and Canadian Academy of Pathology, Inc, 31*(10), 1502−1512.

Bengio, Y. (2009). Learning deep architectures for AI. *Found. Trends® Mach. Learn., 2*(1), Art. no. 1.

Bengio, Y., Courville, A., & Vincent, P. (2013). Representation learning: A review and new perspectives. *IEEE Transactions on Pattern Analysis and Machine Intelligence, 35*(8), 1798−1828.

Blomgren, K., & Pitkäranta, A. (2003). Is it possible to diagnose acute otitis media accurately in primary health care? *Family Practice*, *20*(5), 524−527.

Booth, T. C., Nathan, M., Waldman, A. D., Quigley, A.-M., Schapira, A. H., & Buscombe, J. (2015). The role of functional dopamine-transporter SPECT imaging in Parkinsonian syndromes, Part 1. *American Journal of Neuroradiology*, *36*(2), 229. Available from https://doi.org/10.3174/ajnr.A3970.

Bottou L. (2010). Large-scale machine learning with stochastic gradient descent. In *Proceedings of COMPSTAT'2010*, Y. Lechevallier and G. Saporta, (Eds.) Heidelberg: Physica-Verlag HD, pp. 177−186.

Breiman, L. (2001). Random forests. *Machine Learning*, *45*(1), 5−32.

Bringas, S., Salomón, S., Duque, R., Lage, C., & Montaña, J. L. (2020). Alzheimer's Disease stage identification using deep learning models. *Journal of Biomedical Informatics*, *109*, 103514.

Canayaz, M. (2020). "MH-COVIDNet: Diagnosis of COVID-19 using deep neural networks and *meta*-heuristic-based feature selection on X-ray images. *Biomedical Signal Processing and Control*, *64*, 102257.

Cha, D., Pae, C., Seong, S.-B., Choi, J. Y., & Park, H.-J. (2019). Automated diagnosis of ear disease using ensemble deep learning with a big otoendoscopy image database. *EBioMedicine*, *45*, 606−614.

Chengsheng T., Huacheng L., & Bing X. (2017). *AdaBoost typical algorithm and its application research*, vol. 139, p. 00222.

Choi, H., Ha, S., Im, H. J., Paek, S. H., & Lee, D. S. (2017). Refining diagnosis of Parkinson's disease with deep learning-based interpretation of dopamine transporter imaging. *NeuroImage: Clinical*, *16*, 586−594. Available from https://doi.org/10.1016/j.nicl.2017.09.010, Jan.

Chollet F. (2017). *Xception: Deep learning with depthwise separable convolutions*, pp. 1251−1258.

Cruz-Roa, A., Gilmore, H., Basavanhally, A., Feldman, M., Ganesan, S., Shih, N. N. C., ... Madabhushi, A. (2017). Accurate and reproducible invasive breast cancer detection in whole-slide images: A deep learning approach for quantifying tumor extent. *Science Report*, *7*(1), 1−14.

Cummings, J. L., Henchcliffe, C., Schaier, S., Simuni, T., Waxman, A., & Kemp, P. (2011). The role of dopaminergic imaging in patients with symptoms of dopaminergic system neurodegeneration. *Brain*, awr177.

Duan L., Xu D., & Tsang I. (2012). Learning with augmented features for heterogeneous domain adaptation. *arXiv12064660*.

Durga, P., Jebakumari, V. S., & Shanthi, D. (2016). Diagnosis and classification of parkinsons disease using data mining techniques. *International Journal of Advanced Research Trends in Engineering and Technology*, *3*, 86−90.

Elaziz, M. A., Hosny, K. M., Salah, A., Darwish, M. M., Lu, S., & Sahlol, A. T. (2020). New machine learning method for image-based diagnosis of COVID-19. *PLoS One*, *15*(6), e0235187. Available from https://doi.org/10.1371/journal.pone.0235187, Jun.

Esteva, A., Kuprel, B., Novoa, R. A., Ko, J., Swetter, S. M., Blau, H. M., & Thrun, S. (2017). Dermatologist-level classification of skin cancer with deep neural networks. *Nature*, *542*(7639), 115−118.

Evgeniou T. & Pontil M. (1999). *Support vector machines: Theory and applications*, pp. 249−257.

Fricker, K. S., Moret, M., Rupp, N., Hermanns, T., Fankhauser, C., Wey, N., ... Claassen, M. (2018). Automated Gleason grading of prostate cancer tissue microarrays via deep learning. *Science Report*, *8*(1), 1−11.

Gao, Y., & Mosalam, K. M. (2018). Deep transfer learning for image-based structural damage recognition. *Computer-Aided Civil and Infrastructure Engineering*, *33*(9), 748−768. Available from https://doi.org/10.1111/mice.12363.

Ghosh, A., Sufian, A., Sultana, F., Chakrabarti, A., & De, D. (2020). *Fundamental concepts of convolutional neural network. Recent Trends and Advances in Artificial Intelligence and Internet of Things* (pp. 519−567). Springer.

Giger, M. L., Doi, K., & MacMahon, H. (1988). Image feature analysis and computer-aided diagnosis in digital radiography. 3. Automated detection of nodules in peripheral lung fields. *Medical Physics*, *15*(2), 158−166.

Goodfellow, I., Bengio, Y., Courville, A., & Bengio, Y. (2016). no. 2 *Deep learning* (1). Cambridge: MIT press.

Goodfellow, I., Pouget-Abadie, J., Mirza, M., Xu, B., Warde-Farley, D., Ozair, S., & Bengio, Y. (2014). Generative adversarial nets. *Proceedings of the twenty-seventh international conference on neural information processing systems, 2*, 2672–2680.

Granter, S. R. (2016). Reports of the death of the microscope have been greatly exaggerated. *Archives of Pathology & Laboratory Medicine, 140*(8), 744–745. Available from https://doi.org/10.5858/arpa.2016-0046-(ED.).

Granter, S. R., Beck, A. H., & Papke, D. J., Jr (2017). AlphaGo, deep learning, and the future of the human microscopist. *Archives of Pathology & Laboratory Medicine, 141*(5), 619–621. Available from https://doi.org/10.5858/arpa.2016-0471-(ED.).

Greenspan, H., Van Ginneken, B., & Summers, R. M. (2016). Guest editorial deep learning in medical imaging: Overview and future promise of an exciting new technique. *IEEE Transactions on Medical Imaging, 35*(5), 1153–1159.

Gulshan, V., Peng, L., Coram, M., Stumpe, M. C., Wu, D., Narayanaswamy, A., ... Webster, D. R. (2016). Development and validation of a deep learning algorithm for detection of diabetic retinopathy in retinal fundus photographs. *JAMA: The Journal of the American Medical Association, 316*(22), 2402–2410.

Gutman, D. A., Cooper, L. A. D., Hwang, S. N., Holder, C. A., Gao, J., Aurora, T. D., ... Brat, D. J. (2013). MR imaging predictors of molecular profile and survival: Multi-institutional study of the TCGA glioblastoma data set. *Radiology, 267*(2), 560–569.

Hall M., Witten I., & Frank E. (2011). *Data mining: Practical machine learning tools and techniques.* Kaufmann Burlingt.

Han, J., Pei, J., & Kamber, M. (2011). *Data mining: Concepts and techniques.* Elsevier.

Harel, M. & Mannor, S. (2010). Learning from multiple outlooks. *arXiv10050027.*

Hatem, A., Mohamed, S., Elhassan, U. E. A., Ismael, E. A. M., Rizk, M. S., El-Kholy, A., & El-Harras, M. (2019). Clinical characteristics and outcomes of patients with severe acute respiratory infections (SARI): Results from the Egyptian surveillance study 2010–2014. *Multidisciplinary Respiratory Medicine, 14*(1), 11. Available from https://doi.org/10.1186/s40248-019-0174-7.

He, K., Zhang, X., Ren, S., & Sun, J. (2016a). *Deep residual learning for image recognition*, pp. 770–778.

He, K., Zhang, X., Ren, S., & Sun, J. (2016b). Identity mappings in deep residual networks. *arXiv160305027 Cs.* <http://arxiv.org/abs/1603.05027>. Accessed 07.08.20.

Hemdan, E. E.-D., Shouman, M. A., & Karar, M. E. (2020). COVIDX-Net: A framework of deep learning classifiers to diagnose COVID-19 in X-ray images. *arXiv200311055 Cs Eess.* <http://arxiv.org/abs/2003.11055>. Accessed 07.08.20.

Hoheisel, M., Lawaczeck, R., Pietsch, H., & Arkadiev, V. (2005). Advantages of monochromatic X-rays for imaging. *Proceedings of SPIE - The International Society for Optical Engineering, 5745.* Available from https://doi.org/10.1117/12.593398, Apr.

Hosny, A., Parmar, C., Quackenbush, J., Schwartz, L. H., & Aerts, H. J. W. L. (2018). Artificial intelligence in radiology. *Nature Reviews. Cancer, 18*(8), 500–510. Available from https://doi.org/10.1038/s41568-018-0016-5.

Howard, A.G., Zhu, M., Chen, B., Kalenichenko, D., Wang, W., Weyand, T., ..., Adam, H. (2017). Mobilenets: Efficient convolutional neural networks for mobile vision applications. *arXiv170404861.*

Huang, G., Liu, Z., Van Der Maaten, L., & Weinberger, K. Q. (2017). *Densely connected convolutional networks*, pp. 4700–4708.

Hussain, M., Bird, J., & Faria, D. (2018). *A study on CNN transfer learning for image classification.*

Jaderberg, M., Simonyan, K., & Zisserman, A. (2015). Spatial transformer networks. *Advances in Neural Information Processing Systems, 28*, 2017–2025.

Jamaludin, A., Lootus, M., Kadir, T., Zisserman, A., Urban, J., Battié, M. C.... The Genodisc Consortium. (2017). Automation of reading of radiological features from magnetic resonance images (MRIs) of the lumbar spine without human intervention is comparable with an expert radiologist. *European Spine*

Journal: Official Publication of the European Spine Society, the European Spinal Deformity Society, and the European Section of the Cervical Spine Research Society, 26(5), 1374−1383.

Jha, S., & Topol, E. J. (2016). Adapting to artificial intelligence: Radiologists and pathologists as information specialists. *JAMA: The Journal of the American Medical Association, 316*(22), 2353−2354. Available from https://doi.org/10.1001/jama.2016.17438.

Jiang, F., Jiang, Y., Zhi, H., Dong, Y., Li, H., Ma, S., . . . Wang, Y. (2017). Artificial intelligence in healthcare: Past, present and future. *Stroke and Vascular Neurology, 2*(4).

Kalender, W. A. (2006). X-ray computed tomography. *Physics in Medicine and Biology, 51*(13), R29−R43. Available from https://doi.org/10.1088/0031-9155/51/13/R03, Jul.

Karlo, C. A., Paolo, P. L. D., Chaim, J., Hakimi, A. A., Ostrovnaya, I., Russo, P., . . . Akin, O. (2014). Radiogenomics of clear cell renal cell carcinoma: associations between CT imaging features and mutations. *Radiology, 270*(2), 464−471.

Kawata Y., Niki, N., Ohmatsu, H., Kusumoto, M., Kakinuma, R., Mori, K., . . . Moriyama, N. (1999). *Computer aided differential diagnosis of pulmonary nodules using curvature based analysis*, pp. 470−475.

Kelleher, J. D. (2019). *Deep learning*. Mit Press.

Kermany, D. S., Goldbaum, M., Cai, W., Valentim, C. C. S., Liang, H., Baxter, S. L., . . . Zhang, K. (2018). Identifying medical diagnoses and treatable diseases by image-based deep learning. *Cell, 172*(5), 1122−1131. Available from https://doi.org/10.1016/j.cell.2018.02.010, e9, Feb.

Koitka, S. & Friedrich, C. M. (2016). *Traditional feature engineering and deep learning approaches at medical classification task of ImageCLEF 2016*. pp. 304−317.

Kooi, T., Litjens, G., van Ginneken, B., Gubern-Mérida, A., Sánchez, C. I., Mann, R., . . . Karssemeijer, N. (2017). Large scale deep learning for computer aided detection of mammographic lesions. *Medical Image Analysis, 35*, 303−312. Available from https://doi.org/10.1016/j.media.2016.07.007, Jan.

Krizhevsky, A., Sutskever, I., & Hinton, G. E. (2012). Imagenet classification with deep convolutional neural networks. *Advances in Neural Information Processing Systems, 25*, 1097−1105.

Krizhevsky, A., Sutskever, I., & Hinton, G. E. (2017). ImageNet classification with deep convolutional neural networks. *Communications of the ACM, 60*(6), 84−90. Available from https://doi.org/10.1145/3065386.

Kulis, B., Saenko, K., & Darrell, T. (2011). *What you saw is not what you get: Domain adaptation using asymmetric kernel transforms*. pp. 1785−1792.

Kumar, A., Kim, J., Lyndon, D., Fulham, M., & Feng, D. (2017). An ensemble of fine-tuned convolutional neural networks for medical image classification. *IEEE Journal of Biomedical and Health Informatics, 21*(1), 31−40. Available from https://doi.org/10.1109/JBHI.2016.2635663, Jan.

LeCun, Y., Bengio, Y., & Hinton, G. (2015). Deep learning. *Nature, 521*(7553), 436−444.

Leger, S., Zwanenburg, A., Pilz, K., Lohaus, F., Linge, A., Zöphel, K., . . . Richter, C. (2017). A comparative study of machine learning methods for time-to-event survival data for radiomics risk modelling. *Science Report, 7*(1), 1−11.

Levine, A. B., Schlosser, C., Grewal, J., Coope, R., Jones, S. J., & Yip, S. (2019). "Rise of the machines: Advances in deep learning for cancer diagnosis. *Trends Cancer, 5*(3), 157−169.

Li, R., Pei, S., Chen, B., Song, Y., Zhang, T, Yang, W., . . . Shaman, J. (2020). Substantial undocumented infection facilitates the rapid dissemination of novel coronavirus (COVID-19). *Infectious Diseases (except HIV/AIDS), preprint*. Available from https://doi.org/10.1101/2020.02.14.20023127, Feb.

Liang, W., Liang, H., Ou, L., Chen, B., Chen, A., & Li, C.. . . for the China Medical Treatment Expert Group for COVID-19. (2020). Development and validation of a clinical risk score to predict the occurrence of critical illness in hospitalized patients with COVID-19. *JAMA Internal Medicine, 180*(8), 1081−1089. Available from https://doi.org/10.1001/jamainternmed.2020.2033.

Liew, C. (2018). The future of radiology augmented with artificial intelligence: A strategy for success. *European Journal of Radiology, 102*, 152−156.

Litjens, G., Kooi, T., Bejnordi, B. E., Setio, A. A. A., Ciompi, F., Ghafoorian, M., . . . Sánchez, C. I. (2017). A survey on deep learning in medical image analysis. *Medical Image Analysis*, *42*, 60−88.

Loey, M., Smarandache, F., & Khalifa, N.E.M. (2020). *Within the lack of COVID-19 benchmark dataset: A novel gan with deep transfer learning for corona-virus detection in chest x-ray images*. No April.

Lundström, C. F., Gilmore, H. L., & Ros, P. R. (2017). Integrated diagnostics: The computational revolution catalyzing cross-disciplinary practices in radiology, pathology, and genomics. *Radiology*, *285*(1), 12−15. Available from https://doi.org/10.1148/radiol.2017170062.

Makaju, S., Prasad, P. W. C., Alsadoon, A., Singh, A. K., & Elchouemi, A. (2018). Lung cancer detection using CT scan images. *Procedia Computer Science*, *125*, 107−114. Available from https://doi.org/10.1016/j.procs.2017.12.016, Jan.

Makris, A., Kontopoulos, I., & Tserpes, K. (2020). COVID-19 detection from chest X-ray images using deep learning and convolutional neural networks. *Radiology and Imaging*. Available from https://doi.org/10.1101/2020.05.22.20110817, May.

Mandic, D., & Chambers, J. (2001). *Recurrent neural networks for prediction: Learning algorithms, architectures and stability*. Wiley.

Martinez-Murcia F.J., Ortiz, A., Gorriz, J.M., Ramírez, J., Segovia, F., Salas-Gonzalez, D., ... Illan, I.A. (2017). *A 3D convolutional neural network approach for the diagnosis of Parkinson's disease*, pp. 324−333.

Mazurowski, M. A., Zhang, J., Grimm, L. J., Yoon, S. C., & Silber, J. I. (2014). Radiogenomic analysis of breast cancer: Luminal B molecular subtype is associated with enhancement dynamics at MR imaging. *Radiology*, *273*(2), 365−372.

Merkow, J., Lufkin, R., Nguyen, K., Soatto, S., Tu, Z., & Vedaldi, A. (2017). DeepRadiologyNet: Radiologist level pathology detection in CT head images. *arXiv171109313*.

Mobadersany, P., Yousefi, S., Amgad, M., Gutman, D. A., Barnholtz-Sloan, J. S., Vega, J. E. V., . . . Cooper, L. A. D. (2018). Predicting cancer outcomes from histology and genomics using convolutional networks. *Proceedings of the National Academy of Sciences of the United States of America*, *115*(13), E2970−E2979.

Mohammed, F., He, X., & Lin, Y. (2020). An easy-to-use deep-learning model for highly accurate diagnosis of Parkinson's disease using SPECT images. *Computerized Medical Imaging and Graphics: The Official Journal of the Computerized Medical Imaging Society*, *87*, 101810.

Motlagh, M. H., Jannesari, M., Aboulkheyr, H., Khosravi, P., Elemento, O., Totonchi, M., & Hajirasouliha, I. (2018). Breast cancer histopathological image classification: A deep learning approach. *BioRxiv*, 242818.

Myszczynska, M. A., Ojamies, P. N., Lacoste, A. M. B., Neil, D., Saffari, A., Mead, R., . . . Ferraiuolo, L. (2020). Applications of machine learning to diagnosis and treatment of neurodegenerative diseases. *Nature Reviews Neurology*, *16*(8). Available from https://doi.org/10.1038/s41582-020-0377-8, Art. no. 8.

Nagpal, K., Foote, D., Liu, Y., Chen, P.-H. C., Wulczyn, E., Tan, F., . . . Stumpe, M. C. (2019). Development and validation of a deep learning algorithm for improving Gleason scoring of prostate cancer. *NPJ Digital Medicine*, *2*(1), 1−10.

Nanni, L., Ghidoni, S., & Brahnam, S. (2017). Handcrafted vs. non-handcrafted features for computer vision classification. *Pattern Recognition*, *71*, 158−172.

Narin, A., Kaya, C., & Pamuk, Z. (2020). Automatic detection of coronavirus disease (COVID-19) using X-ray images and deep convolutional neural networks. *arXiv200310849 Cs Eess*. <http://arxiv.org/abs/20030.10849>. Accessed: 08.08.20.

Ning, Z., Luo, J., Li, Y., Han, S., Feng, Q., Xu, Y., . . . Zhang, Y. (2018). Pattern classification for gastrointestinal stromal tumors by integration of radiomics and deep convolutional features. *IEEE Journal of Biomedical and Health Informatics*, *23*(3), 1181−1191.

Olczak, J., Fahlberg, N., Maki, A., Razavian, A. S., Jilert, A., Stark, A., . . . Gordon, M. (2017). Artificial intelligence for analyzing orthopedic trauma radiographs. *Acta Orthopaedica*, *88*(6), 581−586. Available from https://doi.org/10.1080/17453674.2017.1344459.

Oliveira, F. P., Faria, D. B., Costa, D. C., Castelo-Branco, M., & Tavares, J. M. R. (2018). Extraction, selection and comparison of features for an effective automated computer-aided diagnosis of Parkinson's disease based on [123 I] FP-CIT SPECT images. *European Journal of Nuclear Medicine and Molecular Imaging, 45*(6), 1052−1062.

Orru, G., Pettersson-Yeo, W., Marquand, A. F., Sartori, G., & Mechelli, A. (2012). Using support vector machine to identify imaging biomarkers of neurological and psychiatric disease: a critical review. *Neuroscience and Biobehavioral Reviews, 36*(4), 1140−1152.

Pacal, I., Karaboga, D., Basturk, A., Akay, B., & Nalbantoglu, U. (2020). A comprehensive review of deep learning in colon cancer. *Computers in Biology and Medicine*, 104003.

Pan, S. J., & Yang, Q. (2010). A survey on transfer learning. *IEEE Transactions on Knowledge and Data Engineering, 22*(10), 1345−1359. Available from https://doi.org/10.1109/TKDE.2009.191.

Phan, T. V., Sultana, S., Nguyen, T. G., & Bauschert, T. (2020). *Q-TRANSFER: A novel framework for efficient deep transfer learning in networking.* pp. 146−151.

Poostchi, M., Silamut, K., Maude, R. J., Jaeger, S., & Thoma, G. (2018). Image analysis and machine learning for detecting malaria. *Translational Research: the Journal of Laboratory and Clinical Medicine, 194*, 36−55. Available from https://doi.org/10.1016/j.trsl.2017.12.004, Apr.

Prashanth, R., Roy, S. D., Mandal, P. K., & Ghosh, S. (2014). Automatic classification and prediction models for early Parkinson's disease diagnosis from SPECT imaging. *Expert Systems with Applications, 41*(7), Art. no. 7.

Prashanth, R., Roy, S. D., Mandal, P. K., & Ghosh, S. (2016). High-accuracy classification of parkinson's disease through shape analysis and surface fitting in 123I-Ioflupane SPECT imaging. *IEEE Journal of Biomedical and Health Informatics, 21*(3), 794−802.

Rajpurkar, P., Irvin, J., Zhu, K., Yang, B., Mehta, H., Duan, T., ..., Ng, A.Y. (2017). Chexnet: Radiologist-level pneumonia detection on chest x-rays with deep learning. *arXiv171105225.*

Ramraj, S., Uzir, N., Sunil, R., & Banerjee, S. (2016). Experimenting XGBoost algorithm for prediction and classification of different datasets. *International Journal of Control Theory and Applications, 9*, 651−662.

Sak, H., Senior, A. W., & Beaufays, F. (2014). *Long short-term memory recurrent neural network architectures for large scale acoustic modeling.*

Saltz, J., Gupta, R., Hou, L., Kurc, T., Singh, P., Nguyen, V., ... Thorsson, V. (2018). Spatial organization and molecular correlation of tumor-infiltrating lymphocytes using deep learning on pathology images. *Cell Reports, 23*(1), 181−193.

Santosh, T. S., Parmar, R., Anand, H., Srikanth, K., & Saritha, M. (2020). A review of salivary diagnostics and its potential implication in detection of Covid-19. *Cureus, 12*(4).

Sawada, S., Kuklane, K., Wakatsuki, K., & Morikawa, H. (2017). New development of research on personal protective equipment (PPE) for occupational safety and health. *Industrial Health, 55*(6), 471−472. Available from https://doi.org/10.2486/indhealth.55-471.

Schalekamp, S., Huisman, M., van Dijk, R. A., Boomsma, M. F., Jorge, P. J. F., de Boer, W. S., ... Schaefer-Prokop, C. M. (2020). Model-based prediction of critical illness in hospitalized patients with COVID-19. *Radiology, 298*(1), E46−E54. Available from https://doi.org/10.1148/radiol.2020202723.

Schaumberg, A. J., Rubin, M. A., & Fuchs, T. J. (2018). H&E-stained whole slide image deep learning predicts SPOP mutation state in prostate cancer. *BioRxiv*, 064279.

Schiffman, J. D., Fisher, P. G., & Gibbs, P. (2015). Early detection of cancer: Past, present, and future. *American Society of Clinical Oncology Educational Book, 35*(1), 57−65.

Schmidhuber, J. (2015). Deep learning in neural networks: An overview. *Neural Networks: The Official Journal of the International Neural Network Society, 61*, 85−117. Available from https://doi.org/10.1016/j.neunet.2014.09.003, Jan.

Seoh R. (2020). Qualitative analysis of Monte Carlo dropout. *arXiv200701720 Cs Stat.* <http://arxiv.org/abs/2007.01720>. Accessed: 07.08.20.

Setio, A. A. A., Traverso, A., De Bei, T., Berens, M. S. N., van den Bogaard, C., Cerello, P., . . . Jacob, C. (2017). Validation, comparison, and combination of algorithms for automatic detection of pulmonary nodules in computed tomography images: The LUNA16 challenge. *Medical Image Analysis, 42,* 1−13. Available from https://doi.org/10.1016/j.media.2017.06.015, Dec.

Sharma, G., & Carter, A. (2017). Artificial intelligence and the pathologist: Future Frenemies? *Archives of Pathology & Laboratory Medicine, 141*(5), 622−623. Available from https://doi.org/10.5858/arpa.2016-0593-(ED.).

Shi, F., Wang, J., Shi, J., Wu, Z., Wang, Q., Tang, Q., . . . Shen, D. (2020). Review of artificial intelligence techniques in imaging data acquisition, segmentation and diagnosis for COVID-19. *IEEE Reviews in Biomedical Engineering,* 1. Available from https://doi.org/10.1109/RBME.2020.2987975, 1.

Shin, H., Roth, H. R, Gao, M., Lu, L., Xu, Z., Nogues, I., . . . Summers, R. M. (2016). Deep convolutional neural networks for computer-aided detection: CNN architectures, dataset characteristics and transfer learning. *IEEE Transactions on Medical Imaging, 35*(5), 1285−1298. Available from https://doi.org/10.1109/TMI.2016.2528162.

Simonyan, K. & Zisserman, A. (2014). Very deep convolutional networks for large-scale image recognition. *arXiv14091556.*

Sufian, A., Ghosh, A., Sadiq, A. S., & Smarandache, F. (2020). A survey on deep transfer learning to edge computing for mitigating the covid-19 pandemic. *Journal of Systems Architecture, 108,* 101830.

Sufian, A., Jat, D. S., & Banerjee, A. (2020). *Insights of artificial intelligence to stop spread of covid-19. Big Data Analytics and Artificial Intelligence Against COVID-19: Innovation Vision and Approach* (pp. 177−190). Springer.

Sultana, F., Sufian, A., & Dutta, P. (2018). *Advancements in image classification using convolutional neural network,* pp. 122−129.

Szegedy, C., Liu, W., Jia, Y., Sermanet, P., Reed, S., Anguelov, D., & Rabinovich, A. (2015). *Going deeper with convolutions* (pp. 1−9).

Szegedy, C., Ioffe, S., Vanhoucke, V., & Alemi, A. (2017). *Inception-v4, inception-resnet and the impact of residual connections on learning.* 31, 1.

Tajbakhsh, N., Shin, J. Y., Gurudu, S. R., Hurst, T., Kendall, C. B., Gotway, M. B., & Liang, J. (2016). Convolutional neural networks for medical image analysis: Full training or fine tuning? *IEEE Transactions on Medical Imaging, 35*(5), 1299−1312.

Talo, M., Baloglu, U. B., Yıldırım, Ö., & Rajendra Acharya, U. (2019). Application of deep transfer learning for automated brain abnormality classification using MR images. *Cognitive Systems Research, 54,* 176−188. Available from https://doi.org/10.1016/j.cogsys.2018.12.007, May.

Titano, J. J., Badgeley, M., Schefflein, J., Pain, M., Su, A., Cai, M., . . . Oermann, E. K. (2018). "Automated deep-neural-network surveillance of cranial images for acute neurologic events. *Nature Medicine, 24*(9), 1337−1341.

Topol, E. J. (2019). High-performance medicine: the convergence of human and artificial intelligence. *Nature Medicine, 25*(1), 44−56. Available from https://doi.org/10.1038/s41591-018-0300-7.

Vasuki, A., & Govindaraju, S. (2017). *Deep neural networks for image classification,* . *Deep Learning for Image Processing Applications* (vol. 31, p. 27). IOS Press.

Waite, S., Scott, J. M., Legasto, A., Kolla, S., Gale, B., & Krupinski, E. A. (2017). Systemic error in radiology. *American Journal of Roentgenology, 209*(3), 629−639.

Wang, C. & Mahadevan, S. (2011). *Heterogeneous domain adaptation using manifold alignment.* 22, 1, p. 1541.

Wang, L. & Wong, A. (2020). COVID-Net: A tailored deep convolutional neural network design for detection of COVID-19 cases from chest X-ray images. *arXiv200309871 Cs Eess.* <http://arxiv.org/abs/2003.09871>. Accessed 07.08.20.

Wang, L., Lin, Z. Q., & Wong, A. (2020). Covid-net: A tailored deep convolutional neural network design for detection of covid-19 cases from chest x-ray images. *Science Reports, 10*(1), 1−12.

Wang, S., Kang, B., Ma, J., Zeng, X., Xiao, M., Guo, J., . . . Xu, B. (2020). A deep learning algorithm using CT images to screen for Corona Virus disease (COVID-19). *MedRxiv*.

Wang, X., Yang, W., Weinreb, J., Han, J., Li, Q., Kong, X., . . . Wang, L. (2017). Searching for prostate cancer by fully automated magnetic resonance imaging classification: deep learning vs non-deep learning. *Science Reports*, *7*(1), 15415. Available from https://doi.org/10.1038/s41598-017-15720-y.

Wong, K. K. L., Fortino, G., & Abbott, D. (2020). Deep learning-based cardiovascular image diagnosis: A promising challenge. *Future Generation Computer Systems*, *110*, 802−811. Available from https://doi.org/10.1016/j.future.2019.09.047, Sep.

Wu, S. Y., Yau, H. S., Yu, M. Y., Tsang, H. F., Chan, L. W. C., Cho, W. C. S., . . . Wong, S. C. C. (2020). The diagnostic methods in the COVID-19 pandemic, today and in the future. *Expert Review of Molecular Diagnostics*, *20*(9), 985−993. Available from https://doi.org/10.1080/14737159.2020.1816171, Sep.

Wu, X., Kumar, V., Quinlan, J. R., Ghosh, J., Yang, Q., Motoda, H., . . . Steinberg, D. (2008). Top 10 algorithms in data mining. *Knowledge and Information Systems*, *14*(1), 1−37.

Yassin, N. I. R., Omran, S., El Houby, E. M. F., & Allam, H. (2018). Machine learning techniques for breast cancer computer aided diagnosis using different image modalities: A systematic review. *Computer Methods and Programs in Biomedicine*, *156*, 25−45. Available from https://doi.org/10.1016/j.cmpb.2017.12.012, Mar.

Yosinski, J., Clune, J., Bengio, Y., & Lipson, H. (2014). How transferable are features in deep neural networks? *arXiv14111792*.

Zamir, A. R., Sax, A., Shen, W., Guibas, L., Malik, J., & Savarese, S. (2018). Taskonomy: Disentangling task transfer learning. In *Proceedings of the* IEEE/CVF *conference on computer vision and pattern recog*nition, June, pp. 3712−3722, doi: 10.1109/CVPR.2018.00391.

Zeiler, M. D. & Fergus, R. (2014). *Visualizing and understanding convolutional networks*, pp. 818−833.

Zhang, J., Xie, Y., Pang, G., Liao, Z., Verjans, J., Li, W., . . . Xia, Y. (2020). Viral pneumonia screening on chest x-ray images using confidence-aw are anomaly detection. *arXiv200312338*, vol. 3.

Zhu, Y., Chen, Y., Lu, Z., Pan, S. J., Xue, G.-R., Yu, Y., & Yang, Q. (2011). *Heterogeneous transfer learning for image classification*, *25*(1).

Zhu, Z., Albadawy, E., Saha, A., Zhang, J., Harowicz, M. R., & Mazurowski, M. A. (2019). Deep learning for identifying radiogenomic associations in breast cancer. *Computers in Biology and Medicine*, *109*, 85−90.

Zintgraf, L. M., Cohen, T. S., Adel, T., & Welling, M. (2017). Visualizing deep neural network decisions: Prediction difference analysis. *arXiv170204595*.

Brain-computer interface in Internet of Things environment

9

Vijay Jeyakumar[1], Palani Thanaraj Krishnan[2], Prema Sundaram[3] and Alex Noel Joseph Raj[4]

[1]Department of Biomedical Engineering, Sri Sivasubramaniya Nadar College of Engineering, Chennai, Tamil Nadu, India [2]Department of Electronics and Instrumentation Engineering, St. Joseph's College of Engineering, Chennai, Tamil Nadu, India [3]Department of Biomedical Engineering, RVS Educational Trust's Group of Institutions, Dindigul, Tamil Nadu, India [4]Key Laboratory of Digital Signal and Image Processing of Guangdong Province, Department of Electronic Engineering, College of Engineering, Shantou University, Shantou, P.R. China

9.1 Introduction

Brain-computer interface (BCI) popularly called brain-machine interface (BMI) is a communication system involving hardware and software components. It acquires brain signals generated by the central nervous system, interprets, translates them into commands, converts them and relays them into an artificial output device to carry out the desired action, thereby changing the interaction of the user with their external and internal environment.

BCI provides direct communication between a wired brain of a person and a device attached externally to translate the neuronal activity of the brain into signals to assist, augment or repair vision, hearing, mobility, communication, human cognitive or sensory-motor functions without involving peripheral nerves and muscles to achieve a therapeutic effect. BCI records and decodes brain signals and develops a new nonmuscular channel to communicate the intention of a person to systems, speech producers, assistive appliances and neuroprostheses which is due to the neuroplasticity of the brain. The user and the interface work together in which the trained user generates signals from prostheses which can be handled by the user's brain, similar to natural sensor channels by encoding intention, whereas the BCI decodes signals and converts them into commands to an external device by the user intention.

The advancement of technology has allowed humans to employ the electrical signals produced from the brain to interrelate with, control or modify one's environment. The upcoming era of BCI technology will soon improve the quality of lives of many disabled individuals who are not able to speak, utilize their limbs to operate assistive devices and reduce the cost of their medical expenses.

BCI technology is an upcoming research and development area which is of great interest to scientists, clinicians and engineers. Research on BCI is a young multidisciplinary field and hails researchers from different areas such as neuroscience, engineering, physiology, psychology, computer science, mathematics and rehabilitation.

Hans Berger detected neural activity and recorded electroencephalography (EEG) in 1924. The term BCI was first introduced and coined by Dr. J. Vidal in the early 1970s and the field has been evolving at a fast pace and reached many milestones for creating a pathway for severely disabled patients and in further helping humans for control of external devices. The principal aim of BCI is to rehabilitate the normal physiology of patients affected by neuromuscular disorders like cerebral palsy, stroke, sclerosis or injury in the spine. Research on BCI is growing extremely at a rapid rate over the past decade and enables patients to check a simulated e-mail, control television and to perform activities with the robotic arm. In principle, BCI uses brain signals (electroencephalogram—EEG) to control and to gather data on user intention based on polarity changes generated by neuronal postsynaptic membrane due to activation of voltage or ion gated channels. In BCI, electrodes placed on the scalp, on the surface or in the cortex are used to measure electrical signals produced from brain activity and are the widely studied signals in BCI. Thus, brain activity is measured and is translated into tractable electrical signals using a BCI (Table 9.1).

Usually, BCI monitors two types of brain activities:

1. Electrophysiological activity

   It is produced by neurons exchanging electrochemical neurotransmitters generating ionic currents within and across neurons traveling through the dendritic trunk called primary currents. This primary current is enclosed by secondary currents and this activity can be measured using techniques such as EEG, electrocorticography (ECoG), magnetoencephalography (MEG) and signal acquisition in individual neurons.

2. Hemodynamic response

   It involves the release of glucose by blood at a greater rate at the active neurons than in the inactive ones leading to a change in the proportion of oxyhemoglobin to de-oxyhemoglobin levels in the veins of the active area, which can be estimated by indirect methods such as functional magnetic resonance (fMRI) and near-infrared spectroscopy (NIRS).

9.1.1 Components of BCI

The working component of a BCI consists of:

1. Sensors to measure brain signals (usually electrodes)
2. Amplifier to strengthen signals and
3. A computer system to transmit signals into commands to control external devices which can be portable or wearable.

Table 9.1 Electroencephalography frequency bands with properties.

Band	Frequency (Hz)	Amplitude (μV)	Location	Activity
Delta	0.5–4	100–200	Frontal	Deep sleep
Theta	4–8	5–10	Various	Drowsiness, light sleep
Alpha	8–13	20–80	Posterior region of the head	Relaxed
Beta	13–30	1–5	Left and right side, symmetrical distribution, most evident frontally	Active thinking, alert
Gamma	>30	0.5–2	Somatosensory cortex	Hyperactivity

9.1.2 **Types of BCI**

Based on the nature of the input signals, BCI can be categorized as

1. Exogenous BCI

   This involves the use of an external stimulus to elicit neuron activity within the brain and does not require extensive training due to easy and quick setup, which shows a fast information transfer rate of up to 60 bits/min and could be acquired with only one EEG channel. This mode of BCI requires permanent attention to the applied stimulus and sometimes causes tiredness in users.

2. Endogenous BCI

   Involves self-regulation of brain potential without the use of an external stimulus. Through training, the users whose sensory organs are affected, learn to produce brain patterns decoded by BCI that can act as a two-dimensional space, suitable for cursor control applications. Multichannel EEG recordings are required for best performance at a lower bit rate of 20−30 bits/min and are a tiresome process where all users are not able to gain control.

Based on the input data processing module, BCI systems can be categorized as

1. Synchronous (or cue paced)

   In this type of BCI, brain signals in a predefined time window are analyzed whereas those outside the window are rejected. Hence the user is permitted to send commands within the stipulated period represented by the BCI system. Thus the time of mental activity is known beforehand and is linked with a specific cue. Also, the patient can blink and do all other eye movements but may not show a usual mode of interaction.

2. Asynchronous (or self-paced)

   In asynchronous mode, brain signals are analyzed continually without considering the user's actions. This is the natural way of BCI involving complex computation and difficult evaluation.

9.1.3 **How does BCI work?**

A BCI documents and elucidates brain signals. The electrical signals are transmitted by the neurons communicating with each other which could be acquired through advanced electrical sensors.

A precise communication pathway between the brain and the muscle is needed for a normal person to control body movements. Medical conditions such as stroke or neuromuscular disorders can alter or cease the communication between brain and body and can cause paralysis or cerebral palsy. However, the brain is still able to bring about activity for desired movements and thus, BCI can use those brain signals to control assistive devices (Christian, 2018).

Various techniques are made available to compute brain signals. The most commonly used among them is Electroencephalography which implies electrodes to be placed on the scale. Certain other techniques employ electrodes to be placed directly under the scalp or in the brain tissue itself involving surgical intervention without damaging the brain. The quality of the electrical signals obtained directly from the brain is better than signals recorded from the scalp and thus implantable BCIs are preferred and developed for paralyzed people.

Other procedures followed for measuring brain activity includes functional MRI (fMRI), using MRI scanner and MEG with the MEG-scanner not suitable for home use, large and expensive infrastructure. An advanced technique of NIRS is used to measure brain activity through the skull employing near-infrared light and does not require surgery.

Various centers in the brain can control different parts of the body accountable for a variety of movements. The different techniques employed to measure brain activity can analyze when the different control centers in the brain are active, thereby allowing BCIs to recognize body movements from brain activity. A unique characteristic of the brain is that these control centers are active when thinking about moving without actually doing it. The distinct control centers in the brain can be purposefully turned on or off by doing mental or physical tasks even in paralyzed patients, thus making BCI a realistic and promising technology.

By incorporating electrodes on specific control centers in the brain, the obtained signals are detected and could be converted to a command to perform a specific task or function on a device. An example is to make a computer mouse "click" by paralyzed patients when they lift their eyebrows. Thus BCI can be used as a "button" to control devices.

9.1.4 Key features of BCI

The essential key features of BCI to be

- Effectiveness—The BCI system should be really helpful to users.
- Robustness—The system must be stable during regular use and robust concerning anomalies.
- Quick operation—Task execution time should be as low as possible.
- More functionality—The system must allow the user to do as many tasks as feasible to increase the autonomy of the user.
- Safety—The system must pose no danger to user-health and comfort. The EEG cap should be comfortable enough to be worn for several hours.
- Mobility—To ensure mobility, the BCI system should be wireless, lightweight, and compact.
- User-friendliness—The system should be simple to operate and need no expert help for daily use.
- Cost-effectiveness—The price should be affordable to all kinds of users.

9.1.5 Applications

BCI exerts its applications in different fields of:

- Medicine
- Neuroergonomics and smart environment
- Neuromarketing and advertisement
- Educational and self-regulation
- Games and entertainment
- Security and authentication

Being a dynamic and growing field, BCI finds many useful applications in medicine for-

1. Disease prevention
   a. Smoking
   b. Alcoholism
   c. Motion sickness

2. Detection and diagnosis of diseases
    a. Tumors
    b. Brain disorders
    c. Sleep disorders
3. Rehabilitation and restoration
    a. Human-computer interaction using EEG data for classifying various mental states such as Relaxed, neutral or concentrating and mental-emotional states namely negative, neutral or positive, provides assistive technology for disabled individuals having neuromuscular problems such as schizophrenia or depression.
    b. In locked-in state patients, to allow communication within them and to restore normal function and who have no control over motor functions to improve sensory processing.
    c. To restore social interaction for patients by bringing out cursor control, robotic arms, prosthetic devices and wheelchairs.
    d. To improve rehabilitation of patients after stroke, head trauma and other disorders.
    e. Augments natural motor outputs for the improvement in the achievement of skilled pilots, surgeons and other professionals.

9.2 Brain-computer interface classification

Generally, BCI can be classified based on dependability, method of recording and operating mode as shown in Fig. 9.1. Regarding dependability, it can be dependent or independent. In the case of dependent BCI, the motor control by the user is required and vice versa for the independent BCI. MI-based BCI is the best example of independent BCI and are exclusively useful for stroke and impaired patients. Whereas in the case of the method of recording, BCI can be classified as invasive and noninvasive. Usually, a signal to acquire elements like sensors, electrodes are used primarily to record EEG. If any needle or implant type (usually microelectrode arrays are often used) sensing elements are used to acquire EEG, then it is an invasive BCI and surface electrodes are used in most of the noninvasive BCI applications.

9.2.1 Noninvasive BCI

In noninvasive types of BCI, the sensors are located to quantify brain activity by measuring electrical potentials on the scalp using EEG or the magnetic field using MEG. Various advanced techniques such as fMRI, positron emission tomography (PET), NIRS are also used in noninvasive BCI to measure brain activity which depends on changes in the rate of flow of blood called a hemodynamic response.

In severely and partially paralyzed patients, noninvasive BCI seems to be successful to reacquire communication and to control prosthetic devices, but motor recovery is of limited use.

1. EEG - Electroencephalography
2. ECoG - Electrocorticography
3. MEG - Magnetoencephalography
4. fMRI - Functional Magnetic Resonance Imaging
5. PET - Positron Emission Tomography
6. FNIRS - Functional Near Infrared Spectroscopy
7. SSEP - Steady State Evoked Potentials
8. SSVEP - Steady State Visual Evoked Potentials
9. SCP - Slow Cortical Potential
10. ErrP - Error-related Potential

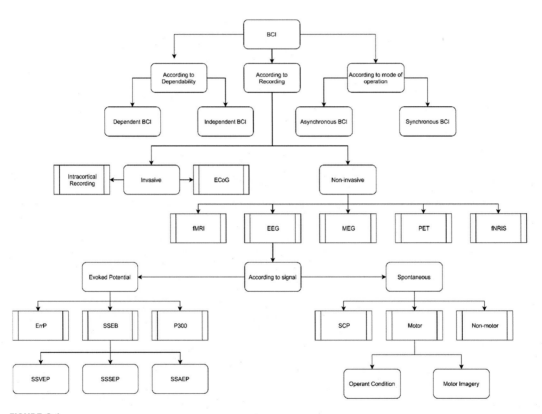

FIGURE 9.1

Classification of BCI systems based on dependability, method of recording and operating mode.

Courtesy Rashid, M., Sulaiman, N., Abdul Majeed, A. P. P., Mosa, R. M., Nasir, A. F. A., Bari, B. S., & Khatun, S. (2020). Current status, challenges, and possible solutions of EEG-based brain-computer interface: A comprehensive review. Frontiers in Neurorobotics, 14. https://doi.org/10.3389/fnbot.2020.00025.

9.2.2 Semiinvasive or partially invasive BCI

In this approach, BCI devices are implanted on the outer surface of the brain, inside the skull, but not within the gray matter to acquire brain electric potentials from the cerebral cortex using ECoG. This method requires a craniotomy to implant the electrodes and is used only when surgery is mandatory especially in epileptic patients.

Being a very promising modality, semiinvasive BCI can provide greater resolution signals than the noninvasive type and shows a reduced level of formation of scar tissue in the brain than in invasive BCI. ECoG based BCI recordings and performance proves to be stable over several months and is best suited for long-term use.

9.2.3 Invasive BCI

In this technique, microelectrodes are directly implanted into the intraparenchymal cortex, and the signals obtained from a single neuron are measured. Invasive BCI enables straight communication between the brain and a device but involves placing electrodes directly into the gray matter of the brain through neurosurgery. They can be a single unit BCI to find a signal from a single region of brain cells or a multiunit BCI to detect from multiple regions. These devices provide more accurate and best-quality signals rather than are prone to the formation of scar tissue making brain signals weaker as the body responds to a foreign object placed in the brain. Because of the expensiveness and high risk, the usual main targets of direct brain implants are noncongenital blind (acquired) and paralyzed patients.

Thus, shortly, direct brain implants used for neuronal recordings in humans may permit valuable insights into the live human brain to improve the quality of life and to regain autonomy.

9.3 Key elements of BCI

The brain is constantly producing electrical signals that can be recorded by keeping electrodes on the scalp or skull which act as good electrical insulators. In humans, every nerve is linked to nearly 10,000 nerves through dendrites. So, when the neurons make contact, currents occur and an electrical signal is passed along an axon or dendrite by the release of neurotransmitters, which transmit through the synapse to the dendrite of another neuron and are reconverted to electrical signals.

When the electric current leaves the neuron, it creates positive polarity and vice versa. These primary currents in the brain tissue reach the skull and the scalp and these voltage differences generated or trapped by EEG electrodes. Millions of parallelly oriented adjacent dendrites have to be active synchronously to have a quantifiable signal.

BCI aims to recognize and measure brain activity to denote the user objective and to interpret the features into device commands to fulfill the user intention in real-time (Fig. 9.2).

To achieve this, BCI being an artificial intelligence system is designed to detect a set of patterns in brain activity in six sequential stages:

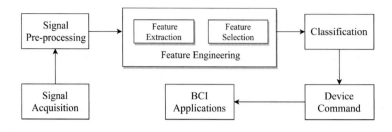

FIGURE 9.2

The general architecture of brain-computer interface.

9.3.1 Signal acquisition

Using any one of the particular sensors, for example, scalp or intracranial electrodes, brain signals are measured. The signals obtained are amplified and are made suitable for electronic processing. Numerous wired and wireless devices are available in the market to acquire EEG signals without the use of conductive gels or pastes. A list of devices with their specifications are available in Table 9.2.

9.3.2 Preprocessing or signal enhancement

This is a multistep process involving the removal of noise and artifacts from the data. High and low pass filters are applied to get rid of undesired power line interference. Other methods employed are used to cut away artifacts such as eyeball movements or eye blinking. After which the recording of signals, the significant features are generated for further processing and analysis and also for computer transmission and communication.

9.3.3 Feature extraction

Being a very challenging task, it analyzes brain signals and extracts information from them. EEG is very complex and needs processing algorithms to differentiate important signal characteristics from extra content and to map them onto vectors in a ready form suitable for interpretation into output commands. The extracted features must have strong relevance to the user intention as the most strongly correlated brain activity can be transient or oscillatory. In current BCI systems, the most prominently extracted features are response amplitudes and frequency ranges in EEG or ECoG and firing rates of individual cortical neurons. To guarantee precise computation of brain signals, environmental and physiologic artifacts are avoided or eliminated without relevant information loss. Many methods are available for feature extraction and include band powers and cross-correlation.

9.3.4 Classification stage

This step classifies the signals considering feature vectors with the help of classification algorithms. Using different techniques, a classifier is trained to recognize different features into different

Table 9.2 Electroencephalography devices.

Device name	No. of channels	Sampling frequency	Communication
NeuroScan	SynAmps: 64 Grael: 32 NuAmps: 40 Siesta: 32	SynAmps: 20 kHz Grael: 4096 Hz NuAmps: 1000 Hz Siesta: 1024 Hz	Wired
Brain Products	LiveAmp: 8/16/32	Between 250, 500, and 1000 Hz	Wireless
BioSemi	16, 32, or 64	2/4/8/16 kHz	Wired
Emotiv	INSIGHT: 5 EPOC + : 14 EPOC FLEX: 32	128 Hz	Wireless
NeuroSky	1	512 Hz	Wireless
Advanced brain monitoring	ABM B-Alert X24: 24	256 Hz	Wireless
G.tec nautilus	64	500 Hz	Wireless
AntNeuro EEG	64	2048 Hz	Wireless
Neuroelectrics Enobio 32	32	500 Hz	Wireless
Muse	4	256 Hz	Wireless
OpenBCI	Up to 16	256 Hz	Wireless
Cognionics Mobile	72	500–1000 Hz	Wireless
mBrainTrain	24	250–500 Hz	Wireless
MyndBand EEG headset	3	512 Hz	Wireless
Enobio	8, 20, or 32	500 Hz	Wireless

classes mathematically to achieve effective pattern recognition and to decrypt the user's intention in performing mental tasks.

9.3.5 Feature translation or control interface stage

The signals classified are fed to the feature translation algorithm that translates the features into respective meaningful commands to the external device such as a wheelchair or a computer to accomplish the user's intent. The algorithm used must be dynamic enough to assist or adapt to the natural or acquired alterations in the signal features and must contribute to a complete range of device control.

9.3.6 Device output or feedback stage

The translated commands from the algorithms could be able to operate external devices and provide operations such as letter selection, a robotic arm to allow movements, cursor control in a computer and so on. The operating device will send feedback to the user, thus terminating the process.

9.4 Modalities of BCI

With the advancement of functional neuroimaging technology, various potential novel modalities are employed based on the following factors.

9.4.1 Electrical and magnetic signals

Recording of electrical signals and generation of magnetic signals to control a BCI involves the use of:

9.4.1.1 Intracortical electrode array

Intracortical neuron recording technology is the most invasive method because it implants a single or array of electrodes into the cortex of the brain to detect the action signals out of individual neurons. The electrodes need to be implanted near to the signal source to ensure their stability over long periods but may face long-term signal variability due to neuron death or high tissue resistance and significant noise effect.

This technology is used for some severely disabled people suffering from a neurodegenerative disease whose neurons in the brain and spinal cord are affected, for example, amyotrophic lateral sclerosis (ALS). These direct brain implant electrodes are implanted into the visual cortex of blind people to restore sight and to synthesize phosphenes, the sensation of seeing light. Also, motor neuroprosthetics are developed to restore movements in paralytic patients to use computer cursors, robotic prosthetic arms, lights and television.

9.4.1.2 Electrocorticography

ECoG is measured by placing electrodes on the cortex of the brain, outside the dura mater called epidural ECoG or below the dura mater termed subdural ECoG, requiring craniotomy to implant electrodes onto the cortex involving 4−256 electrodes to perform a diverse range of studies. It is a less invasive modality, however, preserves the benefits of the invasive approach. It provides high spatial resolution and signals fidelity with the reduction in noise and artifacts, lesser clinical risk and better long-term stability with minimal training, rapid control, wider frequency and low technical difficulties. Thus making ECoG a good candidate for seizure localization in epileptic patients for carrying out motor tasks such as the finger, hand and arm flexion to control cursor movements or prosthetic hands in paralyzed and nonparalyzed epileptic patients and to provide speech and language processing for speech synthesis.

9.4.1.3 Electroencephalography

It provides a recording of the electrical activity of the tuft of neurons along the surface of the scalp by detecting voltage changes accompanying neurotransmission with the neurons. Multiple electrodes of up to 500 can be mounted on cap-like devices allowing data collection from the same scalp region taking thousands of snapshots of brain activity from different sensors per second. EEG data contains rhythmic activity reflecting neural oscillations explained by frequency, phase and power to show an association between rhythms and different states of brain activity. EEG provides unique usability benefits and makes them suitable for commercial use. It is the most widely preferred approach due to minimal risk, portability, inexpensiveness, ease of convenience in wearing without the need for surgery, relative ease of implementation and performance. Brain activity acquired through BCI can control a cursor, spelling device, assistive devices and prosthetic devices at a faster response time but show poor spatial resolution, signal to noise ratio and cannot be effectively used for high-frequency signals as the skull dampens the signal produced by the neurons and also requires training for each session. Spatial resolution can be enhanced by employing suitable filters or by integrating EEG with other modalities.

9.4.1.4 Magnetoencephalography

Using highly sensitive magnetometers, magnetic fields created by electric currents occurring in the brain are utilized for mapping brain activity. Superconducting quantum interference devices (SQUID) is used for acquiring magnetic signals outside the head. This recording method involves lab configuration with shields and special equipment so that MEG signals did not interfere with the earth's magnetic field. The received signals are less disturbed by the skull and cerebrospinal fluid (CSF). It offers better spatial resolution and is highly sensitive to tangential sources. They are better in detecting high-frequency signals as the magnetic field could easily penetrate through the skull and scalp, however, show portability and cost issues.

9.4.2 Metabolic signals

Acquiring metabolic signals based on oxygen level-dependent changes from the brain to control a BCI involves the use of:

9.4.2.1 Positron emission tomography

It is the nuclear imaging technique used to analyze blood flow, metabolism and neurotransmitter activity within brain cells. A specific amount of a radioactive tracer is injected into the bloodstream to reach brain cells. Radiotracer attaches to the glucose molecule and forms fluorodeoxyglucose. The principal fuel of the brain is glucose. So, the regions of the brain which are active will use glucose at a higher rate than inactive regions. By performing a PET scan, it allows scientists to find the working pattern of the brain and to diagnose diseases like dementia, Alzheimer's, Parkinson's and epilepsy by color codings showing highly active regions in warmer colors of yellow and red.

9.4.2.2 Functional magnetic resonance imaging

fMRI underlays MRI Technology that analyzes blood flow changes associated with brain activity. fMRI works on the principle that cerebral blood flow is closely coupled to neuronal activation and thus helps in mapping activity to the respective used area of the brain. It relies on the fact that when an area of the brain is in the active state, it requires increased blood flow which is termed blood-oxygen-level-dependent contrast, dependent on hemodynamic response.

Hemoglobin, the oxygen-carrying protein present in the red blood cells that distribute oxygen to the brain tissue exists in two forms-oxyhemoglobin and deoxyhemoglobin and the magnetic properties of the two states of hemoglobin change based on their oxygen levels. A more active state of the brain increases the demand for more oxygen leading to an increased blood flow to that part of brain tissue thereby changing the magnetic characteristics which would detect the more active areas of the brain at a specified time. Though it provides low temporal resolution, fMRI proves to be a safe and convenient technique with high spatial resolution as it could acquire details from deep parts of the brain than any other technique.

9.4.2.3 Functional near-infrared spectroscopy

NIRS is employed for functional neuroimaging to evaluate brain activity through blood dynamics associated with neuron activity using a near-infrared range of light. fNIRS could collect only 10 samples per second with low spatial resolution and cannot detect electrical signals past the cortex. However, this technology uses devices that are portable, more accessible, less expensive and less sensible to artifacts.

The upcoming field of BCI relies on the progress in three critical areas of research- to develop a stable acquisition system to function in all environments, to validate and disseminate in the long-term and to improve moment-to-moment reliability of BCI performance for different user populations.

9.5 Computational intelligence methods in BCI/BMI

The main objective of this chapter is to audit a wide choice of EEG signal handling procedures utilized in BCI-based EEG frameworks, with a specific spotlight on the cutting edge with respect to feature extraction and selection. It also examines the difficulties and impediments faced during the plan and execution of signal processing procedures. In the vast majority of the EEG-BCI frameworks these days, essential EEG information is pre-prepared to dispose of noise, however not all frameworks pre-process information. Features are then extracted from the EEG information and the most notable highlights for classification might be chosen. Each part of this outline will be examined in more prominent detail in this section.

9.5.1 State of the prior art

9.5.1.1 Preprocessing

EEG signals are ordinarily utilized for BCIs because of their high resolution in time and the cost-adequacy of brain signal acquisition when contrasted with different strategies like fMRI and MEG. Notwithstanding, EEG signals present processing challenges; since they are nonfixed, they can experience the ill effects of external noise and are inclined to signal distortion.

Jafarifarmand et al. (2017) suggested an automatic artifact removal in EEG recording is possible by independent component analysis (ICA) involving five methods for identifying trials consisting of artifacts. They divided trials into six epochs and applied ICA for them. The illegal epochs were marked and if the percentage of unmarked epochs is below 30% in each trial, then they were deleted from the training and testing dataset. ICA can distinguish between physiological noise from multichannel EEG signal which enables restoration to a noise-free physiological signal. The independent components are determined by their spectral density. Eltaf et al. (2018) utilized a temporal filter for preprocessing which is a bandpass filter to remove artifacts and has a range of 0.5−30 Hz.

A filter bank approach can be used for extracting alpha and beta activity since these waves are related to mental images by Serdar Bascil et al. (2015). Khan and Hong (2017) demonstrated that many studies utilized channel averaging, value-based channel selection and vector phase analysis as preprocessing tools to remove artifacts. In the channel averaging technique all channels related to brain activity are averaged. The channel selection technique allows the removal of a redundant channel.

9.5.1.2 Feature extraction

The extracted features should seize salient signal characteristics which may be used as a foundation for the differentiation between specific brain states. For feature extraction, many different methods are available in time Domain and Frequency Domain. Time-frequency evaluation is strong because it allows spectral data about an EEG signal to be mapped to the temporal domain, which is beneficial for BCI technology given that spectral brain interest varies at some point of the duration of use of the device as different tasks are executed. Jafarifarmand et al. (2017) used wavelet transform (WT) as a feature extraction technique because they are powerful decomposition techniques to derive dynamic features from the nonstationary EEG signals. Wavelet packet decomposition (WPD) is utilized to split the EEG signals into various frequency bands. Haar mother wavelet is used for decomposition and the extracted signal features are namely power spectrum, variance and mean.

Eltaf et al. (2018) utilized intrinsic time scale decomposition (ITD) which is an updated version of empirical mode decomposition (EMD) for analyzing the EEG signal into various proper rotation components (PRC). ITD is advantageous since it can provide exact time, frequency and energy localization. Serdar et al. proposed an approach combining the average signal power method for feature extraction and the principal component analysis (PCA) algorithm for the selection of features. PCA enables a reduction in redundancy in terms of Complexity. The main purpose of PCA is for dimension reduction. EEG and fNIRS are the two primary BCI modalities requiring mobility and trends have demonstrated that these two frameworks can be coordinated to improve the BCI execution. Hong et al. (2017) have briefly discussed the various feature extraction schemes such as power spectrum density based on Welch's method.

Logarithmic band power is another approach in which the logarithmic power of different frequency bands is calculated. For EEG feature extraction, common spatial pattern (CSP) is used to map multichannel EEG signals into a less dimensional subspace. Chakladar and Chakraborty (2018) documented that the fast Fourier transform (FFT) is implemented to obtain a frequency spectrum for extracting features from the EEG signal. Since EEGs are nonstationary signals, short-time Fourier transform (SSFT) is applied to get needed features. An extensive feature extraction category has also been explained. PCA is used for reducing the dimension by removing artifacts. Autoregressive components (AR) is a time-domain feature extraction technique. It uses a parametric approach. AR is not suitable with nonstationary EEG signals hence to overcome this disadvantage multivariate adaptive AR (MVAAR) is used.

9.5.1.3 Feature classification

This subsection intends to offer a short precis of the various classification techniques used. Support vector machines (SVMs) and linear discriminant analysis (LDA) have been determined to be the broadly used linear classifiers and artificial neural network (ANN), K−nearest neighbor (KNN) is mostly used as a nonlinear classifier. Jafarifarmand et al. (2017) employed SVM and artificial neural types [multilayer perceptron, probabilistic neural network (PNN)] as classifiers. Among which

SVM uses optimum hyperplane to distinguish between two classes and gives better results. ANN approaches such as feed-forward back propagation neural networks (FFBPNN) were designed, trained and investigated with different activation functions by Eltaf et al. (2018). The mean classification accuracy of 92.20% was obtained.

Serdar Bascil et al. (2015) implemented three classifiers: learning vector quantization (LVQ), multilayer neural network (MLNN) and PNN. LVQ is a supervised competitive learning algorithm used for classifying patterns. They have used the Levenberg—Marquardt algorithm for training the MLNN which gives a faster convergence. PNN is used which is formulated on a probability density function for pattern recognition. The most common classifier as per Khan and Hong (2017) used for hybrid EEG-fNIRS systems is LDA, SVM, extreme machine learning (ELM). ELM is a form of a feed-forward neural network, with high generalization performance at a fast learning speed.

It utilizes Moore-Penrose generalized inverse to set weights. ELMs aren't as correct as conventional neural networks, but they may be used while managing issues that require retraining of the network in real-time.

Convolutional neural networks (CNN) are the other schemes of ANN-based on the visual cortex. It can understand features from the processed input data by forward and backward propagation for optimization of weight parameters to minimize the classification error. One benefit of using a deep learning (DL) method is that it calls for minimum preprocessing considering that most desirable settings are discovered automatically. Feature extraction and classification are integrated into a single entity and optimized automatically by CNN. In Chakladar and Chakraborty (2018), several approaches for classification have been discussed. The Bayesian statistical classifier uses prior probabilities and calculates the posterior probability for classification. Enhanced versions of LDA namely Fisher LDA (FLDA) and Bayesian LDA (BLDA) provide better classification accuracy in disabled subjects. Quadratic discriminant analysis (QDA) gives better classification accuracy compared to LDA and KNN. Hidden Markov Model is a classification technique for classifying features of EEG signals responsible for finger movements. Ensemble classifiers deal with training multiple classifiers to solve the same problems. Ensemble methods usually employ three techniques: Bagging, boosting and random forest. Bagging is used to decrease variance by making use of additional data from the training dataset. Boosting is an iterative method wherein a strong classifier is built from several weak classifiers. Random forest is based on decision trees. The oddity of the technique employed by Ewan et al. is that it requires no predecided features to perform grouping. All things considered; the neural time series is allowed straightforwardly into an ANN that has a structure that is obliged by an arbitrary search (genetic algorithm). In this way, a broadly useful BCI arrangement calculation is made that mitigates dependence on a priori assumptions about the information.

9.5.1.4 Performance evaluation of BCI systems

The BCI systems efficiency relies on the performance of a classifier. Hence, the sensitivity-specificity pair, and precision are the commonly estimating parameters. The most important performance measure is accuracy which is usually calculated from the confusion matrix. Receiver operating characteristic curve, and area under the curve (AUC) are very often utilized when the classification variables are continuous. Few researchers reported information rate (ITR), the practical bit rate, task completion duration, and the count of successful trials in their results. All these parameters do not emphasize much about the overall performance of BCI systems.

9.6 Online and offline BCI applications

It is at most important to categorize BCI and its applications. The buzz word "BCI" means a system that can record, perform analysis, and transforms the inputs (i.e., Brain signal) into device instructions. Based on the commands generated by BCI algorithms, the applications of BCI can be categorized as clinical or nonclinical. In recent days, the BCI has focused on non-clinical applications such as automotive vehicle control, ambient control, communication, gaming, home control, and many more. A list of BCI applications with their associated steps is given in Table 9.3.

9.7 BCI for the Internet of Things

In a few applications, the BCI and Internet of Things (IoT) systems are combined to bring a new technology called brain—to—thing (communication) system, particularly used for healthcare applications. The main objective is to connect small objects like wheelchairs with the BCI environment to assist the end-users who are affected by paraplegics, ALS, and brain stroke. BTC offers features like smart environments, mobility, transparency for the system, user-aware interfaces and confirmations. The simple architecture of BTC is shown in Fig. 9.3.

From the Fig. 9.3, the signals evolving from the brain can be used to control the target items like home appliances through the IoT features. If the commands and the data being used for the control mechanism are simple, then the traditional machine learning algorithms including neural networks, pattern recognition, and SVM like classifiers are sufficed. The application of BCI can not be limited to control only a few objects. Since the IoT has proliferated rapidly, it is expected to extend its impact in other environments including industry, manufacturing, and transportation. Though several algorithms are proposed for cognitive recognition systems, the accuracy attained is between 70% and 80% which is not sufficient for real-time applications. Also, they consume much time when the data increases.

To make a reliable BCI system with the IoT, a DL framework can be proposed to deal with huge dimensional data. Such frameworks create a great impact on BCI users to connect multiple devices rather than a single task. The availability of EEGs large datasets resulted in the use of DL architecture to reveal the potential information which was not recognized by the traditional approaches (Jeyakumar, Nirmala, & Sarate, 2022). BCI systems have been explored by various DL models like CNN, Boltzmann machine, recurrent neural network (RNN), long short term memory (LSTM), spatial-temporal neural network (STNN), multilayer LSTM-RNN, and generative adversarial network (GAN). An autoencoder (AE) being a DL method focused on unsupervised learning, consisting of phases of encoding and decoding. In the encoding phase, the input signals are mapped with a constructive feature (low dimensional feature space). In the decoding phase, the actual features are revived from the lower features (Kurian & Jeyakumar, 2020). There are many architectures, that is, stacked encoders, variational autoencoder, event-related potential (ERP) encoder network, stacked sparse encoders and subject-specific multivariate empirical mode decomposition, which can be used for various applications like control of a

Table 9.3 Brain-computer interface applications.

BCI wheelchair

References	No. of subjects	EEG control signal	EEG features	Classification algorithm	Performance evaluation
Cao et al. (2014)	3	MI + SSVEP	CSP for MI; CCA for SSVEP	RBF SVM	ITR: 295.20; Time required: 370
Mara et al. (2013)	9	SSVEP	PSD	Decision tree	Success rate: 83% ± 15%, ITR: 70.3 ± 28.8 bits/min
Li et al. (2013)	3	MI	CSP	SVM	Success rate: 82.56%

BCI cursor control studies

References	No. of subjects	EEG control signal	EEG features	Classification algorithm	Performance evaluation
Serdar Bascil et al. (2015)	2	MI	BP	PNN	CA: PNN: 93.05%
Bascil et al. (2016)	5	MI	PSD	SVM	CA: 81.22%
Chakladar and Chakraborty (2018)	1	MI	PSD	DBSCAN	Execution time: 4.663 min, Success rate: 70.36%

Gaming and VR studies

References	No. of subjects	EEG control signal	EEG features	Classification algorithm	Performance evaluation
Djamal et al. (2017)	10	MI	FFT	LVQ	Success rate: 70%
Kreilinger et al. (2016)	10	MI	BP	LDA	"Upper 10%" MI detection rates: >70%

Robotic arm studies

References	No. of subjects	EEG control signal	EEG features	Classification algorithm	Performance evaluation
Yang et al. (2018)	2	SSVEP	FFT	CCA	Five tasks performed
Bhattacharyya et al. (2015)	11	MI	MFDFA	ANFIS	Success rate: >60%
Boussetta et al. (2018)	4	MI	FFT	RBF SVM	Success rate: 85.45%

BCI speller studies

References	No. of subjects	EEG control signal	Method	Classification algorithm	Typing speed	Success rate
Cao et al. (2014)	3	MI	Oct-O-Spell	SVM	67.33 bits/min	98.23%
Ansari and Singla (2016)	20	SSVEP	Multiphase spellers	SVM	13 chars/min	96.04%
Chang et al. (2016)	10	SSVEP + P300	Hybrid speller	CCA, SWLDA	31.8 bits/min	93%

(Continued)

Table 9.3   Brain-computer interface applications. *Continued*

BCI biometrics research

References	No. of subjects	EEG features	Classification algorithm	Performance evaluation	Applications
Hu (2018)	28	Fuzzy Entropy	Vote classifier	Accuracy—99.8%	Gender recognition
Nguyen et al. (2017)	125	PSD	Enroll and keygen	99	Cryptographic key Generation
Bashar et al. (2016)	9	MSD, WPS, WPES	SVM	94.44	Human identification
Blondet et al. (2016)	50	Average ERP	SVM	100	Biometric identification

Environmental control

References	No. of subjects	EEG features	Classification algorithm	Performance evaluation	Applications
Aydin et al. (2018)	10	P300	LDA	93.71	Environmental control system
Zhang et al. (2017)	3	ERP	BLDA	89.2%	Environmental control system
Shyu et al. (2013)	15	SSVEP	FPGA	Accuracy: 92.5%	Hospital bed nursing system

Emotion recognition studies

References	No. of subjects	Stimulation	Emotion types	EEG features	Classification algorithm	Performance evaluation
Wei et al. (2017)	12	Pictures	Positive and negative	PSD, CP, CSP	LDA	86.83%
Liu et al. (2017)	30	Movie clips	Joy, anger, fear, sadness, disgust, and neutrality	PSD	LIBSVM	89.45%
Kaur et al. (2018)	10	Video clips	Calm, angry, and happy	FD	RBF SVM	CA: 60%
Murugappan (2011)	20	Video clips	Disgust, happiness, fear, surprise, and neutral	Entropy	K-NN	CA: 82.87%

robot, environment control, figuring out the workload, driver fatigue monitoring, virtual reality, biometric systems, emotion recognition and gaming. An IoT connected with different scenarios through a DL framework is shown in Fig. 9.4.

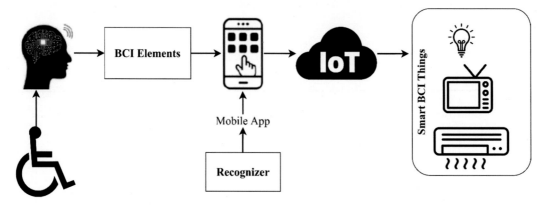

FIGURE 9.3

Elements of BTC communication.

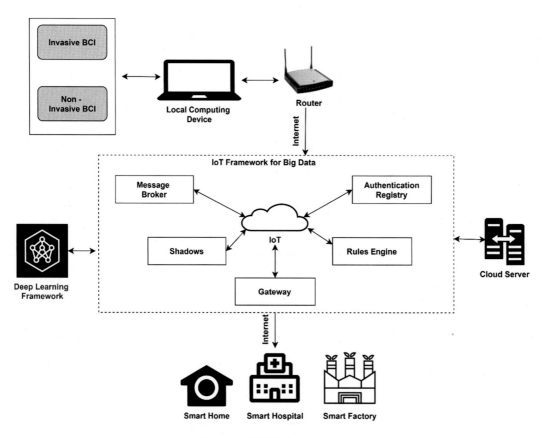

Smart BCT Scenarios

FIGURE 9.4

IoT with deep learning framework.

9.8 Secure brain-brain communication

The brain-brain communication termed the "Internet of Brains" is a technology-mediated communication system connecting two brains without the concern of the peripheral nervous system. It consists of two components:

1. BCI finds neural signals from one brain and converts them to computer commands, and
2. computer-brain interface carries computer commands to another brain.

The networking of human brains results in a networked biological supercomputer to function across language barriers. Hu (2018) explained their experiment as the "first organic computer" with the connected brains together. These systems were experimented with for their capability to differentiate two types of electrical stimuli patterns and were better than individual animals. After so many trials, recently they achieved 80% accuracy by implementing "brain Net" to decode an activity similar to a Tetris game. EEG signals from two senders were obtained, recorded, decoded and translated and sent to a receiver person in the network (Jiang et al., 2019). The decision of the sender on whether to move or not to move a block in the Tetris-like game was conveyed as signals of transcranial magnetic stimulation (TMS) to the occipital cortex of a receiver. If the response is "yes," then a flash of light is recognized by the receiver, that is, a phosphene. Based on this experience, using an EEG interface, a decision was made by the receiver whether or not to move the block. A feedback control was included in the experimental study to allow senders to give feedback on whether they accepted the decision taken by the receiver. Based on brain-to-brain communication, the senders' information reliability is varied to validate who is more reliable by the receiver (Fig. 9.5).

Department of Defense Advanced Research Projects Agency (DARPA), USA has recently sanctioned $8 billion to Prof. Jacob Rabinson, Rice University for his Proof of principle toward a wireless brain link in 2018.

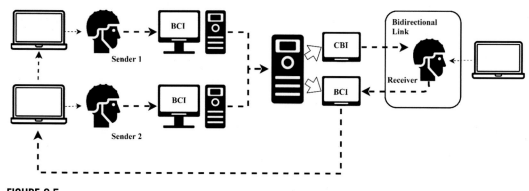

FIGURE 9.5

Brain—to—brain architecture.

9.8.1 Edge computing for brain–to–things

Edge computing is becoming more popular rather than cloud computing due to its processing of time-sensitive data while cloud computing is used to process the data which is not. Also, it is preferred in remote locations where connectivity is limited or not available. The devices connected in an edge computing environment are designed to gather and process onsite data very quickly instead of revolving around large, centralized servers stored in data centers. It does not focus on storing data. There are various limitations such as security threats, issues in performance, and higher cost of operations in traditional cloud computing. Since the information stored in the cloud is not frequently utilized, it leads to wastage of resources and storage space. Thus, edge computing comes in to ease and localize data processing and to reduce cloud dependence.

Edge computing has many benefits namely,

- Enhancing data security and privacy
- Betterment of application performance in a more responsive and robust manner
- Lowering the cost of operation
- Ameliorating efficiency and reliability
- Boundless scalability
- Preserving network and computing resources
- Minimizing latency

Rajesh et al. (2020) proposed a stable brain-to-brain interface with edge computing for helping poststroke paralyzed patients. A paralyzed person's thought based message transmission to the caretaker has been developed using a lightweight tiny symmetric algorithm (NTSA). IF a person is connected to multiple objects to share his thoughts to activate the object or give alerts to the end-user. During the brain-to-devices or brain-to-brain communication, the device authentication, secure channel establishment, device management, data privacy and analysis must be ensured by the proposed methods and algorithms. As given in Fig. 9.6, The framework is divided into four parts, including the consortium blockchain network, Internet of Medical Things environment, trusted edge computing layer and cloud service layer. As the foundation of other parts of the framework, the consortium blockchain network is composed of multiple medical and health institutions.

The Certificate Authority (CA) in the blockchain network is responsible for the authentication of externally connecting BCI interfacing devices, and the smart contract in the blockchain manages the data generated by the BCI interfacing devices, including metadata and access policies. The BTC environment plays the role of the data source in the framework. Once the data generated by the user is over, the data is encrypted by the device itself or the gateway and pushed to the edge computing node for further processing. The edge computing layer connects the BTC environment and cloud services to provide data preprocessing and storage services. In particular, in this framework, edge computing integrates Software Guard Extensions-based trusted computing services to ensure the integrity and confidentiality of BTC data and protect the privacy of data owners. The final cloud service layer is responsible for further processing of the BTC data preprocessed by the edge computing layer, such as big data analysis, machine learning prediction, etc. (Abdulkader et al., 2015). Finally, the command will be issued to the respective devices to be activated.

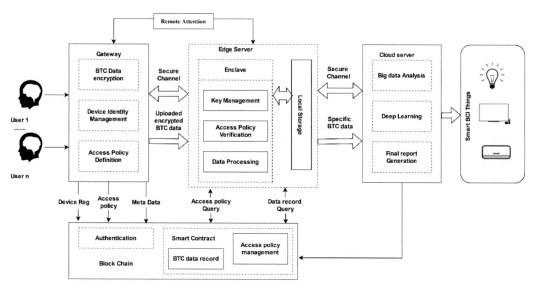

FIGURE 9.6

Brain—to—things (BTC) communication with Edge computing with the secure Blockchain mechanism.

9.9 Summary and conclusion

There are many limitations in the current BCI applications. For example, the application using P300 has poor ITR. If the system expects external visual stimuli, it is another drawback of the system when the person is suffering from sore eyes due to continuous exposure. Though there are many advantages of BCI controlled wheelchairs, the unexpected impediments, collision detection, unflatten areas, staircase climbing, pathfinding and planning along with GPS navigation systems must be addressed. In most cursor control and emotion recognition applications, the accuracy rate is less than 90% which is not sufficient for real-time scenarios. Though many data preprocessing, feature extraction and classification models are proposed for signal acquisition, deriving an optimum method is the need of the hour. A good deal between the BCI hardware and software must happen to have a significant impact on the improvement of BCI system performance. Brain activity recognition in the usage of the Biometric system is very crucial due to its sensitivity to external factors like emotions. As the recording of EEG signals may vary from 1 to 5 weeks, the stability of EEG is questionable. In most of the researcher's findings, the stability issues were not addressed.

9.10 Future research directions and challenges

The conventional kind of feature extraction and selection from the EEG signal does not provide a common solution for various BCI applications. This is the right time to utilize the advantage of Deep learning models while dealing with real-time EEG signals in huge dimensions. As the BCI market is expected to be a trillion-dollar business in 2025, BCI interfacing devices in the Internet of Things would be a promising platform for any disabled person. The researchers working in machine vision need to derive a common approach from the varied evaluation metrics thereby an efficient and standard method will be followed. The data privacy and authentication must be ensured while more number of objects are connected in a BTC environment. If the above limitations and scope for improvements are considered by the researchers working in BCI, human−machine interaction will emerge to new great heights soon.

Abbreviations

ALS	amyotrophic lateral sclerosis
ANFIS	adaptive neuro fuzzy inference system
ANN	artificial neural network
AR	auto regression
BCI	brain-computer interface
BLDA	Bayesian LDA
BMI	brain-machine interface
BP	band power
BTC	brain−to−thing communication system
CCA	Canonical correlation analysis
CNN	convolutional neural networks
CSF	cerebrospinal fluid
CSP	common spatial pattern
DARPA	Defense Advanced Research Projects Agency
DBSCAN	density-based spatial clustering of applications with noise
DL	deep learning
ECoG	electrocorticography
EEG	electroencephalography
EMD	empirical mode decomposition
ERP	event-related potential fMRI
fMRI	functional magnetic resonance imaging
fNIRS	functional near-infrared spectroscopy
FFBPNN	feed-forward back propagation neural networks
FFT	fast Fourier transform
FPGA	field programmable gate array
GAN	generative adversarial network
GPS	global positioning system
Hz	hertz
ICA	independent component analysis
IoT	Internet of Things
ITD	intrinsic time scale decomposition
ITR	information rate
kHz	kilohertz
KNN	K−nearest neighbor
LDA	linear discriminant analysis

LSTM	long short term memory
LVQ	learning vector quantization
MEG	magnetoencephalography
MI	Mental Imagery
MLNN	multilayer neural network
MRI	magnetic resonance imaging
MVAAR	multivariate adaptive AR
NIRS	near-infrared spectroscopy
NTSA	Novel Tiny Symmetric Encryption Algorithm
PCA	principal component analysis
PET	positron emission tomography
PNN	probabilistic neural network
PSD	power spectral density
PRC	proper rotation components
RBF	radial basis function
RNN	recurrent neural network
SQUID	superconducting quantum interference devices
SSFT	short-time Fourier transform
SSVEP	steady state visually evoked potential
STNN	spatial-temporal neural network
SVM	support vector machines
VR	virtual reality
WPD	wavelet packet decomposition
WT	wavelet transform

References

Abdulkader, S. N., Atia, A., & Mostafa, M.-S. M. (2015). Brain computer interfacing: Applications and challenges. *Egyptian Informatics Journal*, *16*(2), 213–230, ISSN 1110-8665. Available from https://doi.org/10.1016/j.eij.2015.06.002.

Ansari, I. A., & Singla, R. (2016). BCI: An optimised speller using SSVEP. *International Journal of Biomedical Engineering and Technology*, *22*, 31. Available from https://doi.org/10.1504/IJBET.2016.078988.

Aydin, E. A., Bay, O. F., & Guler, I. (2018). P300-based asynchronous brain-computer interface for environmental control system. *IEEE Journal of Biomedical and Health Informatics*, *22*, 653–663. Available from https://doi.org/10.1109/JBHI.2017.2690801.

Bascil, M. S., Tesneli, A. Y., & Temurtas, F. (2016). Spectral feature extraction of EEG signals and pattern recognition during mental tasks of 2-D cursor movements for BCI using SVM and ANN. *Australasian Physical & Engineering Sciences in Medicine/Supported by the Australasian College of Physical Scientists in Medicine and the Australasian Association of Physical Sciences in Medicine*, *39*, 665–676. Available from https://doi.org/10.1007/s13246-016-0462-x.

Bashar, M. K., Chiaki, I., & Yoshida, H. (2016). *Human identification from brain EEG signals using advanced machine learning method EEG-based biometrics. 2016 IEEE EMBS conference on biomedical engineering and sciences (IECBES)* (pp. 475–479). Kuala Lumpur: IEEE. Available from http://doi.org/10.1109/IECBES.2016.7843496.

Bhattacharyya, S., Basu, D., Konar, A., & Tibarewala, D. N. (2015). Interval type-2 fuzzy logic based multi-class ANFIS algorithm for real-time EEG based movement control of a robot arm. *Robotics and Autonomous Systems*, *68*, 104−115. Available from https://doi.org/10.1016/J.ROBOT.2015.01.007.

Blondet, M., Jin, Z., & Laszlo, S. (2016). CEREBRE: A novel method for very high accuracy event-related potential biometric identification. *IEEE Transactions on Information Forensics and Security*, *11*, 1. Available from https://doi.org/10.1109/TIFS.2016.2543524.

Boussetta, R., Elouai kouak, I., Gharbi, M., & Regragui, F. (2018). EEG based brain computer interface for controlling a robot arm movement through thought. *IRBM*, *39*, 129−135. Available from https://doi.org/10.1016/J.IRBM.2018.02.001.

Cao, L., Li, J., Ji, H., & Jiang, C. (2014). A hybrid brain computer interface system based on the neurophysiological protocol and brain-actuated switch for wheelchair control. *Journal of Neuroscience Methods*, *229*, 33−43. Available from https://doi.org/10.1016/j.jneumeth.2014.03.011.

Chakladar, D. D., & Chakraborty, S. (2018). Multi-target way of cursor movement in brain computer interface using unsupervised learning. *Biologically Inspired Cognitive Architectures*, *25*, 88−100. Available from https://doi.org/10.1016/J.BICA.2018.06.001.

Chang, M. H., Lee, J. S., Heo, J., & Park, K. S. (2016). Eliciting dual-frequency SSVEP using a hybrid SSVEP-P300 BCI. *Journal of Neuroscience Methods*, *258*, 104−113. Available from https://doi.org/10.1016/j.jneumeth.2015.11.001.

Christian, K. (2018). *Invasive brain-computer interfaces and neural recordings from humans. Handbook of behavioral neuroscience* (28, pp. 527−539). Elsevier ISSN 1569-7339, ISBN 9780128120286. Available from https://doi.org/10.1016/B978-0-12-812028-6.00028-8.

Djamal, E. C., Abdullah, M. Y., & Renaldi, F. (2017). Brain computer interface game controlling using fast fourier transform and learning vector quantization. *Journal of Telecommunication, Electronic and Computer Engineering*, *9*, 71−74.

Eltaf, A., Mohd Zuki, Y., Ibrahim, A., Abdalla, H., Ali, E., Fares, A.-S., & Muhammad, M. (2018). Enhancing EEG signals in brain computer interface using intrinsic time-scale decomposition. *Journal of Physics: Conference Series*, *1123*, 012004. Available from https://doi.org/10.1088/1742-6596/1123/1/012004.

Hong, X., Lu, Z. K., Teh, I., Nasrallah, F. A., Teo, W. P., Ang, K. K., Phua, K. S., Guan, C., Chew, E., & Chuang, K.-H. (2017). Brain plasticity following MI-BCI training combined with tDCS in a randomized trial in chronic subcortical stroke subjects: A preliminary study. *Scientific Reports*, *7*, 9222. Available from https://doi.org/10.1038/s41598-017-08928-5.

Hu, J. (2018). An approach to EEG-based gender recognition using entropy measurement methods. *Knowledge-Based Systems*, *140*, 134−141. Available from https://doi.org/10.1016/J.KNOSYS.2017.10.032.

Jafarifarmand, A., Badamchizadeh, M. A., Khanmohammadi, S., Nazari, M. A., & Tazehkand, B. M. (2017). A new self-regulated neuro-fuzzy frame-work for classification of EEG signals in motor imagery BCI. *IEEE Transactions on Fuzzy Systems*, *26*(3), 1485−1497.

Jeyakumar, V., Nirmala, K., & Sarate, G. S. (2022). Non-contact measurement system for COVID-19 vital signs to aid mass screening—An alternate approach. In Ramesh Chandra Poonia, et al. (Eds.), *Cyber-Physical Systems: AI and COVID-19* (pp. 75−92). Academic Press.

Jiang, L., Stocco, A., Losey, D. M., Abernethy, J. A., Prat, C. S., & Rao, R. P. N. (2019). BrainNet: A multi-person brain-to-brain interface for direct collaboration between brains. *Scientific Reports*, *9*, 6115. Available from https://doi.org/10.1038/s41598-019-41895-7.

Kaur, B., Singh, D., & Roy, P. P. (2018). EEG based emotion classification mechanism in BCI. *Procedia Computer Science*, *132*, 752−758. Available from https://doi.org/10.1016/J.PROCS.2018.05.087.

Khan, M. J., & Hong, K.-S. (2017). Hybrid EEG−fNIRS-based eight-command decoding for BCI: Application to quadcopter control. *Frontiers in Neurorobotics*, *11*, 1−6. Available from https://doi.org/10.3389/fnbot.2017.00006.

Kreilinger, A., Hiebel, H., & Muller-Putz, G. R. (2016). Single vs multiple events error potential detection in a BCI-controlled car game with continuous and discrete feedback. *IEEE Transactions on Bio-medical Engineering, 63*, 519−529. Available from https://doi.org/10.1109/TBME.2015.2465866.

Kurian, P., & Jeyakumar, V. (2020). Multimodality medical image retrieval using convolutional neural network. In *Deep Learning Techniques for Biomedical and Health Informatics*, (pp. 53−95). Academic Press.

Li, J., Liang, J., Zhao, Q., Li, J., Hong, K., & Zhang, L. (2013). Design of an assistive wheelchair system directly steered by human thoughts. *International Journal of Neural Systems, 23*, 1350013. Available from https://doi.org/10.1142/S0129065713500135.

Liu, Y., Yu, M., Zhao, G., Song, J., Ge, Y., & Shi, Y. (2017). Real-time movie-induced discrete emotion recognition from EEG signals. *IEEE Transactions on Affective Computing, 9*, 2660485. Available from https://doi.org/10.1109/TAFFC.2017.2660485.

Mara, S., Müller, T., Freire, T., Mário, B., & Filho, S. (2013). Proposal of a SSVEP BCI to command a robotic wheelchair. *Journal of Control, Automation and Electrical Systems, 24*, 97−105. Available from https://doi.org/10.1007/s40313-013-0002-9.

Murugappan, M. (2011). *Human emotion classification using wavelet transform and KNN. Proceedings of the 2011 international conference pattern analysis. intelligence robot ICPAIR 2011* (Vol. 1, pp. 148−153). Putrajaya: ICPAIR. Available from http://doi.org/10.1109/ICPAIR.2011.5976886.

Nguyen, D., Tran, D., Sharma, D., & Ma, W. (2017). On the study of EEG-based cryptographic key generation. *Procedia Computer Science, 112*, 936−945. Available from https://doi.org/10.1016/JPROCS.2017.08.126.

Rajesh, S., Paul, V., Menon, V. G., Jacob, S., & Vinod, P. (2020). Secure brain-to-brain communication with edge computing for assisting post-stroke paralyzed patients. *IEEE Internet of Things Journal, 7*(4), 2531−2538. Available from https://doi.org/10.1109/JIOT.2019.2951405.

Rashid, M., Sulaiman, N., Abdul Majeed, A. P. P., Mosa, R. M., Nasir, A. F. A., Bari, B. S., & Khatun, S. (2020). Current status, challenges, and possible solutions of EEG-based brain-computer interface: A comprehensive review. *Frontiers in Neurorobotics, 14*. Available from https://doi.org/10.3389/fnbot.2020.00025.

Serdar Bascil, M., Tesneli, A. Y., & Temurtas, F. (2015). Multi-channel EEG signal feature extraction and pattern recognition on horizontal mental imagination task of 1-D cursor movement for brain computer interface. *Australasian Physical & Engineering Sciences in Medicine/Supported by the Australasian College of Physical Scientists in Medicine and the Australasian Association of Physical Sciences in Medicine, 38*, 229−239. Available from https://doi.org/10.1007/s13246-015-0345-6.

Shyu, K.-K., Chiu, Y.-J., Lee, P.-L., Lee, M.-H., Sie, J.-J., Wu, C.-H., et al. (2013). Total design of an FPGA-based brain−computer interface control hospital bed nursing system. *IEEE Transactions on Industrial Electronics, 60*, 2731−2739. Available from https://doi.org/10.1109/TIE.2012.2196897.

Wei, Y., Wu, Y., & Tudor, J. (2017). A real-time wearable emotion detection headband based on EEG measurement. *Sensors and Actuators A: Physical, 263*, 614−621. Available from https://doi.org/10.1016/J.SNA.2017.07.012.

Yang, C., Wu, H., Li, Z., He, W., Wang, N., & Su, C.-Y. (2018). Mind control of a robotic arm with visual fusion technology. *IEEE Transactions on Industrial Informatics, 14*, 3822−3830. Available from https://doi.org/10.1109/TII.2017.2785415.

Zhang, R., Wang, Q., Li, K., He, S., Qin, S., Feng, Z., et al. (2017). A BCI-based environmental control system for patients with severe spinal cord injuries. *IEEE Transactions on Bio-medical Engineering, 64*, 1959−1971. Available from https://doi.org/10.1109/TBME.2016.2628861.

Early detection of COVID-19 pneumonia based on ground-glass opacity (GGO) features of computerized tomography (CT) angiography

H.M.K.K.M.B. Herath, G.M.K.B. Karunasena and B.G.D.A. Madhusanka

Faculty of Engineering Technology, The Open University of Sri Lanka, Nugegoda, Sri Lanka

10.1 Introduction

Novel COVID-19 has caused a coronavirus strain termed Severe Acute Respiratory Syndrome Coronavirus 2 or SARS-CoV-2 and the Middle East Respiratory Syndrome Coronavirus (MERS-CoV) (Herath et al., 2020). It has also been described as an acute respiratory infection caused by a novel coronavirus (CoV-2), later referred to as COVID-19 by the World Health Organization (Zhu et al., 2020). As of December 2020, Fig. 10.1 indicates COVID-19 recovered, and death cases worldwide. In December 2019, a pneumonia case of uncertain cause has been identified in Wuhan (Eurosurveillance Editorial Team, 2020). Since then, COVID-19 has spread rapidly in China and other countries, rendering it a significant global public health threat (Jin et al., 2020). The disease spread quickly across the globe after it was first identified and has become an international concern. The symptoms of infection with COVID-19 are similar to previously reported infections with SARS (Severe Acute Respiratory Syndrome), MERS (Middle East Respiratory Syndrome) (Herath, 2021; Yan, Chang, & Wang, 2020; Herath, Karunasena, & Herath, 2021), influenza (Potter, 2001) with fever, sore throat, nasal inflammation, pain in the chest, and dry cough.

A valuable noninvasive technique for identifying and diagnosing a range of diseases is medical image processing. Deep learning algorithms are now gaining a lot of interest for solving numerous medical imaging challenges. Latest advancements in medical image processing have focused on the detection of brain tumors (Kadkhodaei et al., 2016), skin lesions (Jafari et al., 2016), heart ventricle monitoring (Nasr-Esfahani et al., 2018), and liver diagnosis (Rafiei et al., 2018). Traditional medical imaging modalities (Llovet, Schwartz, & Mazzaferro, 2005) used in modern healthcare industries include X-rays (X-radiation), Ultrasound, Computed Tomography (CT), and Magnetic Resonance Imaging (MRI) (Katti, Ara, & Shireen, 2011).

Lung-related diseases have emerged worldwide as one of the most common medical disorders in humans. Early detection for timely treatment of patients with COVID-19 pneumonia is crucial

5G IoT and Edge Computing for Smart Healthcare. DOI: https://doi.org/10.1016/B978-0-323-90548-0.00013-9

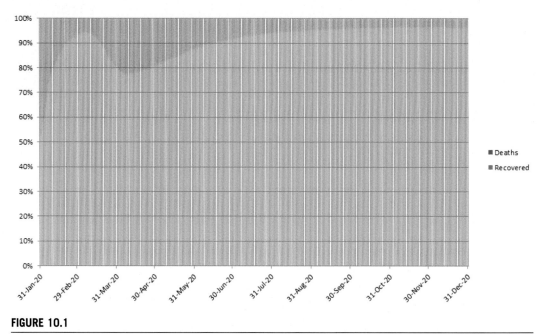

FIGURE 10.1

COVID-19 global recovered and death cases as of 31st December 2020 (Zhu et al., 2020).

for preventing spread, particularly in epidemic regions. Real-time Reverse Transcription Polymerase Chain Reaction or RT-PCR (Gibson, Heid, & Williams, 1996) is the most recent technique used to make a definitive diagnosis of SARS-CoV-2 infection (Jin et al., 2020). Computerized tomography, or CT scan, is a type of computer-assisted imaging constructed from multiple X-rays. Doctors and researchers can acquire precise, highly accurate, 3D photographs of a body with solid mass depth through CT scans. This facilitates visual diagnosis of the interior of patients without actual intrusion into the body. In diagnosing and evaluating internal fractures, diseases, cancers, blood clots, or excess fluid and conditions in the heart, kidneys, lungs, bones, and joints, CT scans are used. X-ray and CT photographs of a Chinese citizen infected by COVID-19 revealed the harm done to the human lungs, according to the evidence shared by the RSNA (Radiological Society of North America).

In COVID-19 patients with severe symptoms, including pneumonia, GGOs were observed in both lungs. It is important to remember that GGOs may suggest the emergence of lung fibrosis and maybe an indicative characteristic of lung cancer growth in high-risk individuals (Sadhukhan, Ugurlu, & Hoque, 2020). Galiatsatos, a lung disease specialist at Johns Hopkins Bayview Medical Center, treats COVID-19 patients and has identified some of the short-term and long-term lung complications caused by the novel coronavirus. COVID-19, a disease caused by the new coronavirus, may cause lung complications such as Acute Respiratory Distress Syndrome or ARDS (Umbrello, Formenti, Bolgiaghi, & Chiumello, 2017), pneumonia, and in the most severe cases. pneumonia allows the lungs to swell with fluid and become inflamed, resulting in respiratory complications. Individual patients' respiratory conditions can become serious enough to necessitate hospitalization for oxygen or ventilator help. COVID-19 induced pneumonia begins to take place in both lungs. In the lungs, air sacs fill with fluid, restricting their capacity to draw in oxygen and causing symptoms such as shortness of breath, coughing, and fatigue. While most people recover from pneumonia without permanent lung injury, COVID-19-associated pneumonia can be serious. Lung damage can lead to respiratory problems long after the illness has passed, which may take months to recover. Considering the lung damages caused by COVID-19 pneumonia, there is a high requirement for early detection of patients with COVID-19, particularly the elderly. This study aims to develop a system that uses machine intelligence (MI) with CT angiography to diagnose COVID-19 pneumonia patients.

Rest of this chapter is composed of four sections. The theoretical background and previous research related to medical imaging and deep learning techniques are listed in Section 10.2. The materials and methods of the proposed schemes are discussed in Section 10.3. The results and analysis of the experiments are listed in Section 10.4. Finally, Section 10.5 concludes the proposed system.

10.2 Background

Recently, the use of image recognition in machine learning (ML) has been growing increasingly. Medical image recognition (Jiang, Trundle, & Ren, 2010), image segmentation (Haralick & Shapiro, 1985), computer-aided diagnostics (Duggirala, Paine, Comaniciu, Krishnan, & Zhou, 2007), image transformation (Acharya & Ray, 2005), image fusion (Sahu & Parsai, 2012) associated with artificial intelligence (AI) plays a vital role in the field of healthcare. This section discussed the theoretical background related to deep learning techniques and characteristics of COVID-19 pneumonia by leveraging medical imaging data.

10.2.1 Ground-glass opacity (GGO)

COVID-19 can affect different organ systems, with symptoms most frequently seen in the respiratory tract. In Wuhan, a prospective study reported bilateral lung opacities in 40 of 41 (98%) chest CTs in COVID-19 patients and identified their most typical findings as lobular and subsegmental consolidation areas (Huang et al., 2020). Ground-glass opacity (GGO) (Collins & Stern, 1997) describes a finding on high-resolution CT (HRCT) of the lungs in which is seen hazy increased attenuation of the lung. Previous studies have shown that the existence of multifocal GGOs is the most common CT characteristic of COVID-19 pneumonia. GGOs are commonly detected and found to be present in 77%−100% of confirmed COVID-19 cases (Basler et al., 2020; Caruso et al., 2020; Song et al., 2020). Pure GGO lesions could be seen in the early stages of COVID-19 pneumonia (Zu et al., 2020), and were confirmed to be the critical finding after symptom onset (Wang et al., 2020). Schmitta and Marchiorib (Schmitt & Marchiori, 2020) have identified two characteristics of GGOs that may suggest the diagnosis of COVID-19 in the sense of the current pandemic. Fig. 10.2

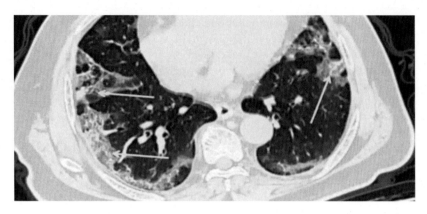

FIGURE 10.2

CT scan image of a patient with severe COVID-19 (Schmitt & Marchiori, 2020).

depicts GGO regions on a COVID-19 patient's CT scan image. The arrowheads show the discernible hazy region on the lungs' outer edges.

10.2.2 Support vector machine (SVM)

Support vector machine (SVM) is a supervised algorithm for machine learning that can be used for classification or regression problems. In classification problems, however, it is mostly used. The SVM is originally a binary classification method developed by Vladimir N. Vapnik and Alexey Ya at Bell laboratories (Burges, 1998; Vapnik, 2013), with further algorithm improvements by others (Joachims, 1998). For a binary problem, training data points are described, as shown in Eq. (10.1):

$$\{x_i, y_i\}, i = 1, ..., l, y_i \in \{-1, 1\}, x_i \in R^d \tag{10.1}$$

SVMs are based on statistical learning systems or VC theory suggested by Vapnik and Chervonenkis (1974) and Vapnik (1982, 1995), which are one of the most robust prediction methods. In the SVM algorithm, each data object is interpreted as a point in n-dimensional space (where n is the number of characteristics). The value of each characteristic is the value of a certain coordinate. In this study, we have used the SVM algorithm to detect GGO regions of the lungs.

10.2.3 Histogram of oriented gradients (HOG) algorithm

The histogram of oriented gradients (HOG) algorithm is a feature descriptor used for object detection purposes in computer vision and digital image processing. The HOG algorithm's basic principle is to use each pixel's gradient information to extract discriminating object detection features. In 1986, Robert K. McConnell of Wayland Research Inc. defined the ideas underlying HOG without using the term HOG in a patent application (McConnell, 1986). Mitsubishi Electric Research

Laboratories used the concepts in 1994 (Freeman & Roth, 1995). Navneet Dalal and Bill Triggs, researchers for the French National Institute for Research in Computer Science and Automation (INRIA), presented their additional work on HOG descriptors at the Conference on Computer Vision and Pattern Recognition (CVPR) in 2005.

HOG characteristics in the image are typically derived from different window sizes. The image window in the original HOG algorithm is split into several blocks, and each block is divided into several cells. The extraction algorithm for the HOG function is defined based on the following parameters:

1. Size of the ROI image
2. Type of the gradient ("signed" or "unsigned")
3. Cell size
4. Number of bins of cell histogram
5. Size of the block
6. The overlapping ratio of the block

We decided to use the HOG feature as a descriptive element for the GGOs, introduced by Dalal and Triggs (Dalal & Triggs, 2005) in 2005. The features were inspired by the previous work of Lowe et al. Lowe (1999), who in the late nineties, invented the SIFT (Scale-invariant feature transform) key point method. Binning the magnitude of the gradients into a histogram according to their orientation is the core concept of this function.

10.2.4 Convolutional neural network (CNN)

The convolutional neural network (CNN) has recently been highlighted as part of deep learning in machine vision for supervised and unsupervised learning tasks (Krizhevsky, Sutskever, & Hinton, 2017). Over the past decade, the convolutional neural network (CNN) has seen groundbreaking results in various fields related to pattern recognition, from image processing to voice recognition. The compositions of CNNs are convolutional, pooling, and fully connected layers. The primary roles of the convolutional layer are the identification of patterns, lines, and edges. Each CNN hidden layer consists of convolution layers that combine the input array of convolution kernels parameterized by weight. The multiple kernels produce multiple feature images and have succeeded in numerous vision tasks, such as segmentation and classification. Feature maps are locally progressive and spatially pooled pooling layers between the convolutional layers. The pooling layer transfers the maximum or average value, and therefore the size of feature maps is reduced. This approach captures an image's features with a strong position and shape.

10.2.4.1 Rectified Linear Unit (ReLU) activation function

One of the most prominent nonsaturated activation functions is the Rectified Linear Unit (ReLU). Eq. (10.2) defines the ReLU activation function, where $Z_{i,j,k}$ is the input of the activation function at the location (i,j) on the kth channel.

$$a_{i,j,k} = \max(Z_{i,j,k}, 0) \qquad (10.2)$$

ReLU is a piecewise linear function that prunes the negative portion to zero and preserves the positive portion. ReLU's simple *max (·)* operation helps it to compute much faster than the activation functions of "*Sigmoid*" or "*Tanh*," and it also induces sparsity in the hidden units and makes it possible for the network to obtain sparse representations.

10.2.5 Literature

To prevent the COVID-19 pandemic, science, technical knowledge, and resources were widely used. There is a daily increase in the number of studies related to the novel COVID-19. To detect infection with COVID-19, researchers have recently used imaging patterns on chest CT. In recent years, deep learning has led to excellent success in many research fields, including visual recognition, speech recognition, and natural language processing. A broad survey of the recent developments in convolutional neural networks (CNN) has been provided by Jiuxiang et al. (Gu et al., 2018). They have also addressed the enhancements to CNNs in terms of layer architecture, activation function, loss function, regularization, optimization, and fast computing. Brosch et al. (Brosch et al., 2015; Brosch et al., 2016) have suggested and applied a novel segmentation method focused on deep convolutional encoder networks to the segmentation of multiple sclerosis lesions in magnetic resonance photographs. Lee et al. (2017) have reviewed perspectives on deep learning technology's history, development, and implementations, especially concerning its medical imaging applications. Mishra, Tripathy, and Acharya (2020) have discussed the different levels of deep learning design, image classifications, and medical image segmentation optimization. They have also presented a thorough review of the deep learning algorithms used in the clinical image concerning recent works and their potential approaches. Das, Kumar, Kaur, Kumar, and Singh (2020) have developed an artificial deep transfer learning-based approach for detecting COVID-19 infection in chest X-rays using the extreme variant of the inception (Xception) model.

The aim of this study is to develop a COVID-19 pneumonia early detection system focused on CT angiography by referencing theoretical background and previous research work discussed here. The material and process used for constructing the proposed system are described in this chapter's next section.

10.3 Materials and methods

This section presents the materials and methodology used for the development of the COVID-19 pneumonia early detection system. Section 10.3.1 lists the datasets used for the study. In Section 10.3.2, the methodology of the proposed algorithm is present.

10.3.1 Dataset description

The CT scan image dataset from Kaggle (Ning et al., 2020) was used in this study and is freely accessible to researchers. Three forms of CT images collected from Union Hospital (HUST-UH), and Liyuan Hospital (HUST-LH) was included in the dataset. The CT scan image dataset is

Table 10.1 Description of the COVID-19 positive and negative CT scan image dataset.

CT scan image type	Data used	Description
Positive CT (pCT) images	27% of data from the original dataset	Imaging characteristics are related to COVID-19 pneumonia
Negative CT (nCT) images	11% of data from the original dataset	Imaging characteristics are irrelevant to COVID-19 pneumonia in both lungs.

composed of noninformative CT (NiCT) images, positive CT (pCT) images, and negative CT (nCT) images. 27% of pCT and 11% of nCT data were extracted randomly for this study. In the testing process, 120 of the nCT and pCT data were used to validate the algorithm. Table 10.1 depicts an overview of the datasets used in the study. Initially, the images in the datasets were scaled at 512 × 512 pixels.

10.3.2 Methodology

In the past few decades, computer vision technology has learned, changed, and evolved dramatically by being developed on the roots of AI and deep learning techniques (Mittal et al., 2019). The precise identification of natural characteristics from medical photographs greatly influences accurate inference (Pandey, Pallavi, & Pandey, 2020). Recent efforts have been made to improve the performance of artificial intelligence algorithms. The introduction of hybridized AI techniques is a prominent and essential feature in this area (Pham, Xu, & Prince, 2000). This section describes the methodology of the proposed system. The architecture of the proposed machine learning algorithm is represented in Fig. 10.3.

All the tests and evaluations were carried out on a personal computer (PC) with an Intel(R) Core (TM) i7−3.07 GHz CPU, 4GB RAM, running a Windows 7 Professional 64-bit operating system (OS).

10.3.2.1 Data preprocessing

As the first stage of this study, chest CT scan images of COVID-19 subjects and normal healthy subjects were taken and saved on a computer. We performed multiple image preprocessing (Bhattacharyya, 2011) techniques, such as image cropping and image resizing, before using the dataset to extract usable pulmonary regions. MATLAB image processing tools were used to crop and resize the images. The preprocessed images were manually verified for the training process. Images that were not adequately preprocessed were withdrawn from the training set but not replaced.

10.3.2.2 The development of convolution neural network (CNN) model

A well-known deep learning architecture influenced by living organisms' natural visual perception mechanism is the convolution neural network (CNN) (O'Shea & Nash, 2015). CNN is a scalable technology that is widely used to classify images. CNN is a complex image classification model

Dataset

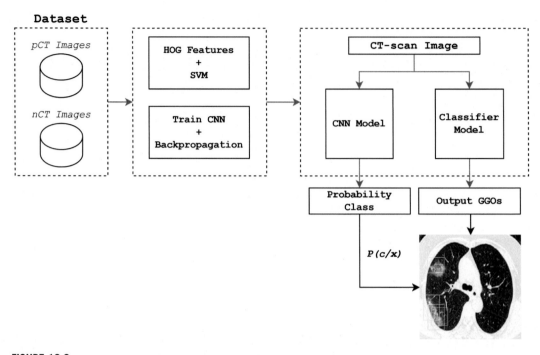

FIGURE 10.3

Architecture of the proposed machine learning algorithm.

due to its hierarchical structure and efficient image extraction features. The proposed CNN architecture comprises two stages: a feature learning stage and a classification stage, as shown in Fig. 10.4.

The developed feature learning stage consists of two convolutional layers and two pooling layers. A 3×3 convolutional filter for initial feature extraction was used in the first convolution layer. The resulting characteristics were then transferred to the first pooling layer composed of 2×2 max-pooling filters. Then, from the first convolution and pooling, the extracted features transferred to the second convolution and pooling layers.

The second convolution layer consists of 3×3 convolutional filters, and the pooling layer consists of 2×2 max-pooling filters. The feature score matrix passed into a fully connected layer in the classification stage, consisting of three neural layers fully connected. Each layer includes 500, 100, and 2 artificial neurons. Finally, by obtaining probability and classify input samples, the softmax layer was used to determine whether the test subjects were positive or negative for COVID-19 infection. We also used an input layer of 200×200 sizes in this model. The Model Hyperparameters (Ali, Caruana, & Kapoor, 2014) are properties that govern the entire process of training. This includes the variables that decide the network structure and the variables that determine how to train the network. As the training network solver, the

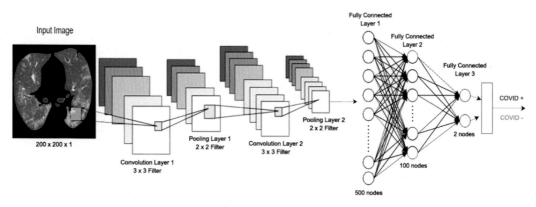

FIGURE 10.4

Proposed CNN architecture of the COVID-19 detection algorithm.

Stochastic Gradient Descent with Momentum (SGDM) (Dogo, Afolabi, Nwulu, Twala, & Aigbavboa, 2018) optimizer was used. For a particular task, a proper activation function significantly enhances the efficiency of CNN. To activate the nodes, the ReLU (Schmidt-Hieber, 2020) activation function was used.

10.3.2.3 The development of cascade classifier model

This section aims to develop a cascade classifier model to detect GGO regions on the lungs. The proposed cascade classifier model consists of the HOG feature extraction process and feature classification process. For the training process of the cascade classifier, we used 200 positive sample CT scan images and 500 negative sample CT scan images. The trained model was then used to distinguish infected regions by classifying extracted HOG features from the model. By splitting the picture into tiny cells, the HOG descriptor technique was carried out. Each cell evaluated a gradient direction histogram for the pixel inside the cell. For the object extraction, the HOG features extraction was completed in four phases. An overview of the proposed HOG/SVM architecture is shown in Fig. 10.5.

Gradient values are calculated in the first step by using the horizontal and vertical directions of the derivative mask using Eqs. (10.3) and (10.4).

$$D_x = \begin{bmatrix} -1 & 0 & 1 \end{bmatrix} \tag{10.3}$$

$$D_y = \begin{bmatrix} 1 \\ 0 \\ -1 \end{bmatrix} \tag{10.4}$$

The convolution procedure was used to obtain the x and y derivatives of object Image I, as defined by Eq. (10.5).

$$I_x = I_x \times D_x, \; I_y = I_y \times D_y \tag{10.5}$$

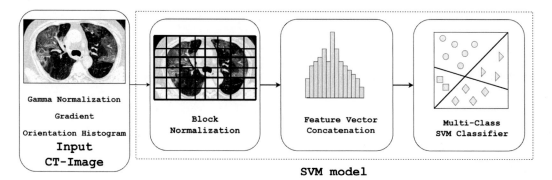

FIGURE 10.5

Overview of the proposed HOG/SVM architecture.

The gradient magnitude was estimated using Eq. (10.6),

$$|G| = \sqrt{I_x^2 \times I_y^2}$$ (10.6)

The gradient orientation was estimated using Eq. (10.7),

$$\theta = \arctan \frac{I_x}{I_y}$$ (10.7)

The second stage was the spatial orientation binning method. This stage aims to give the polling method for the product of a cell histogram. Based on the direction, each pixel of the CT scan image inside the cell was divided. The nearest bin in the range of 0 to 180 degree was allocated by the pixels. In the next phase, the cell and histogram normalized to be a vector shape using the HOG descriptor. In the final phase, the block's standardization was achieved using the L2 norm (Luo, Chang, & Ban, 2016) as described in Eq. (10.8).

$$b = \frac{b}{\sqrt{\|b\|^2 + \varepsilon^2}}$$ (10.8)

All of the block descriptors were translated into vector form during the HOG normalization procedure. For the identification process, the descriptor was used a detection window, which includes a 52×32 pixel window. This detection window was divided into 16×16 pixel blocks. Each block contains four cells, each cell contains an 8×8 pixel size with a 9-bin histogram, and each block includes 36 values. Finally, all the vector results of the classification process have been stored in the XML database.

10.4 Results and analysis

The main objective of this study is to develop an algorithm to early detection of COVID-19 pneumonia using CT scan images. In this section, results and analysis are presented.

Section 10.4.1 describes the results of the COVID-19 pneumonia detection system. In Section 10.4.2, Analysis is presented.

10.4.1 Test results of the COVID-19 pneumonia detection system

A classification model was developed for the training process. A total of 20 epochs and 2000 iterations were performed during the data training phase to achieve maximal model parameters. The accuracy curve of the training phase is shown in Fig. 10.6.

Based on the training data, each mini-batch classification's average accuracy was 94.25%, and the accuracy of the classification for each mini-batch has reached the maximum at epoch 8. The mini-batch loss for multiclass classifications decreased from 0.695 to 0.0977 at the end of the 20 epochs, according to the mini-batch loss curve shown in Fig. 10.7. Fig. 10.8 depicts the sample of healthy test subjects used in this study.

According to the images, lungs are observed to be clear and detect fewer gray spots. Detection of fewer gray spots suggested that the test subject is negative from COVID-19. Clinical trials have determined that these healthy test subjects are COVID-19 negative.

The CNN test results of 30 test subjects are shown in Table 10.2. (67% of COVID-19 positive subjects and 33% of negative subjects). Each test results describe the COVID-19 detection through the number of GGO regions of the patients. The test findings showed that positive subjects for COVID-19 were graded with a range of 0.69−0.99 probabilities, while stable subjects (COVID-19 negative subjects) ranged from 0.05−0.48 probabilities. Fig. 10.9

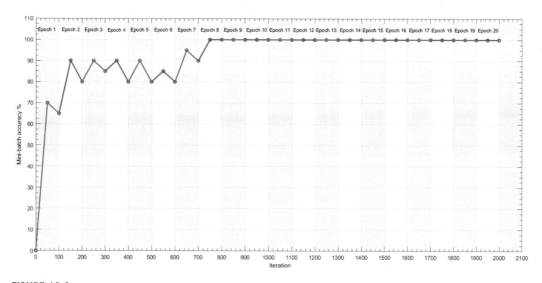

FIGURE 10.6

Mini-batch accuracy curve of the model training for 20 epochs.

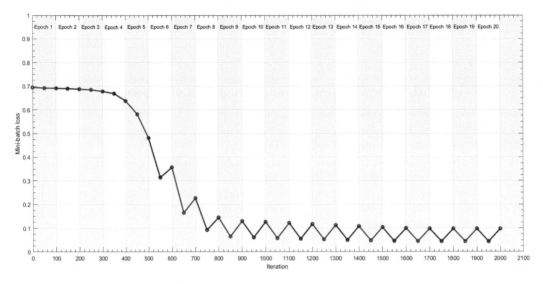

FIGURE 10.7

Mini-batch loss curve of the model training for 20 epochs.

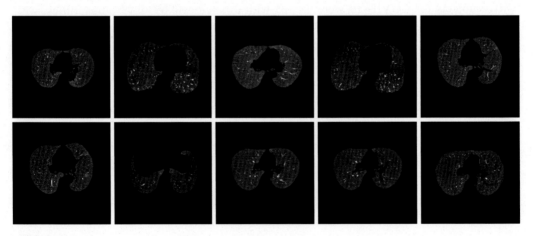

FIGURE 10.8

CT chest scan images for healthy test subjects (COVID-19 negative test subjects).

illustrates the experiment results for 10 test subjects of COVID-19 positive cases in this study.

Fig. 10.10 displays the test outcomes for 200 test subjects. The positive test subjects for COVID-19 describe the range between 0.69 and 0.99. A probability ranging from 0.05 to 0.48 is defined for healthy test subjects.

Table 10.2 Sample test results of the COVID-19 detection system.

Test no	Probability		COVID-19 detection		GGO Count
	Positive (+)	Negative (−)	Real	Predicted	
1	0.8213	0.1787	Positive	82.1% Positive	3
2	0.8014	0.1986	Positive	80.1% Positive	1
3	0.9116	0.0884	Positive	91.1% Positive	4
4	0.8323	0.1677	Positive	83.2% Positive	3
5	0.2453	0.7547	Negative	75.4% Negative	0
6	0.1569	0.8431	Negative	84.3% Negative	0
7	0.9214	0.0786	Positive	92.1% Positive	5
8	0.0187	0.9813	Negative	98.1% Negative	0
9	0.8473	0.1527	Positive	84.7% Positive	4
10	0.9256	0.0744	Positive	92.56% Positive	5
11	0.3048	0.6952	Negative	69.5% Negative	1
12	0.8823	0.1177	Positive	88.2% Positive	4
13	0.8015	0.1985	Positive	80.1% Positive	2
14	0.2541	0.7459	Negative	74.5% Negative	0
15	0.8341	0.1659	Positive	83.4% Positive	3
16	0.9084	0.0916	Positive	90.8% Positive	2
17	0.2209	0.7791	Negative	77.9% Negative	0
18	0.7913	0.2087	Positive	79.1% Positive	2
19	0.7215	0.2785	Positive	72.1% Positive	1
20	0.1057	0.8943	Negative	89.4% Negative	0
21	0.9403	0.0597	Positive	94.0% Positive	4
22	0.3301	0.6699	Negative	66.9% Negative	0
23	0.7012	0.2988	Positive	70.1% Positive	1
24	0.7756	0.2244	Positive	77.5% Positive	2
25	0.2185	0.7815	Negative	78.1% Negative	0
26	0.8702	0.1298	Positive	87.0% Positive	3
27	0.3047	0.6953	Negative	69.5% Negative	0
28	0.9503	0.0497	Positive	95.0% Positive	5
29	0.8981	0.1019	Positive	89.8% Positive	4
30	0.8213	0.1787	Positive	82.1% Positive	3

For ten COVID-19 test subjects, Fig. 10.11 demonstrates the detection of GGO regions. Fig. 10.12 indicates the test results shown in the GUI for the COVID-19 negative test subject. As shown in the figure, GGO regions were not found in the subject's lungs. Therefore, detected GGOs were counted as zero. The possibility of having COVID-19 of this test subject was 6.6%. The interface shown in Fig. 10.13 indicates the test results for positive test subjects with COVID-19. As shown in the CT image, five GGO regions in the lungs were identified by the system. Test findings

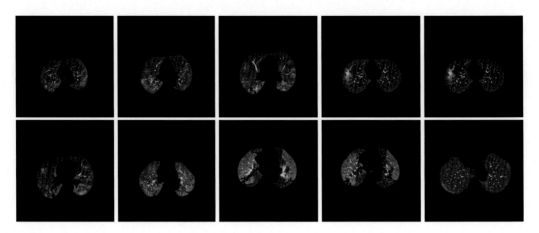

FIGURE 10.9

CT chest scan images for COVID-19 positive test subjects.

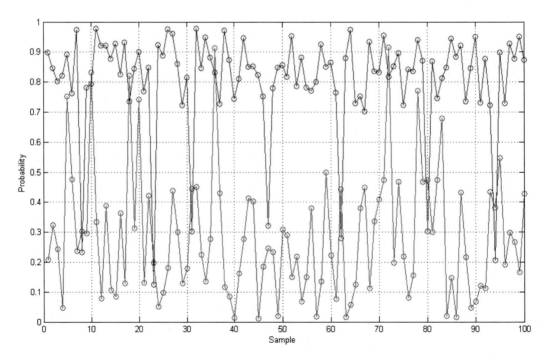

FIGURE 10.10

Model validation results of the COVID-19 positive and healthy test subjects.

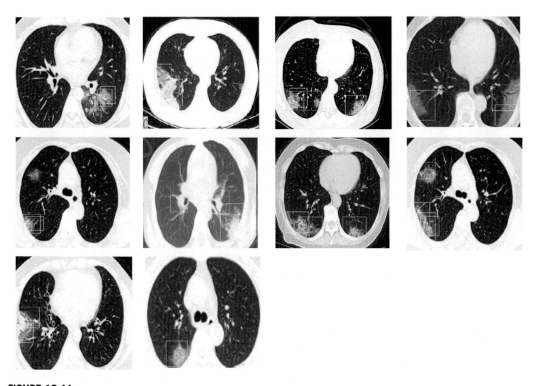

FIGURE 10.11

Ground-glass opacity (GGO) feature identification for ten COVID-19 test subjects.

have identified the patient as positive for 98.32% COVID-19 contamination. Besides, the cascade classifier algorithm identified the gray areas (GGO regions) of the lungs.

10.4.2 Analysis of the test results

A confusion matrix represents the output of an algorithm, also known as an error matrix. In the field of machine learning, usually supervised learning, the confusion matrix is widely used. Using the following assumption (Table 10.3), the entries in the confusion matrix were determined from the coincidence matrix.

The number of points at which the expected label is equal to the true label is defined by the diagonal elements, while off-diagonal elements are mislabeled by the classifier. The higher the diagonal values of the confusion matrix, the better the proposed algorithm's prediction results. Fig. 10.14 illustrates the confusion matrix of the proposed COVID-19 pneumonia detection system. According to the illustration, diagonal values are higher than off-diagonal values. As a result, the proposed algorithm produced higher accurate predictions. The

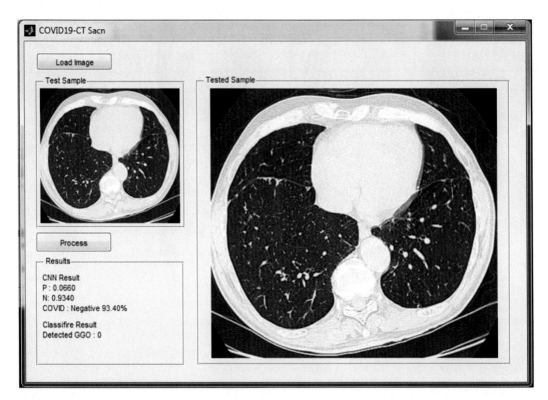

FIGURE 10.12

GUI test results of the COVID-19 detection system for negative test subjects.

accuracy and precision of the test results based on the confusion matrix data are shown in Eqs. (10.9) and (10.10).

$$\text{Accuracy} = \frac{\sum(\text{TP} + \text{TN})}{\sum(\text{TP} + \text{FP} + \text{FN} + \text{TN})} \tag{10.9}$$

Where,

- True Positives (TP): 93
- True Negatives (TN): 91
- False Positives (FP): 07
- False Negatives (FN): 09

$$\text{Precision} = \frac{\sum(\text{TP})}{\sum(\text{TP} + \text{FP})} \tag{10.10}$$

$$\text{MAE} = \frac{\sum_{t=1}^{N} |Y_t - F_t|}{N} \tag{10.11}$$

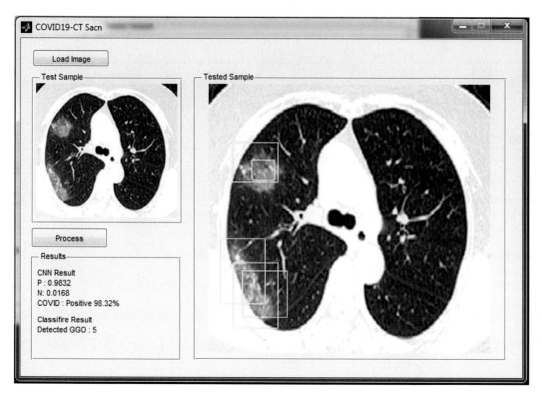

FIGURE 10.13

GUI test results of the COVID-19 detection system for positive test subjects.

Table 10.3 Hypothesis of the coincidence matrix.

Notation	Case	Description
TN	True Negative	The number of correct predictions of a negative case
TP	True Positive	The number of correct predictions of a positive case
FP	False Positive	The number of incorrect predictions of a positive case
FN	False Negative	The number of incorrect predictions of a negative case

Mean Absolute Error (MAE) (Chai & Draxler, 2014) calculates errors which describing the same phenomenon between paired measurements. Eq. (10.11) represents the relationship between actual data and expected data. Less than 0.200 is known to be the best MAE of a system, and MAE of the proposed system was 0.095. Therefore, using chest CT scan images, lower MAE validates the proposed model's accuracy for the COVID-19 pneumonia detection. Root Mean Square Error

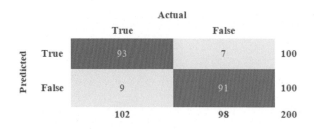

FIGURE 10.14

Confusion matrix of the proposed COVID-19 detection system.

Table 10.4 Accuracy, precision, MAE, and F1-score of the proposed system.	
Measure	**Observation**
Accuracy	92.8%
Precision	0.931
F1 Score	0.921
MAE	0.095

(RMSE) (Chai & Draxler, 2014) of the system was 0.149 calculated. Lower RMSE suggested the higher accuracy of the proposed algorithm (Table 10.4).

10.5 Conclusion

The novel coronavirus (COVID-19) has caused world alarm since its emergence in Wuhan, China, in late 2019. Infection can lead to severe pneumonia with clusters of diseases. It is essential to clarify the clinical features of other pneumonia due to the impact on public health. Real-time Reverse Transcription Polymerase Chain Reaction (RT-PCR) is the most commonly used novel coronavirus (COVID-19) detection technique. However, to confirm COVID-19 infection in the patient, RT-PCR kits are expensive and require 6−9 h to get results. It provides higher false negative outcomes due to less RT-PCR sensitivity.

This chapter suggested a possible way of using medical imaging (CT scan) to early detection of COVID-19 pneumonia. A GUI was developed using the MATLAB development environment for the identification of GGO regions of the lungs. GGOs were detected using the cascade classifier model, while CNN detected the probability of infection with COVID-19. The convolutional neural network and cascade classifier algorithm for COVID-19 pneumonia detection achieved 92.8% accuracy at 0.931 precision. Moreover, as positive cases of COVID-19, patients with lung diseases were reported. As a response, in the other lung condition with hazy regions of the lungs, an error could be developed. The probability of COVID-19 recognition indicates the stage of COVID-19 infection in the patient. A higher probability implies the existence of larger hazy spots in the lungs, showing that the individual is in a higher level of the COVID-19 virus.

In the future, we plan to use transfer learning to expand the algorithm to quantify the severity of other pneumonia or bacterial infections and to verify the findings using bio-imaging data obtained from various geographical regions.

References

Acharya, T., & Ray, A. K. (2005). *Image processing: Principles and applications*. John Wiley & Sons.

Ali, A., Caruana, R., & Kapoor, A. (2014, June). Active learning with model selection. In *Proceedings of the AAAI conference on artificial intelligence* (Vol. 28, No. 1).

Basler, L., Gabryś, H. S., Hogan, S. A., Pavic, M., Bogowicz, M., Vuong, D., & Levesque, M. P. (2020). Radiomics, tumor volume, and blood biomarkers for early prediction of pseudoprogression in patients with metastatic melanoma treated with immune checkpoint inhibition. *Clinical Cancer Research*, 26(16), 4414−4425.

Bhattacharyya, S. (2011). A brief survey of color image preprocessing and segmentation techniques. *Journal of Pattern Recognition Research*, 1(1), 120−129.

Brosch, T., Tang, L. Y., Yoo, Y., Li, D. K., Traboulsee, A., & Tam, R. (2016). Deep 3D convolutional encoder networks with shortcuts for multiscale feature integration applied to multiple sclerosis lesion segmentation. *IEEE Transactions on Medical Imaging*, 35(5), 1229−1239.

Brosch, T., Yoo, Y., Tang, L.Y., Li, D.K., Traboulsee, A., & Tam, R. (2015, October). Deep convolutional encoder networks for multiple sclerosis lesion segmentation. In *International conference on medical image computing and computer-assisted intervention* (pp. 3−11). Springer, Cham.

Burges, C. J. (1998). A tutorial on support vector machines for pattern recognition. *Data Mining and Knowledge Discovery*, 2(2), 121−167.

Caruso, D., Zerunian, M., Polici, M., Pucciarelli, F., Polidori, T., Rucci, C., & Laghi, A. (2020). Chest CT features of COVID-19 in Rome, Italy. *Radiology*, 296(2), E79−E85.

Chai, T., & Draxler, R. R. (2014). Root mean square error (RMSE) or mean absolute error (MAE)?− Arguments against avoiding RMSE in the literature. *Geoscientific Model Development*, 7(3), 1247−1250.

Collins, J., & Stern, E. J. (1997). Ground-glass opacity at CT: The ABCs. *American Journal of Roentgenology*, 169(2), 355−367.

Dalal, N., & Triggs, B. (2005, June). Histograms of oriented gradients for human detection. In *2005 IEEE computer society conference on computer vision and pattern recognition (CVPR'05)* (Vol. 1, pp. 886−893). IEEE.

Das, N.N., Kumar, N., Kaur, M., Kumar, V., & Singh, D. (2020). https://www.sciencedirect.com/science/article/pii/S1959031820301172.

Dogo, E.M., Afolabi, O.J., Nwulu, N.I., Twala, B., & Aigbavboa, C.O. (2018, December). A comparative analysis of gradient descent-based optimization algorithms on convolutional neural networks. In *2018 International conference on computational techniques, electronics and mechanical systems (CTEMS)* (pp. 92−99). IEEE.

Duggirala, B., Paine, D.S., Comaniciu, D., Krishnan, A., & Zhou, X. (2007). U.S. Patent No. 7,244,230. Washington, DC: U.S. Patent and Trademark Office.

Eurosurveillance Editorial Team. (2020). Note from the editors: World Health Organization declares novel Coronavirus (2019-nCoV) sixth public health emergency of international concern. *Eurosurveillance*, 25(5), 200131e.

Freeman, W.T., & Roth, M. (1995, June). Orientation histograms for hand gesture recognition. In *International workshop on automatic face and gesture recognition* (Vol. 12, pp. 296−301).

Gibson, U. E., Heid, C. A., & Williams, P. M. (1996). A novel method for real time quantitative RT-PCR. *Genome Research*, 6(10), 995−1001.

Gu, J., Wang, Z., Kuen, J., Ma, L., Shahroudy, A., Shuai, B., & Chen, T. (2018). Recent Advances in Convolutional Neural Networks. *Pattern Recognition, 77*, 354–377.

Haralick, R. M., & Shapiro, L. G. (1985). Image segmentation techniques. *Computer Vision, Graphics, and Image Processing, 29*(1), 100–132.

Herath, H.M.K.K.M.B. (2021). Internet of Things (IoT) enable designs for identify and control the COVID-19 pandemic. In D. Oliva, S. A. Hassan, & A. Mohamed (Eds.), *Artificial intelligence for COVID-19* (Studies in systems, decision and cntrol (ed.), Vol. 358, pp. 1–14). https://doi.org/10.1007/978-3-030-69744-0_24.

Herath, H.M.K.K.M.B., Karunasena, G.M.K.B., & Herath, H.M.W.T. (2021). Development of an IoT based systems to mitigate the impact of COVID-19 pandemic in smart cities. In U. Ghosh, Y. Maleh, M. Alazab, & A.-S. Khan Pathan (Eds.), *Machine intelligence and data analytics for sustainable future smart cities* (Vol. 971, pp. 1–23). https://doi.org/10.1007/978-3-030-72065-0_16.

Herath, H.M.K.K.M.B., Karunasena, G.M.K.B., Ariyathunge, S.V.A.S.H., Priyankara, H.D.N.S., Madhusanka, B.G.D.A., Herath, H.M.W.T., & Nimanthi, U.D.C. (2020, December). Deep learning approach to recognition of novel COVID-19 using CT scans and digital image processing. In *4th SLAAI - international conference on artificial intelligence* (pp. 1–12). SLAAI.

Huang, C., Wang, Y., Li, X., Ren, L., Zhao, J., Hu, Y., & Cao, B. (2020). Clinical features of patients infected with 2019 novel coronavirus in Wuhan, China. *The lancet, 395*(10223), 497–506.

Jafari, M.H., Karimi, N., Nasr-Esfahani, E., Samavi, S., Soroushmehr, S.M.R., Ward, K., & Najarian, K. (2016, December). Skin lesion segmentation in clinical images using deep learning. In *2016 23rd International conference on pattern recognition (ICPR)* (pp. 337–342). IEEE.

Jiang, J., Trundle, P., & Ren, J. (2010). Medical image analysis with artificial neural networks. *Computerized Medical Imaging and Graphics, 34*(8), 617–631.

Jin, Y. H., Cai, L., Cheng, Z. S., Cheng, H., Deng, T., Fan, Y. P., & Wang, X. H. (2020). A Rapid Advice Guideline for the diagnosis and treatment of 2019 Novel Coronavirus (2019-nCoV) infected pneumonia (standard version). *Military Medical Research, 7*(1), 1–23.

Joachims, T. (1998). *Making large-scale SVM learning practical (no. 1998, 28). Technical report.*

Kadkhodaei, M., Samavi, S., Karimi, N., Mohaghegh, H., Soroushmehr, S.M.R., Ward, K., & Najarian, K. (2016, August). Automatic segmentation of multimodal brain tumor images based on classification of super-voxels. In *2016 38th Annual international conference of the IEEE Engineering in Medicine and Biology Society (EMBC)* (pp. 5945–5948). IEEE.

Katti, G., Ara, S. A., & Shireen, A. (2011). Magnetic Resonance Imaging (MRI)-A review. *International Journal of Dental Clinics, 3*(1), 65–70.

Krizhevsky, A., Sutskever, I., & Hinton, G. E. (2017). ImageNet classification with deep convolutional neural networks. *Communications of the ACM, 60*(6), 84–90.

Lee, J. G., Jun, S., Cho, Y. W., Lee, H., Kim, G. B., Seo, J. B., & Kim, N. (2017). Deep learning in medical imaging: General overview. *Korean Journal of Radiology, 18*(4), 570.

Llovet, J.M., Schwartz, M., & Mazzaferro, V. (2005, May). Resection and liver transplantation for hepatocellular carcinoma. In *Seminars in liver disease* (Vol. 25, No. 02, pp. 181–200). New York, NY 10001, USA: Thieme Medical Publishers, Inc., 333 Seventh Avenue.

Lowe, D.G. (1999, September). Object recognition from local scale-invariant features. In *Proceedings of the seventh IEEE international conference on computer vision* (Vol. 2, pp. 1150–1157). IEEE.

Luo, X., Chang, X., & Ban, X. (2016). Regression and Classification Using Extreme Learning Machine Based on L1-norm and L2-norm. *Neurocomputing, 174*, 179–186.

McConnell, R.K. (1986). U.S. Patent No. 4,567,610. Washington, DC: U.S. Patent and Trademark Office.

Mishra, S., Tripathy, H. K., & Acharya, B. (2020). *A precise analysis of deep learning for medical image processing. Bio-inspired neurocomputing* (pp. 25–41). Singapore: Springer.

Mittal, M., Verma, A., Kaur, I., Kaur, B., Sharma, M., Goyal, L. M., & Kim, T. H. (2019). An efficient edge detection approach to provide better edge connectivity for image analysis. *IEEE Access, 7*, 33240−33255.

Nasr-Esfahani, M., Mohrekesh, M., Akbari, M., Soroushmehr, S.R., Nasr-Esfahani, E., Karimi, N., & Najarian, K. (2018, July). Left ventricle segmentation in cardiac MR images using fully convolutional network. *In 2018 40th Annual international conference of the IEEE Engineering in Medicine and Biology Society (EMBC)* (pp. 1275−1278). IEEE.

Ning, W., Lei, S., Yang, J., Cao, Y., Jiang, P., Yang, Q., & Wang, Z. (2020). iCTCF: An integrative resource of chest computed tomography images and clinical features of patients with COVID-19 pneumonia.

O'Shea, K., & Nash, R. (2015). An introduction to convolutional neural networks. Available online: https://white.stanford.edu/teach/index.php/An_Introduction_to_Convolutional_Neural_Networks (accessed on 28 December 2021).

Pandey, P., Pallavi, S., & Pandey, S. C. (2020). *Pragmatic medical image analysis and deep learning: An emerging trend. Advancement of machine intelligence in interactive medical image analysis* (pp. 1−18). Singapore: Springer.

Pham, D. L., Xu, C., & Prince, J. L. (2000). Current methods in medical image segmentation. *Annual Review of Biomedical Engineering, 2*(1), 315−337.

Potter, C. W. (2001). A history of influenza. *Journal of Applied Microbiology, 91*(4), 572−579.

Rafiei, S., Nasr-Esfahani, E., Najarian, K., Karimi, N., Samavi, S., & Soroushmehr, S.R. (2018, October). Liver Segmentation in CT Images using Three Dimensional to Two Dimensional Fully Convolutional Network. In *2018 25th IEEE international conference on image processing (ICIP)* (pp. 2067−2071). IEEE.

Sadhukhan, P., Ugurlu, M. T., & Hoque, M. O. (2020). Effect of COVID-19 on lungs: Focusing on prospective malignant phenotypes. *Cancers, 12*(12), 3822.

Sahu, D. K., & Parsai, M. P. (2012). Different image fusion techniques - A critical review. *International Journal of Modern Engineering Research (IJMER), 2*(5), 4298−4301.

Schmidt-Hieber, J. (2020). Nonparametric regression using deep neural networks with ReLU activation function. *Annals of Statistics, 48*(4), 1875−1897.

Schmitt, W., & Marchiori, E. (2020). Covid-19: Round and oval areas of ground-glassopacity. *Pulmonology, 26*(4), 246.

Song, F., Shi, N., Shan, F., Zhang, Z., Shen, J., Lu, H., & Shi, Y. (2020). Emerging 2019 novel Coronavirus (2019-nCoV) Pneumonia. *Radiology, 295*(1), 210−217.

Umbrello, M., Formenti, P., Bolgiaghi, L., & Chiumello, D. (2017). Current concepts of ARDS: A narrative review. *International Journal of Molecular Sciences, 18*(1), 64.

Vapnik, V. (2013). *The nature of Statistical Learning Theory*. Springer Science & Business Media.

Vapnik, V. N. (1982). *Estimation of dependencies based on empirical data*. New York: Springer-Verlag.

Vapnik, V. N. (1995). *The nature of Statistical Learning Theory*. New York: Springer-Verlag.

Vapnik, V. N., & Chervonenkis, A. Ya. (1974). *Theory of Pattern Recognition*. Moscow: Nauka.

Wang, Y., Dong, C., Hu, Y., Li, C., Ren, Q., Zhang, X., & Zhou, M. (2020). Temporal changes of CT findings in 90 patients with COVID-19 pneumonia: A longitudinal study. *Radiology, 296*(2), E55−E64.

Yan, Y., Chang, L., & Wang, L. (2020). Laboratory testing of SARS-CoV, MERS-CoV, and SARS-CoV-2 (2019-nCoV): Current status, challenges, and countermeasures. *Reviews in Medical Virology, 30*(3), e2106.

Zhu, N., Zhang, D., Wang, W., Li, X., Yang, B., Song, J., & Tan, W. (2020). A Novel Coronavirus from patients with pneumonia in China, 2019. *New England Journal of Medicine, 382*(8), 727−733. Available from https://doi.org/10.1056/nejmoa2001017.

Zu, Z. Y., Jiang, M. D., Xu, P. P., Chen, W., Ni, Q. Q., Lu, G. M., & Zhang, L. J. (2020). Coronavirus Disease 2019 (COVID-19): A perspective from China. *Radiology, 296*(2), E15−E25.

Applications of wearable technologies in healthcare: an analytical study

Hiren Kumar Thakkar[1], Shamit Roy Chowdhury[2], Akash Kumar Bhoi[3] and Paolo Barsocchi[4]

[1]*Department of Computer Engineering, Marwadi University, Rajkot, Gujarat, India* [2]*School of Computer Engineering, KIIT Deemed to be University, Bhubaneswar, Odisha, India* [3]*KIET Group of Institutions, Delhi-NCR, Ghaziabad, India* [4]*Institute of Information Science and Technologies, National Research Council, Pisa, Italy*

11.1 Introduction

The first wearable technology was found when eyeglasses were invented, which may be traceable since the 13th century. Simple, robust clocks were developed in the 16th century, named Nuremberg eggs (Evenson, Goto, & Furberg, 2015). They were designed so that they had to be worn around the neck. Before the pocket watches and wristwatches arrived, it became a famous status symbol in Europe. During the 17th century in China, the abacus ring could be referred to as another early example of wearable technology. During the early 1960s, statistics academician Edward Thorp built the first computer using wearable technology. Thorp explained in his book, which was named "Beat the Dealer," that he was successful in designing a tiny computer.

According to Thorp, that computer could fit in a shoe that would help to cheat at roulette. Thorp and co-developer Claude Shannon got 44 probability edge as an outcome because of the timing device. That same timing device helped to predict the landing of the ball on a roulette table. The next couple of decades marked the modernization of wearable technologies. In 1975, the first calculator wristwatch was released. After four years, the Sony Walkman came into the market. The 1980s saw the first release of digital hearing aids. In Fig. 11.1, the chronological order of the wearables has been shown.

The wearable units based on bio-metrics integrated within mobile units were used to collect specific information (Klonoff, 2013; Lee, Kim, & Welk, 2014). It allowed the accurate and precise management of patient's medical status from the healthcare unit's outskirts. The diagnosis and prediction of the patient's outcome could be made easy from the massive amount of information gathered from the wearable units. It also might play a significant role for healthcare professionals to choose the optimal diagnosis for their patients. Recently, there has been a lot of "hype" for artificial intelligence (AI)-based tools. The wearables help the user gain a better understanding of one's body.

5G IoT and Edge Computing for Smart Healthcare. DOI: https://doi.org/10.1016/B978-0-323-90548-0.00001-2

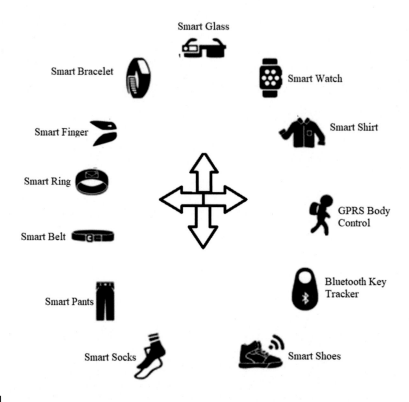

FIGURE 11.1

Types of wearable devices (Sahoo, Thakkar, & Lee, 2017).

There are different types of wearables in the market (Wu, Li, Cheng, & Lin, 2016). Some of them are mentioned below:

- Smart Watches: It helps keep track of time and helps the user by notifying calls, messages, emails, etc.
- Fitness Tracker: Keeps check on the health of the user. The device notes down the number of steps walked by the user.
- Head-Mounted Display: Takes the user to a different virtual reality world and provides virtual information directly to the eyes.
- Sports Watches: These wearable watches are specifically for fitness freaks and sportspersons like swimmers, gymnastics, cyclists, etc.

A GPS tracker is connected to that, which helps record information of the user's pace, heart rate, calories burned, etc. (Bhoi, Sherpa, & Khandelwal, 2018).

- Smart Jewelry: Smartwatches are designed as jewelry, especially for women.
- Smart clothing: The wearable devices are planted in the clothing to give a more fashionable look.
- Implantable: These devices are implanted under the skin of the patient. They help in tracking contraception, insulin level, etc.

In addition to the sensors used in wearable devices to monitor vital health care data, sensor technology advancements greatly influence the diversified domains such as Wireless Sensors Networks (Sahoo & Thakkar, 2019; Sahoo, Thakkar, & Hwang, 2017). Usually, the sensor acquired data are meaningless unless processed and analyzed using intelligent techniques. Several machine learning (ML)-assisted techniques can effectively help analyze healthcare and data acquired for diversified domains. For instance, in Rai, Thakkar, and Rajput (2020), the Seismocardiogram data are analyzed for automatic annotation using ML assisted binary classification. Similarly, robust AI techniques such as Reinforcement Learning can help analyzed health care data in a cloud of useful resource utilization (Thakkar, Dehury, & Sahoo, 2020). Moreover, ML tools can also be used for predicting clinically significant motor function improvements (Thakkar, Liao, Wu, Hsieh, & Lee, 2020). Wearables influence health and medicine, fitness, disability, education, transportation, enterprise, finance, gaming, music, etc. (Mishra, Dash, & Mishra, 2020).

11.2 Application of wearable devices

Wearable technology is an emerging field of innovation that includes designing a portable device worn by people. Wearable devices essentially help monitor vital information related to human health (Bhoi, Sherpa, & Khandelwal, 2015; Mishra, Patel, Panda, & Mishra, 2019). Wearable devices are more or less designed to capture the image and signal data and transfer it to cloud storage via smartphone-enabled wireless technology to monitor vital health parameters. From healthcare to finance wearables, devices have stepped their foot in every domain. Due to the small size of a wearable device, it has become very comfortable for people to use them anywhere, let it be for jogging or while working. By using fitness trackers, people can count the exact number of calories they burned (Mishra, Mishra, Tripathy, & Dutta, 2020), the distance they have walked, and can analyze the sleep trend. Some wearable devices also give an estimation of your heart rate and blood pressure. Wearables help one to live a healthy and balanced life (Mishra, Tripathy, Mishra, & Mohapatra, 2018). The use of wearables can also be seen in the finance sector. Every financial institution wants to create a better experience for its customers (Bhoi, 2017). With the help of fast-growing internet services and India's smartphone market, many private and public sector banks have moved to internet banking, mobile banking, etc. Payments from mobiles have become very common now. Wearables are a revolution in the finance sector. Wearables can also be used for payments and to extend the current mobile banking functions. Many wearables like wristbands, smartwatches (Apple Watch, Samsung Gear s2) jewelry (NFC ring) allows users to connect it with their mobile phones or bank account to make payment for goods and services. Apart from payment, it also helps its customers check the money balance, control money spending, and track cash flow. By the use of wearables, we can ensure the safety of women and elderly people. For example, ATHENA is a wearable technology by the ROAR company that aims to protect women against violence and assaults. It is designed especially for women. It can be easily worn as a pendant or clipped to their bag or dress. It can produce an alarm of 85db and also sends notifications to family members. In Fig. 11.2, the use of wearables in various domains has been described.

The most common wearable technologies include smart jewelry, such as rings, watches, wristbands, and pins (Mishra, Tripathy, Mallick, Bhoi, & Barsocchi, 2020). Specific smaller devices for display and interaction go best with a smartphone app. Smartwatches and fitness trackers have become very common and recognizable as examples of wearable technology. Examples: Garmin, Fitbit, Apple Watch, and Samsung Galaxy Gear. In Fig. 11.3, the different examples of wearables have been shown.

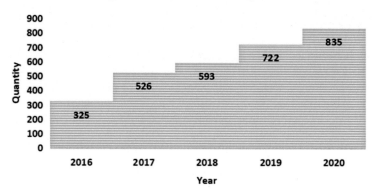

FIGURE 11.2

Number of connected wearable devices worldwide from 2016 to 2020.

Courtesy: https://www.electronic.se/.

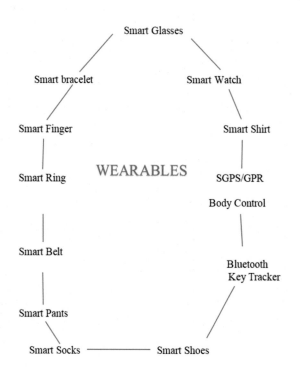

FIGURE 11.3

Examples of wearables.

11.3 **The importance of wearable technology in healthcare**
11.3.1 **Personalization**

In Fig. 11.4, few more examples of wearables have been shown. Inter-sharing norms to monitor fitness and clinical data using wearables and smartphones are to be adopted rather than conventional guidelines were declared in late 2018 by John Hancock, an eminent foreign-based insurer. The company would determine the risks included in every person's lifestyle habits by keeping track of every client's health status. The clients who have the habits included with less time span (e.g., less physical labor, inconsistent diet) must deal with high premiums. This insurance industry took a great step, making their clients adopt healthier lifestyles eventually, which will help reduce the pressure on healthcare.

11.3.2 **Remote patient monitoring**

The main advantage of a small smartwatch is that it can be worn regularly. A smartwatch can provide various events along with measuring the pulse all day long (Mishra, Mishra, & Tripathy, 2020). If these kinds of devices gathered more data, medicare's feasibility to develop a prevention-based prototype would increase, which can interpret the patient's data and alert patients for timely diagnoses.

11.3.3 **Early diagnosis**

Patterns can be established using AI by gathering heaps of information through wearables to track patients' health status. The system will determine the potential problems of the individual much before the problem arrives by using the information aggregated from every person. It will help to adopt much cost-effective and better curative decisions, which is opposite to diagnosing the risks at its full swing.

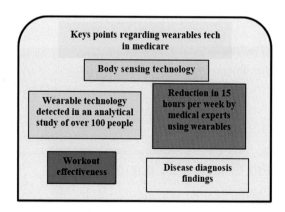

FIGURE 11.4

Use of wearables in healthcare.

11.3.4 Medication adherence

On many occasions, we find patients often forget to take their medications on time. Wearables play a very important role in this aspect. It helps remind the doctors of the patients' medications and remind the doctors whether their patients are taking medications as prescribed.

11.3.5 Complete information

Generally, very limited information is there with the doctors about their cases. The patient's medical analysis regarding risk factors and the beginning of the health discrepancies and the patient's clinical data based on past interactions is everything on which the doctors must rely upon. These kinds of data are often incomplete or imprecise Balas (2021). A wearable device's task is to collect all a patient's data in real-time, mistakes are not made, or an important symptom report never forgets to note. Therefore, the doctor can collect a detailed and precise report about the patient's condition. Hence, it will aid the doctor to establish a more accurate analysis leading to correct decision-making.

11.3.6 Cost savings

Investing in wearables might look like a costly investment, but it will be useful in saving unnecessary costs overall. These units often permit medical staff to keep track of patient's health status. This eliminates the requirement to shift the patient to a medical department. Moreover, with wearable devices, the symptoms could be detected at an early phase, which will allow less expensive treatments (Mishra, Mallick, Jena, & Chae, 2020).

The wearables in health care are still in a developing phase, but in the coming years, it is expected that this market will grow at a rapid pace (Bhoi & Sherpa, 2014; Mallick, Mishra, & Chae, 2020). In Fig. 11.4, some uses of wearables in healthcare have been shown.

Wearables have drastically increased; there are many reasons for the increase in the wearables market. Some of the key points responsible for these change is as follow (Mishra, Sahoo, Mishra, & Satapathy, 2016).

- Rising demand for round the clock inspection of the patients
- The growing culture of health and fitness
- Rising of per capita income
- Growth of technology.

11.4 Current scenario of wearable computing

A flexible distributed mobile platform was developed by the Media Laboratory of MIT, Cambridge. It was named LiveNet. It is primarily aimed at specific health monitoring long-term applications along with data processing in real-time. It also has context classification and streaming (Sung, Marci, & Pentland, 2005).

The EU FP5 IST program financed AMON's project (advanced care and alert portable telemedical monitor) (Anliker et al., 2004). The project's outcome was a wrist-worn device designed to acquire the data related to blood pressure. Moreover, the same wrist-worn device can acquire the blood oxygen saturation along with the skin temperature using the one lead Electrocardiography (ECG). In Lin, Lin, Chou, Chong, and Chen (2006), a real-time wireless physiological monitoring system (RTWPMS) is described. Like AMON, the RTWPMS is primarily designed using the low-power digital 2G cordless phone to acquire data related to the body temperature, pulse rate, and blood pressure. The RTWPMS is a highly custom-made healthcare information retrieval module. In Mundt et al. (2005), a system called LifeGuard is designed on the top of the multiparameter wearables with an application of space and terrestrial. The LifeGuard module comprises a crew physiologic observation device that helps measure the diversified vital health parameters such as oxygen saturation, body movement, respiration rate, heart rate, and blood pressure. In Ren, Chien, and Tai (2006), a prototype of a portable healthcare system is designed. The proposed portable healthcare system is targeted to acquire vital cardiac health-related data such as ECG, body temperature, and phonocardiography.

In addition to the cardiac health-related data, brain injury detection is an important aspect. In Tura, Badanai, Longo, and Guarani (2003), a portable wearable device is designed to detect brain injuries in children. The proposed portable wearable device uses different sensors to acquire health information, such as a pulse oximeter to acquire heart rate and blood oxygen saturation. In addition to that piezoelectric sensor-based belt is designed to acquire the respiration rate from the chest area. Furthermore, the dual-axis thermal accelerometer is used to acquire body movement. Cardiovascular diseases (CVD) are on the rise, and to contain them, the European Commission supported a project named MyHeart. This MyHeart project has engaged nearly 33 partners across ten major developed and developing countries. To make the MyHeart project successful and commercially viable, the European Commission has included several major industry collaborators such as Philips, Nokia, Vodafone, and Medtronic. The primary object of MyHeart is to fight against CVD by early diagnosis and prevention (Habetha, 2006). Under the MyHeart project, smart clothing is designed using the sensing module in garment-integrated or cloth-embedded wearable (Lubrano, Sola, Dasen, Koller, & Chetelat, 2006; Pacelli, Loriga, Tacchini, & Paradiso, 2006). Heart belts were developed by MyHeart, which could be worn across the chest. It could be attached to a standard bra or the waistband of standard underwear (Uhlsteff et al., 2004).

One of the Projects, called Wearable Health Care System (WEALTHY), has designed a wearable garment that covers the upper half of the body, and the garment is designed to be comfortable by wearing it as such the normal clothing. The wearable garment of the WEALTHY project is targeted to acquire physiological signals and biomechanical variables (Paradiso, Loriga, & Tacchini, 2005). A three-lead ECG can be designed using the sensory units combined with conducting items over some textile template (Lymperis & Paradiso, 2008; Scilingo et al., 2005). A washable sensorized vest was developed by researchers in Milan, Italy, named MagIC (Di Rienzo et al., 2005). It involved completely covered textile sensory units for breathing rate and heart rate tracking. It has a robust electronic module for evaluating the movement of the wearer. It performs signal processing and data sharing tasks by Bluetooth to a local system. A very simple and cost-effective stretchable sensory garment was developed in 2006 by the Medical Remote Monitoring of clothes (MERMOTH) project as a part of the European project (Lubrano, 2006; Weber & Porotte, 2006). A physiological tacking module consisting of a vest using several sensors was described by Pandian et al. (2008). It integrates with a fabric of garments that reliably collected various bio-signals.

A wearable body area network was developed in (Milenkovic, Otto, & Jovanov, 2006) consisting of a general wireless interface with transceivers and less powered control units. Harvard University proposed a healthcare sensory model for multipotent tracking developed Code-Blue (Shnayder, Chen, Lorincz, Fulford-Jones, & Welsh, 2005) based on Zigbee protocol. Customized sensory boards were available in the design to monitor pulse rate and heart irregularities. In Monton et al. (2008), a body area network based on a star topology and IEEE 802.15.4 protocol was developed. It constituted two primary units which include the communication unit and the processing unit. Communication unit interfaced with digital and analog sensors while processing unit coordinated with BAN and controlled all SCMs and sharing with external modules.

A customized e-medicare model is presented by Chung et al. Chung, Lee, and Toh (2008). It constitutes several customized interfacing nodes fitted with heart rate and blood pressure sensors. It also consists of an essential mobile phone unit used to display information and extraction features in signals. The ubiquitous sensor nodes hardware is described in detail in Chung, Lee, and Jung (2008) by Chung et al. A wireless body sensor network hardware developed recently is presented by Yuce, Ng, Myo, Khan, and Liu (2007). It makes use of an assigned clinical sharing band. An analytical study termed Human++ proposed a body area network that consists of sensory modules and a sink node (Gyselinckx, Penders, & Vullers, 2007). A heart-related application is presented by Leijdekkers and Gay (Leijdekkers & Gay, 2008), which mainly focuses on a personal healthcare status monitoring model. It involves a traditional cell phone service along with a Bluetooth-based ECG sensor. A wearable device for continual monitoring of ECG is analyzed in work in Fensli, Gunnarson, and Gundersen (2005), Sahoo et al. (2017), Sahoo, Thakkar, Lin, Chang, and Lee (2018), Thakkar and Sahoo (2019). Here the ECG sensor transmits the original and the amplified ECG signal regularly. This band is held device (HHD), which is a general PDA.

11.5 The wearable working procedure

In brief, it lets us know the working of a wearable explicitly. There are video cameras attached to head-mounted displays and glasses that track and record particular objects. Sensors attached can track motion, brain activity, the muscles' activity, and heart rate. They are usually found in health-oriented devices. Wearable devices also contain miniature computers or processors inside them. This helps in interacting with other objects. The data is collected from the different sensors present in the device. After that, the data is passed through algorithms, which helps to make sense of the data. The data collected is stored in the databases and, when required, is displayed in mobile applications and the device's web portals. In Fig. 11.5, the workflow of wearables has been shown.

11.6 Wearables in healthcare

11.6.1 Weight loss

Lose It! It is a kind of application-based service that helps people lose weight and undergo a healthy lifestyle. There are a lot of digital mediums through which Lose It offers several services. It includes the ability to log meals, count the maintenance calories, provide a variety of fitness

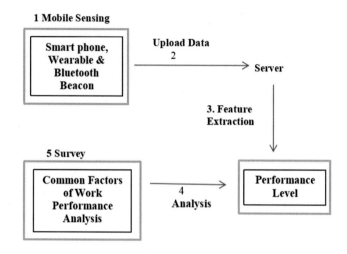

FIGURE 11.5

Workflow of wearables.

challenges. It also offers to provide reminders and reports on daily weight loss goals. It also provides a precise method that analyzes the daily calorie and nutrient intake by the wearable application. Lose it platform works in association with several popular health brands such as Google Fit, Fitbit, Nike, Misfit, MapMyFitness, etc.

11.6.2 Medication tracking

Medisafe is a mobile platform that helps patients to stick to their daily medical regime strictly through wearable technology. There are four key features of medisafe,

- Patients can record their medications on the app by a virtual pillbox.
- Alerts and notifications are there to make the patients aware (remind) about their medications according to the doctor's prescription.
- Medfriends, or third parties can monitor the user and receive medication reminders, which will help track the patient's progress.
- Location-specific coupons are sent to the patients by a partnership that has been formed recently. These medications are imported directly from the hospital records.

Medication notifications and reminders are the focus of the Medisafe wearable application. Its availability can be found on Android and Apple. The wearable interactions include quick finger taps, which indicate whether the medication has been taken.

11.6.3 Virtual doctor consultations

HealthTap, a health management program, provides comprehensive virtual care. Three core components are included in the digital platform.

- An extensive network of doctors creates contents that are present in the information module. Doctors can give real-time answers to the patients according to their questions asked.
- There are doctor consultations 24 × 7 through voice, video, or text by services like There HealthTap Prime and Concierge.
- HealthTap encourages users to remain healthy by providing knowledge and information related to a healthy lifestyle.

Certain health search capabilities are controlled by voice.

11.6.4 Geiger counter for illnesses

Sickweather is a mobile platform. It acts as a Doppler radar for illness. It made its place in the Apple Watch Release in June 2014. From social media and crowdsourcing, the data is mainly collected by Sickweather's mobile platform. Hence the data gets mapped into a real-time map of illness. The advantage of the new platform was taken by Graham Dodge, president, and CEO of Sickweather, to create an innovative, new feature, Sick Score, described by him as a "Geiger counter for contagious illness." There is a feature which contains the top three contagious illness. There is also a handwashing timer. The handwashing timer makes the people alert whenever they are in an area with a high sick score.

11.6.5 Hydration tool

App design and development company Stanfy designed an Apple Watch application named Water balance. It helps users in tracking their water intake, which will improve their hydration habits. The eatable application helps in analyzing the amount of water an individual needs. Height, weight, age, and physical activity level determine the amount of water intake. The Apple Watch was described as a convenient device by Andrew Garkavyi, CEO of Stanfy, which tracks hydration as it is located on the wrist.

11.6.6 Pregnancy and fertility tracking

An independent pregnancy and fertility named BabyMed worked with application design and development company Blue Label Labs to put its mobile application to the Apple Watch. A fertility calendar and a pregnancy calendar feature in the mobile application will help users track, prepare, and calculate due dates or ovulation dates. It allows a user to track these dates in many ways, but the wearable mainly focuses on only two features. Fig. 11.6 shows an example pregnancy and fertility tracking mobile application.

- The user can check the status of her fertility using animations using the fertility tracker.
- The user's present milestone is identified, and the length of time until he/ she reaches the next landmark, is shown by the pregnancy tracker. This, in turn, helps the doctor to monitor the patient easily.

FIGURE 11.6

Pregnancy and fertility tracking functionality (https, mhealthspot).

11.7 State-of-the-art implementation of wearables

We will be discussing three case studies below. Wearables have played a significant role in keeping track of the fitness information and detecting older people's stress levels. It will be more apparent now when we will discuss three case studies now.

11.7.1 Detection of soft fall in disabled or elderly people

According to a World Health Organization report, between 28% and 35% of older people aged above 65 years old fall at least once a year, and these accidents lead to significant injuries. There are already many approaches to solving this particular problem, but most are expensive and time-consuming, leading to delayed results. With the rise of wearables in healthcare, a collection of real-life data samples is possible using wearables. By using the proper ML algorithm, the accuracy of predicting a fall increase. Presently, in wearable devices used to detect falls, the input data is collected from sensors such as accelerometers, altimeters, or gyroscopes. These sensors provide data based on the sudden change in the movement or position of the user. In Fig. 11.7, a flow chart analysis of a specific wearable to detect the soft fall is presented.

- Accelerometer—Detect rapid change in movements.
- Gyroscopes—It is a device used for measuring or maintaining orientation and angular velocity.
- Altimeter—measures the altitude.

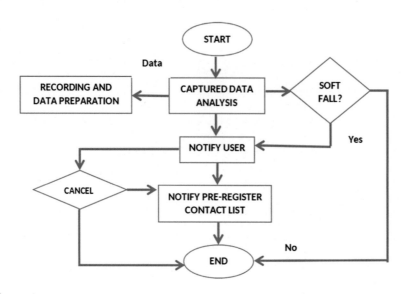

FIGURE 11.7

A flow chart analysis of a specific wearable.

The experiments recorded based on these sensors could majorly detect hard falls but not soft falls. Soft falls are referred to as falls originating from another initial axis rather than the standing coordinates, for example-chair sliding. The present work focuses on soft fall identification. At the end of this work, a prototype will be presented, which will be based on ML. It will be executed on a real-time model using a wearable device running on Android OS. There are several objectives to be achieved to implement a case study in a real-life system. Firstly, choose already available and affordable hardware. Secondly, it should be convenient for an elderly user. Thirdly, provide the user with accurate predictions. Android platform is chosen to implement the model because of its large market size. Using Predictive Model Mark-up Language (PMML), an XML-based language, the predictive model's shareable form is exported. The PMML is generated from the Knime platform, a graphical user interface that integrates ML and data mining through its modular data pipelining concept. Then the consumption of the model is done. The precision of the system is guaranteed using the ML approach as described below.

In the first step, the real-life data collected is divided into two data sets: the training set and the test set. The model is developed with the help of a training set in the training phase and the test phase; with test data, the accuracy of prediction is tested. The data set is divided as 50% randomly allocated training set data and the other half to the test set data; this process is repeated at least 20 times with random allocation of data in the training set and the test set to increase the statistical relevance. As the work is based on soft fall detection, the hard fall data is removed from the dataset, thus leaving only soft fall and other activities in the dataset. The Area Under Curve of a Receiver Operating Characteristic is used to measure the separation of two classes.

Table 11.1 Some records.

Labels	Records	Samples
Soft fall	475	146,743
Others	2651	12,789,643

Table 11.2 Performance of different ML algorithms.

ML algorithms	AUC/Random	AUC/Loo
Decision tree	0.65 ± 0.03	0.69 ± 0.03
Random forest	0.92 ± 0.02	0.92 ± 0.02
KNN	0.85 ± 0.02	0.86 ± 0.02

The data is collected from two wearable devices (LG-G and Moto 360). Table 11.1 presents the characteristics of the collected data. The data is collected and processed with the help of an android application. Twenty persons perform different activities, which include motion and soft fall. Only the data related to the linear accelerometer are kept as the data is sufficient to detect a soft fall and reduces the time required to detect the fall. For stability purpose, the values of three linear accelerator sensors are modified as

$$V_n(i) = \sqrt{((a_x(i))^2 + ((ay(i))^2 + ((a_z(i))^2} \tag{11.1}$$

After a first inspection of the pattern of the soft fall, the inputs of the ML algorithm are needed with records of 200 vectors Vn; the time window is $20*200 = 4000$ Ms.

Different algorithms have been used to detect a soft fall accurately. To counterattack the convergence problem when the ratio of samples between classes is not equilibrated, the simple bootstrap sampling method balances the training data. The platform Knime is used to implement ML models. The algorithms used in the experiment are:

- Decision tree: A decision tree is used to build a classification or regression prototype in the tree-like structure. It splits a dataset into subsets with an increase in the level of the tree.
- DT ensemble: Maximum of 300 trees are used in this algorithm; this algorithm will learn an ensemble of decision trees.
- KNN-50 max neighbors are configuring.

The results observed from the experiment are shown in Table 11.2. Among all other algorithms, the DT ensemble algorithm gives the best accuracy of detecting a soft fall, and the leave-one-out scoring does not show any changes in the result, which shows the robustness of the dataset. With this approach, the disabled and elderly people can be easily monitored by doctors and their family members, and immediate help can be provided when required. This method is very beneficial for someone who lives away from their family. With little modification, the patients' different body conditions, like heart rate, blood pressure, and temperature, can also be shown and immediately alerted if any abnormal conditions occur.

11.7.2 The third case study is based on the detection of stress using a smart wearable band

Mental stress is one of the most common mental phenomena that most people suffer from; stress is not a serious health problem. But if a person goes through stress and anxiety, it can often be the foundation for a more serious health problem. Stress detection may help prevent stress-related medical issues. With the help of a smart band based on skin conductance, stress detection can be done.

In this approach, a smart wearable band is designed to detect stress based on the different conductance levels of the skin. But some activities like running can also produce the same conductance level, so distinguishing between stress and other activities ML algorithm is used. Sleep trends and various activities can also be tracked, and the data collected can thus be transmitted to the user's smartphone via Bluetooth or can be uploaded in a web portal. Apart from many parameters like heart rate, HRV, and blood pressure, this device uses skin conductance to detect stress. The sweat secreted in stress conditions decreases the resistance of the skin and increases skin conductance. This fluctuation in the parameter allows it to become reliable for stress detection. If the biological aspects are seen, the conductance level of skin is denoted to be a marker to detect the degree of stress where the excess level of sweat may be regulated specifically by the sympathetic neuron model can be determined. If an individual is feeling stressed, in that scenario, the neuron model is set into play. The sympathetic nervous system activates the sweat glands. Thus, sweat secretion takes place. The sweat decreases the resistive functionality of the skin and increases the conductance of the skin.

The band's physical structure contains a micro-controller, Bluetooth, skin conductance sensor, and 3-axis accelerometer. The conductance sensor is present at the band's underside. The conductance is measured through a static voltage level provided to silver-coated electrodes that are embedded into the skin layer. To maximize the stress detection efficiency, an accelerometer is also present in the device.

The data was collected by retrieving electric signals among participants with a 23–56 age group. The sampling frequency was taken to be 4 Hz. The data was collected from participants after identifying the psychological state of patients. The information was divided into two groups, namely with efforts and without efforts. Table 11.3 shows the samples of the collected data.

The ML algorithm used for analyzing data is logistic regression. Logistic regression gives the output as 1 or 0. In this case, if the output is 1, the stress is detected; if 0, the individual is not stressed.

The electrodes located at the internal end of the band continuously monitors the skin conductance of the user. The electrodes release a minor voltage to the skin layer while the response is extracted. The data is recorded, and then, using an algorithm, the user's stress level is detected. The band is associated through Bluetooth to the mobile, from where the user can see the results.

The algorithm is found to be 91.66% accurate. The smart band was able to distinguish between a stressed and unstressed individual accurately. It also helps the doctor and family member monitor the user's mental health remotely and take proper preventive measures when required. The design and development of wearables in the medical field hold great attention in the people. The devices provide low-cost, unobtrusive solutions for continuous monitoring of the mental health of the patients. The doctors on the data collected can provide improved treatments. It also helps the user to understand the stress patterns better and know their body better.

Table 11.3 Records of collected data samples.

Name of users	Without efforts	With efforts	Stressed
U1	1.739	1.611	Yes
U2	1.537	1.630	Yes
U3	1.615	1.615	Yes
U4	1.383	1.671	Yes
U5	1.126	1.143	Yes
U6	1.138	1.018	Yes
U7	1.001	1.094	No
U8	0.823	0.958	No
U9	1.101	1.123	No
U10	1.060	1.097	No
U11	0.709	0.793	No
U12	1.052	1.081	No

11.7.3 Use of wearables to reduce cardiovascular risk

The third case study is based on the determining sedentary nature by computational intelligence study of wearables on a day-to-day activities to reduce cardiovascular risk. With continuous developments in wearable research, energy-constrained methodologies, and the data analysis approach, it is important to contain better, effective methods of providing healthcare services. Based on several reports on heart-related diseases, treating and curing such diseases is not an easy task due to the increasing population, inefficient healthcare services, and rising medical expenses. The primary elements for heart-related risk factors include lack of physical activity, improper food habits, regular smoking nature, and high alcohol intake.

The elimination of disease metrics will prevent around 80% of high risks as disclosed during research.

- The evidence in what way to sort the concern of an inactive way of leading life is negatively conceived and act as an issue for further analysis based on the National Institute for Health and Care Excellence guidelines.
- Use of hips, wrists, or even accelerometers mounted on thighs using belts helps analyze static lifestyle study.
- This study persuades elderly people to remain physically active. The target of this study is to prepare a model that automatically recognizes sedentary behavior.
- This depends on the smart shirt equipped with a processor, wearable sensors, power supply, and telemedicine interface.
- There are two phases to this method. Firstly, the usage of sensors comprising two distinct modes is interlinked with a variable vector which is further subjected to detection methods to decode this sedentary nature.

Secondly, A dummy model of a smart shirt is associated with the telemedicine unit responsible for measurement detection and computing relevant data. This way is very helpful for elderly patients who need constant monitoring in the ICUs. The signals that are collected by smart shirts consist of information that recognizes sedentary behavior during day-to-day activities and stored in the telemedical service provider for further analysis.

The experiment was performed with the help of five individuals (three females and two males). The volunteers are required to wear a prototype design of a smart shirt that was embedded with accelerometers, pulse sensors, ambient sensors. The volunteers were asked to perform a set of different activities through which data was generated, and this data was used to test the accuracy of the inactivity classification algorithm. The tasks were conducted in a home environment. The experiments consisted of tasks such as:

- Ambulatory events: lying unmoved, idle sitting, and walking.
- Day-to-day events such as constantly working for hours over a system.
- The battery test functionalities like stand & sit, feet stand, standing at tandem, etc.

The duration of each task was the 60 s, which was followed by a break of 60 s. Over 28,800 samples were collected and lasted 23 min. The shirt prototype is designed so that it is simple to utilize and operates primarily in elderly and physically challenged individuals. Three types of modules present interlink in these three non-disjointed groups.

- Personal Area Network (PAN) is dynamically developed near users.
- Local Area Network (LAN) and the internet.

The shirt is integrated with a body control unit, which creates a PAN network and waits for incoming connections. The network is established when the network coordinator is connected to the BCU. The data is collected, filtered, and encapsulated, and features are extracted with the help of the BCU. The NC and the results obtained to analyze the data packets from BCU are sent to the cloud provider (CP) via LAN. In Fig. 11.8, the design implementation of a sample shirt prototype is presented.

CP does data aggregation, data storage, data analysis for authorized users. The system architecture is shown in Fig. 11.8.

Supervised learning was used to develop the model. States of activity are determined in a regular 1 minute time span. Enough text-based data and coordination for the working of the SPPB protocol were obtained. The sampling rate was predefined at 40 Hz. This approached attribute reduction approach was applied with every time span. Heart rate is estimated with a peak rate of signals beyond a maximum limit of 540(~ 1.74 V) with at least 19 samples. The time-domain feature of accelerometer and intensity computation includes addition, average, standard deviation, variance, and the root mean square is investigated during the experiment. A vector for the optimum intensity for every axis was chosen from the frequency-domain feature of the accelerometer. The standard deviation of bit-to-bit time duration and the square root of the mean of consecutive differences were computed.

In the second step, the 2D feature vector obtained is transferred by the categorization system that differentiates between suitable labels. The first-class events for sedentary nature are lying, sitting, working on the system. The second set of activities is the remaining activities. The data is labeled into the appropriate class and stored in the database. The various ML algorithms used were: Binary decision trees, the Discriminant Analysis model, Naive Bayes, K-nearest neighbors, SVM, and ANN. The validation of the classification model was done by using a 10-fold-cross-validation. The entire process is presented in Fig. 11.9.

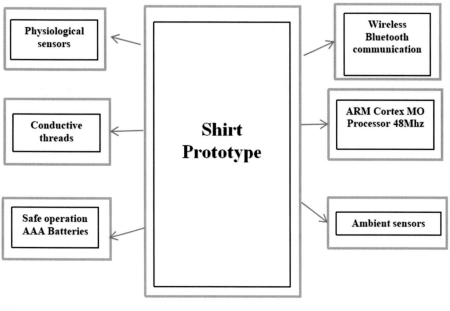

FIGURE 11.8

A look of the shirt prototype.

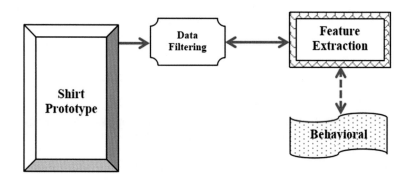

FIGURE 11.9

The functionality of the shirt prototype.

The results are shown in Table 11.4, which have concluded the raw hypothesis regarding the method's feasibility to get and discuss multi-sensors signals for the experimental protocol. The two classifiers: SVM (Support Vector Machines) and Artificial Neural Network (ANN), provide the maximal efficiency with respect to accuracy. The Naive Bayes model delivered a relatively poor performance. With the use of this process, the sedentary behavior in individuals, especially older

Table 11.4 Few details of algorithms.

Methods	Accuracy	Sensitivity	Specificity	Precision	F1
Linear discriminative analysis	95.0	86.6	97.7	92.8	89.6
Support vector machine	96.6	93.3	97.7	93.3	93.3
KNN	93.3	80.0	97.7	92.3	85.7
Binary decision tree	96.6	93.3	97.7	93.3	93.3
Naive Bayes	91.6	100.0	88.8	75.0	85.7
ANN	96.6	93.3	97.7	93.3	93.3
Average	95.0	91.1	96.2	90.0	90.1

people, can be monitored. Using machine learning, high accuracy can be obtained to estimate sedentary behavior. The prototype can be personalized to a shirt-based health technology telemonitoring system, which can provide significant data about the doctors' behavior. This data can be very useful in early diagnosis and detection to the prevention and treatment of cardiovascular disease. It also motivates the user to keep track of their health.

11.8 Future scope and conclusion

The true potential of wearables is not only to wear the screens on the wrist but also about the new, amazing, and interesting sensors incorporated in it. In other words, New sensors signify new data, new data leads to new insights, and new insights take to new applicability of the insights across markets. The beginning was all about accelerometers, gyroscopes, and magnetometers of three axes. Skin Galvanic, glucose levels, and electromyography, and many others can begin to be viewed with the expansion in the scope of better developed wearable devices. It is absolutely, unimaginably fascinating to know the amount of health data that can be gathered with various sensors. This can make humanity much healthier with the possibility of constantly analyzing one's health. Humanity can now know the response of an activity biologically on the human body. These devices' benefits could be as simple as tracking a pushup to early detection of cardiac arrest to various others. The wearable health industry trend is not slowing down any soon as healthcare professionals are excited to invade more possibilities for helping patients with chronic illness and helping people in remote areas. Effective collection of individual health information, more actionable outputs, and clinical validation are the need of the hour. Demand also is increasing for the ability to integrate data from wearables into electronic health records. In conclusion, technology through wearables boosts the healthcare industry by making health analysis of people a much easier task.

References

Anliker, U., Ward, J. A., Lukowicz, P., Tr Oster, G., Dolveck, F., Baer, M., ... Vuckovic, M. (2004). AMON: A wearable multiparameter medical monitoring and alert system. *IEEE Transactions on Information Technology in Biomedicine*, 8(4), 415−427.

Bhoi, A. K. (2017). Classification and Clustering of Parkinson's and healthy control gait dynamics using LDA and K-means. *International Journal Bioautomation*, *21*(1).

Bhoi, A. K., Mallick, P. K., Liu, C. M., & Balas, V. E. (Eds.), (2021). *Bio-inspired neurocomputing*. Springer.

Bhoi, A. K., Sherpa, K. S., & Khandelwal, B. (2018). Ischemia and arrhythmia classification using time-frequency domain features of QRS complex. *Procedia Computer Science*, *132*, 606−613.

Bhoi, A. K., Sherpa, K. S., & Khandelwal, B. (2015). Multidimensional analytical study of heart sounds: A review. *International Journal Bioautomation*, *19*(3), 351−376.

Bhoi, A. K., & Sherpa, K. S. (2014). QRS complex detection and analysis of cardiovascular abnormalities: A review. *International Journal Bioautomation*, *18*(3), 181−194.

Chung, W. Y., Lee, Y. D., & Jung, S. J. (2008). A wireless sensor network compatible wearable u-healthcare monitoring system using integrated ECG, accelerometer and SpO2. In *Proc. 30th ann. int. IEEE EMBS conf.* (pp. 1529−1532).

Chung, W. Y., Lee, S. C., & Toh, S. H. (2008). WSN based mobile u-healthcare system with ECG, blood pressure measurement function. In *Proc. 30th ann. int. IEEE EMBS conf.* (pp: 1533−1536).

Di Rienzo, M., Rizzo, F., Parati, G., Brambilla, G., Ferrarini, M., & Castiglioni, P. (2005). MagIC system: A new textile-based wearable device for biological signal monitoring applicability in daily life and clinical setting. In *Proc. 27th ann. int. IEEE EMBS Conf.* (pp. 7167−7169).

Evenson, K. R., Goto, M. M., & Furberg, R. D. (2015). Systematic review of the validity and reliability of consumer-wearable activity trackers. *International Journal of Behavioral Nutrition and Physical Activity*, *12*(1), 159.

Fensli, R., Gunnarson, E., & Gundersen, T. (2005). A wearable ECG-recording system for continuous arrhythmia monitoring in a wireless tele-home-care situation. In *Proc. 18th IEEE symp. comput.-based med. syst.* (pp. 407−412).

Gyselinckx, B., Penders, J., & Vullers, R. (2007). Potential and challenges of body area networks for cardiac monitoring. *Journal of Electrocardiology*, *40*, S165−S168.

Habetha, J. (2006). The MyHeart project—Fighting cardiovascular diseases prevention and early diagnosis. In *Proc. 28th ann. int. IEEE EMBS con*f. (pp. 6746−6749).

https://mhealthspot.com.

Klonoff, D. C. (2013). Twelve modern digital technologies that are transforming decision making for diabetes and all areas of health care. *Journal of Diabetes Science and Technology*, *7*(2), 291−295.

Lee, J. M., Kim, Y., & Welk, G. J. (2014). Validity of consumer-based physical activity monitors. *Medicine & Science in Sports & Exercise*, *46*(9), 1840−1848.

Leijdekkers, P. & Gay, V. (2008). A self-test to detect a heart attack using a mobile phone and wearable sensors. In *Proc. 21st IEEE CBMS int. symp.* (pp. 93−98).

Lin, B. S., Lin, B. S., Chou, N. K., Chong, F. C., & Chen, S. J. (2006). RTWPMS: A real-time wireless physiological monitoring system. *IEEE Transactions on Information Technology in Biomedicine*, *10*(4), 647−656.

Lubrano, J., Sola, J., Dasen, S., Koller, J. M., & Chetelat, O. (2006). Combination of body sensor networks and on-body signal processing algorithms: The practical case of MyHeart project. In *Proc. int. workshop wearable implantable body sens. netw.* (pp. 76−79).

Lubrano, J. (2006). European projects on smart fabrics, interactive textiles: Sharing opportunities and challenges. *Presented at the workshop wearable technol. intel. textiles*. Helsinki, Finland.

Lymperis, A. & Paradiso, R. (2008). Smart and interactive textile enabling wear-able personal applications: R&D state of the art and future challenges. In *Proc. 30th ann. Int. IEEE EMBS Conf.* (pp. 5270−5273).

Mallick, P. K., Mishra, S., & Chae, G. S. (2020). Digital media news categorization using Bernoulli document model for web content convergence. *Personal and Ubiquitous Computing*. Available from https://doi.org/10.1007/s00779-020-01461-9.

Milenkovic, A., Otto, C., & Jovanov, E. (2006). Wireless sensor networks for personal health monitoring: Issues and an implementation. *Computer Communications*, *29*, 2521−2533.

Mishra, S., Dash, A., & Mishra, B. K. (2020). *An insight of Internet of Things applications in pharmaceutical domain. Emergence of pharmaceutical industry growth with industrial IoT approach* (pp. 245−273). Academic Press.

Mishra, S., Mallick, P. K., Jena, L., & Chae, G. S. (2020). Optimization of skewed data using sampling-based preprocessing approach. *Frontiers in Public Health*, 8, 274. Available from https://doi.org/10.3389/fpubh.2020.00274.

Mishra, S., Mishra, B. K., Tripathy, H. K., & Dutta, A. (2020). *Analysis of the role and scope of big data analytics with IoT in health care domain. Handbook of data science approaches for biomedical engineering* (pp. 1−23). Academic Press.

Mishra, S., Mishra, B. K., & Tripathy, H. K. (2020). *Significance of biologically inspired optimization techniques in real-time applications. Robotic systems: Concepts, methodologies, tools, and applications* (pp. 224−248). IGI Global.

Mishra, S., Patel, S., Panda, A. R., & Mishra, B. K. (2019). Exploring IoT-enabled smart transportation system. In D. Goyal, S. Balamurugan, S. Peng, & D. S. Jat (Eds.), *The IoT and the next revolutions automating the world* (pp. 186−202). IGI Global. Available from http://doi:10.4018/978-1-5225-9246-4.ch012.

Mishra, S., Sahoo, S., Mishra, B. K., & Satapathy, S. (2016, October). A quality-based automated admission system for the educational domain. In *2016 International conference on signal processing, communication, power and embedded system (SCOPES)* (pp. 221−223). IEEE.

Mishra, S., Tripathy, H. K., Mallick, P. K., Bhoi, A. K., & Barsocchi, P. (2020). EAGA-MLP—An enhanced and adaptive hybrid classification model for diabetes diagnosis. *Sensors*, 20(14), 4036.

Mishra, S., Tripathy, H. K., Mishra, B. K., & Mohapatra, S. K. (2018). *A succinct analysis of applications and services provided by IoT. Big data management and the Internet of Things for improved health systems* (pp. 142−162). IGI Global.

Monton, E., Hernandez, J. F., Blasco, J. M., Herve, T., Micallef, J., Grech, I., ... Traver, V. (2008). Body area network for wireless patient monitoring. *IET Communications*, 2, 215−222. Available from https://digital-library.theiet.org/content/journals/10.1049/iet-com_20070046. https://doi.org/10.1049/iet-com:20070046IET DigitalLibrary.

Mundt, C. W., Montgomery, K. N., Udoh, U. E., Barker, V. N., Thornier, G. C., Tellier, A. M., ... Kovacs, G. T. A. (2005). A multiparameter wearable physiological monitoring system for space and terrestrial applications. *IEEE Transactions on Information Technology in Biomedicine: A Publication of the IEEE Engineering in Medicine and Biology Society*, 9(3), 382−391.

Pacelli, M., Loriga, G., Tacchini, N., & Paradiso, R. (2006). Sensing fabrics for monitoring physiological and biomechanical variables: E-textile solutions. In *Proc. 3rd IEEE-EMBS int. summer school symp. med. dev. biosens.* (pp. 1−4).

Pandian, P. S., Mohanavelu, K., Safeer, K. P., Kotresh, T. M., Shakunthala, D. T., Gopal, P., & Padaki, V. C. (2008). Smart vest: Wearable multiparameter remote physiological monitoring system. *Medical Engineering & Physics*, 30, 466−477.

Paradiso, R., Loriga, G., & Tacchini, N. (2005). A wearable health care system based on knitted integral sensors. *IEEE Transactions on Information Technology in Biomedicine: A Publication of the IEEE Engineering in Medicine and Biology Society*, 9(3), 337−344.

Rai, D., Thakkar, H. K., & Rajput, S. S. (2020, October). Performance characterization of binary classifiers for automatic annotation of aortic valve opening in Seismocardiogram signals. In *2020 9th International conference on bioinformatics and biomedical science* (pp. 77−82).

Ren, J., Chien, C., & Tai, C. C. (2006, Dec). A new wireless-type physiological signal measuring system using a PDA and the Bluetooth technology. In *Proc. IEEE int. conf. ind. technol.* (pp. 3026−3031).

Sahoo, P. K., & Thakkar, H. K. (2019). TLS: Traffic load based scheduling protocol for wireless sensor networks. *International Journal of Ad Hoc and Ubiquitous Computing*, 30(3), 150−160.

Sahoo, P. K., Thakkar, H. K., & Hwang, I. (2017). Pre-scheduled and self organized sleep-scheduling algorithms for efficient K-coverage in wireless sensor networks. *Sensors*, 17(12), 2945.

Sahoo, P. K., Thakkar, H. K., & Lee, M. Y. (2017). A cardiac early warning system with multi-channel SCG and ECG monitoring for mobile health. *Sensors, 17*(4), 711.

Sahoo, P. K., Thakkar, H. K., Lin, W. Y., Chang, P. C., & Lee, M. Y. (2018). On the design of an efficient cardiac health monitoring system through combined analysis of ECG and SCG signals. *Sensors, 18*(2), 379.

Scilingo, E. P., Gemignani, A., Paradiso, R., Tacchini, N., Ghelarducci, B., & De Rossi, D. (2005). Performance evaluation of sensing fabrics for monitoring physiological and biomechanical variables. *IEEE Transactions on Information Technology in Biomedicine, 9*(3), 345–352.

Shnayder, V., Chen, B. R., Lorincz, K, Fulford-Jones, T. R. F., & Welsh, M. (2005). *Sensor networks for medical care*. Cambridge, MA: Division Eng. Appl. Sci., Harvard Univ. Tech. Rep. TR-08-05.

Sung, M., Marci, C., & Pentland, A. (2005). Wearable feedback systems for rehabilitation. *Journal of Neuroengineering and Rehabilitation, 2*, 17.

Thakkar, H. K., Dehury, C. K., & Sahoo, P. K. (2020). MUVINE: Multi-stage virtual network embedding in cloud data centers using reinforcement learning-based predictions. *IEEE Journal on Selected Areas in Communications, 38*(6), 1058–1074.

Thakkar, H. K., Liao, W. W., Wu, C. Y., Hsieh, Y. W., & Lee, T. H. (2020). Predicting clinically significant motor function improvement after contemporary task-oriented interventions using machine learning approaches. *Journal of Neuroengineering and Rehabilitation, 17*(1), 1–10.

Thakkar, H. K., & Sahoo, P. K. (2019). Towards automatic and fast annotation of seismocardiogram signals using machine learning. *IEEE Sensors Journal, 20*(5), 2578–2589.

Tura, A., Badanai, M., Longo, D., & Guarani, L. (2003). A medical wearable device with wireless Bluetooth-based data transmission. *Measurement Science Review, 3*, 1–4.

Uhlsteff, J. M., Such, O., Schmidt, R., Perkuhn, M., Reiter, H., Lauter, J., ... Harris, M. (2004). Wearable approach for continuous ECG and activity patient-monitoring. In *Proc. 26th ann. int. IEEE EMBS Conf.* (pp. 2184–2187).

Weber, J. L. & Porotte, F. (2006). Medical remote monitoring with clothes. *Presented at the int. workshop on PHealth*. Luzern, Switzerland.

Wu, J., Li, H., Cheng, S., & Lin, Z. (2016). The promising future of healthcare services: When big data analytics meets wearable technology. *Information & Management, 53*(8), 1020–1033.

Yuce, M. R., Ng, S. W. P., Myo, N. L., Khan, J. Y., & Liu, W. (2007). Wireless body sensor network using medical implant band. *Journal of Medical Systems, 31*, 467–474.

Index

Printed in the United States
by Baker & Taylor Publisher Services